CAMBRIDGE TRACTS IN MATHEMATICS

General Editors

196 The Theory of Hardy's *Z*-Function

CAMBRIDGE TRACTS IN MATHEMATICS

GENERAL EDITORS

A complete list of books in the series can be found at www.cambridge.org/mathematics.
Recent titles include the following:

163. Linear and Projective Representations of Symmetric Groups. By A. Kleshchev
164. The Covering Property Axiom, CPA. By K. Ciesielski and J. Pawlikowski
165. Projective Differential Geometry Old and New. By V. Ovsienko and S. Tabachnikov
166. The Lévy Laplacian. By M. N. Feller
167. Poincaré Duality Algebras, Macaulay's Dual Systems, and Steenrod Operations. By D. Meyer and L. Smith
168. The Cube-A Window to Convex and Discrete Geometry. By C. Zong
169. Quantum Stochastic Processes and Noncommutative Geometry. By K. B. Sinha and D. Goswami
170. Polynomials and Vanishing Cycles. By M. Tibăr
171. Orbifolds and Stringy Topology. By A. Adem, J. Leida, and Y. Ruan
172. Rigid Cohomology. By B. Le Stum
173. Enumeration of Finite Groups. By S. R. Blackburn, P. M. Neumann, and G. Venkataraman
174. Forcing Idealized. By J. Zapletal
175. The Large Sieve and its Applications. By E. Kowalski
176. The Monster Group and Majorana Involutions. By A. A. Ivanov
177. A Higher-Dimensional Sieve Method. By H. G. Diamond, H. Halberstam, and W. F. Galway
178. Analysis in Positive Characteristic. By A. N. Kochubei
179. Dynamics of Linear Operators. By F. Bayart and É. Matheron
180. Synthetic Geometry of Manifolds. By A. Kock
181. Totally Positive Matrices. By A. Pinkus
182. Nonlinear Markov Processes and Kinetic Equations. By V. N. Kolokoltsov
183. Period Domains over Finite and p-adic Fields. By J.-F. Dat, S. Orlik, and M. Rapoport
184. Algebraic Theories. By J. Adámek, J. Rosický, and E. M. Vitale
185. Rigidity in Higher Rank Abelian Group Actions I: Introduction and Cocycle Problem. By A. Katok and V. Niţică
186. Dimensions, Embeddings, and Attractors. By J. C. Robinson
187. Convexity: An Analytic Viewpoint. By B. Simon
188. Modern Approaches to the Invariant Subspace Problem. By I. Chalendar and J. R. Partington
189. Nonlinear Perron Frobenius Theory. By B. Lemmens and R. Nussbaum
190. Jordan Structures in Geometry and Analysis. By C.-H. Chu
191. Malliavin Calculus for Lévy Processes and Infinite-Dimensional Brownian Motion. By H. Osswald
192. Normal Approximations with Malliavin Calculus. By I. Nourdin and G. Peccati
193. Distribution Modulo One and Diophantine Approximation. By Y. Bugeaud
194. Mathematics of Two-Dimensional Turbulence. By S. Kuksin and A. Shirikyan
195. A Universal Construction for R-free Groups. By I. Chiswell and T. Müller
196. The Theory of Hardy's Z-Function. By A. Ivić
197. Induced Representations of Locally Compact Groups. By E. Kaniuth and K. F. Taylor
198. Topics in Critical Point Theory. By K. Perera and M. Schechter
199. Combinatorics of Minuscule Representations. By R. M. Green
200. Singularities of the Minimal Model Program. By J. Kollár

G. H. Hardy, 1877-1947

The Theory of Hardy's Z-Function

ALEKSANDAR IVIĆ
Univerzitet u Beogradu, Serbia

CAMBRIDGE
UNIVERSITY PRESS

CAMBRIDGE UNIVERSITY PRESS
Cambridge, New York, Melbourne, Madrid, Cape Town,
Singapore, São Paulo, Delhi, Mexico City

Cambridge University Press
The Edinburgh Building, Cambridge CB2 8RU, UK

Published in the United States of America by Cambridge University Press, New York

www.cambridge.org
Information on this title: www.cambridge.org/9781107028838

Frontispiece: G. H. Hardy. Kindly supplied by Trinity College, Cambridge

First published 2013

Printed and bound in the United Kingdom by the MPG Books Group

A catalog record for this publication is available from the British Library

Library of Congress Cataloguing in Publication data
Ivic, A., 1949– author.
The theory of Hardy's Z-function / Aleksandar Ivic.
pages cm. – (Cambridge tracts in mathematics ; 196)
Includes bibliographical references and index.
ISBN 978-1-107-02883-8 (hardback)
1. Number theory. I. Title.
QA241.I83 2012
512.7 – dc23 2012024804

ISBN 978-1-107-02883-8 Hardback

Contents

Preface

This text has grown out of a mini-course held at the Arctic Number Theory School, University of Helsinki, May 18-25, 2011. The central topic is Hardy's function $Z(t)$, of great importance in the theory of the Riemann zeta-function $\zeta(s)$. It is named after Godfrey Harold ("G. H.") Hardy FRS (1877-1947), who was a prominent English mathematician, well-known for his achievements in number theory and mathematical analysis. Sometimes by Hardy function(s) one denotes the element(s) of Hardy spaces H^p, which are certain spaces of holomorphic functions on the unit disk or the upper half-plane. In this text, however, Hardy's function $Z(t)$ will always denote the function defined by (0) below. It was chosen as the object of study because of its significance in the theory of $\zeta(s)$ and because, initially, considerable material could be presented on the blackboard within the framework of six lectures. Some results, like Theorem 6.7 and the bounds in (4.25) and (4.26) are new, improving on older ones. It is "Hardy's function" which is the thread that holds this work together. I have thought it is appropriate for a monograph because the topic is not as vast as the topic of the Riemann zeta-function itself. Moreover, specialized monographs such as [Iv4], [Lau5], [Mot5] and [Ram] cover in detail other parts of zeta-function theory, but not $Z(t)$. It appears that one volume cannot suffice today to cover all the relevant material concerning $\zeta(s)$, although some of the most important problems (such as the distribution of zeros, in particular the notorious Riemann hypothesis that all complex zeros of $\zeta(s)$ have real parts 1/2), are not settled yet. On the other hand, there seems now to be enough material on $Z(t)$ to warrant the writing of a monograph dedicated solely to it, especially since there is at present a lot of research concerning $Z(t)$ going on.

In the desire to keep the size of the text compact, I had to avoid going into detail in subjects which are connected with zeta-function theory, but are already adequately covered in the literature. These include spectral theory, random matrix theory and representation theory, among others. Proofs of some well-known results such as the first and second derivative tests for exponential integrals (Lemma 2.2 and Lemma 2.3) are omitted, while the proof of, for example, Lemma 7.2 would lead us too much astray. I have tried to include most of the material of the ongoing research on $Z(t)$, thus there is much material from papers which are just published or in print. Starting with Chapter 2, the theory of Hardy's function is systematically developed, hence the title *The Theory of Hardy's Z-Function* seems appropriate. In developing the theory, some modern tools of analytic number theory are systematically used and discussed, such as approximate functional equations, exponential sums and integrals, and integral transforms, to name just a few.

The text is intended for ambitious graduate students, Ph.D. students, and researchers in the field. The prerequisites are basically standard courses in real and complex analysis, but it will help if the reader is familiar with a general text on the Riemann zeta-function, such as the monographs [Tit3], [KaVo] and [Iv1], although efforts have been made to keep the text self-contained. Besides mathematicians, the book will hopefully appeal to physicists whose work is linked to the applications of $\zeta(s)$ in Physics.

The text offers sufficient material for a course of one semester. Except for the introductory Chapter 1, other chapters are basically independent, and various shorter courses may be given based on disposable time and the interest of students.

The material is organized as follows. In Chapter 1 we present the Riemann zeta-function and discuss some of its basic properties. Hardy's function

$$Z(t) := \zeta(\tfrac{1}{2} + it)\big(\chi(\tfrac{1}{2} + it)\big)^{-1/2}, \qquad \zeta(s) = \chi(s)\zeta(1-s), \tag{0}$$

is introduced and some of its properties are presented. One of the most important aspects of $Z(t)$ is that it is a real-valued function of the real valuable t, and its zeros exactly correspond to the zeros of $\zeta(s)$ on the so-called "critical line" $\mathbb{R}\mathrm{e}\, s = 1/2$. Thus it is very convenient to work with, being real-valued, both theoretically and in conjunction with the calculation of zeta-zeros. It is in Chapter 2 that we discuss the zeros of the zeta-function on the critical line. In particular, we present a variant of Hardy's original proof that there are infinitely many zeta-zeros on $\mathbb{R}\mathrm{e}\, s = 1/2$. Besides that, we discuss Lehmer's phenomenon, namely that the Riemann hypothesis fails if $Z(t)$ has a positive local minimum or a negative local maximum (except at $t = 2.475\,75\ldots$) and the unconditional estimation of gaps between the consecutive zeros of $Z(t)$.

Chapter 3 is devoted to the Selberg class $\mathcal{S}$ of L-functions, of which $\zeta(s)$ can be thought of as a prototype, since the elements of $\mathcal{S}$ inherit the intrinsic properties of $\zeta(s)$, namely the Euler product (see (1.1)) and the functional equation (see (1.5)). Basic notions and properties of functions in $\mathcal{S}$ are presented. The class $\mathcal{S}$ (introduced by A. Selberg in 1989) is a natural generalization of $\zeta(s)$, and the modern theory of L-functions often focuses on families and classes of L-functions containing $\zeta(s)$, such as $\mathcal{S}$.

Chapter 4 is dedicated to AFEs (approximate functional equations) for $\zeta^k(s)$ when $k \in \mathbb{N}$. The AFEs are an essential tool in the theory of L-functions, approximating a Dirichlet series by a finite number of Dirichlet polynomials. For the AFEs that we treat, the accent is on the methods of proof, of which there are several. From an AFE for $\zeta^k(s)$ one then deduces an AFE for $Z^k(t)$.

In Chapter 5 we provide formulas for the derivatives $Z^{(k)}(t)$, and in Chapter 6 we deal with Gram points g_n which satisfy $\theta(g_n) = \pi n$, where $Z(t) = \mathrm{e}^{i\theta(t)}\zeta(\frac{1}{2} + it)$ and $\theta(t)$ is given by (1.19). We also discuss the so-called "Gram's law" related to Gram points.

In Chapter 7 we give formulas for the kth moment of $Z(t)$, that is, for the integral $\int_0^T Z^k(t)\,\mathrm{d}t$, where $k \in \mathbb{N}$. This is achieved by employing an AFE with smooth weights.

M. Jutila's recent formula for the oscillating function $F(T) = \int_0^T Z(t)\,\mathrm{d}t$ is given in detail in Chapter 8. It furnishes the results $F(T) = O(T^{1/4})$ and $F(T) = \Omega_{\pm}(T^{1/4})$, thereby determining the true order of magnitude of $F(T)$. This approach is different from the author's work and from the recent work of M. Korolev [Kor3], [Kor4], who also obtained an explicit formula for $F(T)$.

The Mellin transforms ($k \in \mathbb{N}$)

$$\mathcal{Z}_k(s) = \int_1^\infty |\zeta(\tfrac{1}{2} + ix)|^{2k} x^{-s}\,\mathrm{d}x, \qquad M_k(s) = \int_1^\infty Z^k(x)x^{-s}\,\mathrm{d}x$$

form the subject of Chapter 9 and Chapter 10. In general, classical integral transforms, in particular Fourier, Laplace and Mellin transforms, play an important rôle in analytic number theory. Analytic continuation, pointwise and mean square estimates of $\mathcal{Z}_k(s)$, $\mathcal{M}_k(s)$ are given in detail, by using various techniques of analytic number theory. The importance of these functions is their connections with power moments of $|\zeta(\frac{1}{2} + it)|$, one of the most important topics of zeta-function theory.

The concluding Chapter 11 focuses on some problems involving Hardy's function and zeta moments. I hope that this will be of interest to all who want to do research in areas connected to these topics. This is in the spirit of P. Erdős's motto: "Prove and conjecture!".

Each chapter is followed by "Notes", where many references, comments and remarks are given. This seemed preferable to burdening the body of the text with footnotes, parentheses, etc. On many occasions dates of birth and death are given for the mathematicians mentioned in the text.

A full bibliography of all the relevant works mentioned in the text is to be found at the end of the text.

Acknowledgements. Several experts in the field have read (parts of) the text, and made valuable remarks. I wish to thank all of them. In alphabetical order they are: H. Bui, J. Kaczorowski, M. Korolev, Y. Motohashi, A. Perelli, I. Rezvyakova, T. Trudgian and N. Watt. Special thanks go to M. Jutila.

I also wish to thank the staff of Cambridge University Press for their unfailing courtesy and help.

Notation

Owing to the nature of this text, absolute consistency in notation could not be attained, although whenever possible standard notation is used. Notation used commonly through the text is explained there, while specific notation introduced in the proof of a theorem or lemma is given at the proper place in the body of the text.

$k, l, m, n, \ldots$	Natural numbers (positive integers).
p	A generic prime number.
$\mathbb{N}, \mathbb{Z}, \mathbb{R}, \mathbb{C}$	The sets of natural numbers, integers, real and complex numbers, respectively.
$A, B, C, C_1, \ldots$	Absolute, positive constants (not necessarily the same ones at each occurrence).
ε	An arbitrarily small positive number, not necessarily the same one at each occurrence.
s, z, w	Complex variables ($\mathbb{R}\mathrm{e}\, s$ and $\mathbb{I}\mathrm{m}\, s$ denote the real and imaginary part of s, respectively; common notation is $\sigma = \mathbb{R}\mathrm{e}\, s$ and $t = \mathbb{I}\mathrm{m}\, s$).
t, x, y	Real variables.
$\operatorname{Res}_{s=s_0} F(s)$	Denotes the residue of $F(s)$ at the point $s = s_0$.
$\zeta(s)$	The Riemann zeta-function is defined by $\zeta(s) = \sum_{n=1}^{\infty} n^{-s}$ for $\mathbb{R}\mathrm{e}\, s > 1$ and otherwise by analytic continuation.
$\Gamma(s)$	$= \int_0^\infty x^{s-1} \mathrm{e}^{-x}\, \mathrm{d}x$ for $\mathbb{R}\mathrm{e}\, s > 0$, otherwise by analytic continuation by $s\Gamma(s) = \Gamma(s+1)$. This is the Euler gamma-function.
γ	Euler's constant $\gamma = -\Gamma'(1) = 0.577\,2157\ldots$.

$\chi(s)$	The function defined by $\zeta(s) = \chi(s)\zeta(1-s)$, so that by the functional equation for $\zeta(s)$ we have $\chi(s) = (2\pi)^s/(2\Gamma(s)\cos(\pi s/2))$.
$Z(t)$	Hardy's function, $Z(t) = \zeta(\frac{1}{2}+it)(\chi(\frac{1}{2}+it))^{-1/2}$.
$\theta(t)$	For real t defined as $\theta(t) = \mathbb{I}\mathrm{m}\,\{\log\Gamma(\frac{1}{4}+\frac{1}{2}it)\} - \frac{1}{2}t\log\pi$.
$\rho = \beta + i\gamma$	A complex zero of $\zeta(s)$; $\beta = \mathbb{R}\mathrm{e}\,\rho$, $\gamma = \mathbb{I}\mathrm{m}\,\rho$.
$N(T)$	The number of zeros $\rho = \beta + i\gamma$ of $\zeta(s)$, counted with multiplicities, for which $0 < \gamma \le T$.
$S(T)$	$= \frac{1}{\pi}\arg\zeta(\frac{1}{2}+iT)$.
$\mu(\sigma)$	For real σ defined as $\mu(\sigma) = \limsup_{t\to\infty} \frac{\log\lvert\zeta(\sigma+it)\rvert}{\log t}$.
$\exp(z)$	$= \mathrm{e}^z$.
$\mathrm{e}(z)$	$= \mathrm{e}^{2\pi iz}$.
$\log x$	$= \mathrm{Log}_{\mathrm{e}}\mathrm{x} \equiv \ln \mathrm{x}$.
$[x]$	The greatest integer not exceeding the real number x.
$\{x\}$	$= x - [x]$, the fractional part of x.
$\sum_{n\le x} f(n)$	A sum taken over all natural numbers n not exceeding x; the empty sum is defined to be equal to zero.
$\prod_j$	The product taken over all possible values of the index j; the empty product is defined to be unity.
$d_k(n)$	The number of ways n can be written as a product of $k \ge 2$ fixed factors; $d_2(n) = d(n)$ is the number of divisors of n.
$\Delta(x)$	$= \sum_{n\le x} d(n) - x(\log x + 2\gamma - 1)$, the error term in the Dirichlet divisor problem.
$f(x) \sim g(x)$	Means $\lim_{x\to x_0} f(x)/g(x) = 1$, as $x \to x_0$, with x_0 not necessarily finite.
$f(x) = O(g(x))$	Means $\lvert f(x)\rvert \le Cg(x)$ for $x \ge x_0$ and some constant $C > 0$. Here $f(x)$ is a complex function of the real variable x and $g(x)$ is a positive function of x for $x \ge x_0$. $f(x) = O_{\alpha,\beta,\ldots}(g(x))$ means the implied constant depends on $\alpha, \beta, \ldots$.
$f(x) \ll g(x)$	Means the same as $f(x) = O(g(x))$. Likewise $f(x) \ll_{\alpha,\beta,\ldots} g(x)$ means the implied constant depends on $\alpha, \beta, \ldots$.
$f(x) \gg g(x)$	Means the same as $g(x) = O(f(x))$.
$f(x) \asymp g(x)$	Means that both $f(x) \ll g(x)$ and $g(x) \ll f(x)$ hold.

(a, b)	Means the interval $a < x < b$.
$[a, b]$	Means the interval $a \le x \le b$.
$C^r[a, b]$	The class of functions having a continuous rth derivative in $[a, b]$.
$L^p(a, b)$	The class of measurable functions $f(x)$ such that $\int_a^b \lvert f(x)\rvert^p \,\mathrm{d}x$ is finite.
$f(x) = o(g(x))$ as $x \to x_0$	Means $\lim_{x\to x_0} f(x)/g(x) = 0$, with x_0 possibly infinite.
$f(x) = \Omega\big(g(x)\big)$	Means that $f(x) = o(g(x))$ does not hold when $x \to \infty$.
$f(x) = \Omega_+(g(x))$	Means that there exists a suitable constant $C > 0$ such that $f(x) > Cg(x)$ holds for some arbitrarily large values of x.
$f(x) = \Omega_-(g(x))$	Means that there exists a suitable constant $C > 0$ such that $f(x) < -Cg(x)$ holds for some arbitrarily large values of x.
$f(x) = \Omega_\pm(g(x))$	Means that both $f(x) = \Omega_+(g(x))$ and $f(x) = \Omega_-(g(x))$ hold.
$\int_{(c)} G(s)\,\mathrm{d}s$	$= \lim_{T\to\infty} \int_{c-iT}^{c+iT} G(s)\,\mathrm{d}s$.
$\mathcal{M}_k(s)$	$= \int_1^\infty Z^k(x)x^{-s}\,\mathrm{d}x \quad (k \in \mathbb{N})$.
$\mathcal{Z}_k(s)$	$= \int_1^\infty \lvert\zeta(\frac{1}{2} + ix)\rvert^{2k} x^{-s}\,\mathrm{d}x \quad (k \in \mathbb{N})$.
AFE	Means "approximate functional equation".

1

Definition of $\zeta(s)$, $Z(t)$ and basic notions

1.1 The basic notions

The classical Riemann zeta-function

$$\zeta(s) = \sum_{n=1}^{\infty} n^{-s} = \prod_{p} (1 - p^{-s})^{-1} \qquad (s = \sigma + it,\ \sigma > 1) \tag{1.1}$$

admits analytic continuation to $\mathbb{C}$. It is regular on $\mathbb{C}$ except for a simple pole at $s = 1$. The product representation in (1.1) shows that $\zeta(s)$ does not vanish for $\sigma > 1$. The Laurent expansion of $\zeta(s)$ at $s = 1$ reads

$$\zeta(s) = \frac{1}{s-1} + \gamma_0 + \gamma_1(s-1) + \gamma_2(s-1)^2 + \cdots,$$

where the so-called *Stieltjes constants* γ_k are given by

$$\gamma_k = \frac{(-1)^k}{k!} \lim_{N\to\infty} \left(\sum_{m\le N} \frac{\log^k m}{m} - \frac{\log^{k+1} N}{k+1} \right) \qquad (k = 0, 1, 2, \ldots).$$

In particular

$$\gamma \equiv \gamma_0 = \lim_{N\to\infty} \left(1 + \frac{1}{2} + \cdots + \frac{1}{N} - \log N \right) = -\Gamma'(1) = 0.577\,2157\ldots$$

is the *Euler constant* and

$$\Gamma(s) = \int_0^\infty x^{s-1} \mathrm{e}^{-x}\, \mathrm{d}x \qquad (\mathbb{R}\mathrm{e}\, s > 0) \tag{1.2}$$

is the familiar *Euler gamma-function.*

The product in (1.1) is called the *Euler product*. As usual, p denotes prime numbers, so that by its very essence $\zeta(s)$ represents an important tool for the investigation of prime numbers. This is even more evident from the relation

$$-\frac{\zeta'(s)}{\zeta(s)} = \sum_{n=1}^{\infty} \Lambda(n)n^{-s} \qquad (\sigma > 1),$$

which follows by logarithmic differentiation of (1.1), where *the von Mangoldt function* $\Lambda(n)$ is defined as

$$\Lambda(n) = \begin{cases} \log p & \text{if } n = p^\alpha, \\ 0 & \text{if } n \neq p^\alpha \quad (\alpha \in \mathbb{N}). \end{cases}$$

The zeta-function can be also used to generate many other important arithmetic functions; for example,

$$\frac{\zeta(s)}{\zeta(2s)} \quad (\sigma > 1), \qquad \frac{\zeta(2s)\zeta(3s)}{\zeta(6s)} \quad (\sigma > \tfrac{1}{2})$$

generate the characteristic functions of *squarefree* and *squarefull* numbers, respectively. One also has, for a given $k \in \mathbb{N}$,

$$\zeta^k(s) = \sum_{n=1}^{\infty} d_k(n)n^{-s} \qquad (\sigma > 1), \tag{1.3}$$

where the (general) divisor function $d_k(n)$ represents the number of ways n can be written as a product of k factors, so that in particular $d_1(n) \equiv 1$ and $d_2(n) = \sum_{\delta|n} 1$ is the number of positive divisors of n. The function $d_k(n)$ is a multiplicative function of n (meaning $d_k(mn) = d_k(m)d_k(n)$ if m and n are coprime), and

$$d_k(p^\alpha) = (-1)^\alpha \binom{-k}{\alpha} = \frac{k(k+1)\cdots(k+\alpha-1)}{\alpha!}$$

for primes p and $\alpha \in \mathbb{N}$.

Another significant aspect of $\zeta(s)$ is that it can be generalized to many other similar Dirichlet series (notably to the Selberg class $\mathcal{S}$, which will be discussed in Chapter 3). A vast body of literature exists on many facets of zeta-function theory, such as the distribution of its zeros and power moments of $|\zeta(\frac{1}{2} + it)|$ (see, e.g., the monographs [Iv1], [Iv4], [Mot1], [Ram] and [Tit3]). It is within this framework that the classical Hardy function (see, e.g., [Iv1]) $Z(t)$ $(t \in \mathbb{R})$ arises, and plays an important rôle in the theory of $\zeta(s)$. It is defined as

$$Z(t) := \zeta(\tfrac{1}{2} + it)\big(\chi(\tfrac{1}{2} + it)\big)^{-1/2}, \tag{1.4}$$

where $\chi(s)$ comes from the well-known functional equation for $\zeta(s)$; see (1.5) and (1.6) below. The basic properties of $Z(t)$ will be discussed in Section 1.3.

1.2 The functional equation for $\zeta(s)$

The functional equation is one of the most fundamental tools of zeta-function theory. Therefore we shall, for the sake of completeness, provide a proof which incidentally originated with the great German mathematician B. Riemann (1826-1866), who founded the theory of $\zeta(s)$ in his epoch-making memoir [Rie].

Theorem 1.1 *The function $\zeta(s)$ admits analytic continuation to $\mathbb{C}$, where it satisfies the functional equation*

$$\pi^{-s/2}\zeta(s)\Gamma(\tfrac{1}{2}s) = \pi^{-(1-s)/2}\zeta(1-s)\Gamma(\tfrac{1}{2}(1-s)). \tag{1.5}$$

Remark 1.2 The functional equation (1.5) is in a symmetric form. Alternatively we can write (1.5) as

$$\zeta(s) = \chi(s)\zeta(1-s), \tag{1.6}$$

where we set

$$\chi(s) = \frac{\Gamma(\frac{1}{2}(1-s))}{\Gamma(\frac{1}{2}s)}\pi^{s-1/2}.$$

This expression can be put into other equivalent forms. For example, we have

$$\chi(s) = 2^s\pi^{s-1}\sin(\tfrac{1}{2}\pi s)\Gamma(1-s) = \frac{(2\pi)^s}{2\Gamma(s)\cos(\pi s/2)}, \tag{1.7}$$

where we used the well-known identities

$$\Gamma(s)\Gamma(1-s) = \frac{\pi}{\sin(\pi s)}, \quad \Gamma(s)\Gamma(s+\tfrac{1}{2}) = 2^{1-2s}\sqrt{\pi}\Gamma(2s). \tag{1.8}$$

Remark 1.3 Note that (1.6) gives the identity

$$\chi(s)\chi(1-s) = 1. \tag{1.9}$$

All identities (1.5)-(1.9) hold for $s \in \mathbb{C}$.

Before we proceed to the proof of the functional equation (1.5), we need a result on a transformation formula for *the theta-function* (see (1.14)), embodied in the following lemma.

Lemma 1.4 *We have*

$$\sum_{n=-\infty}^{\infty} \mathrm{e}^{-\pi n^2 t} = \frac{1}{\sqrt{t}}\sum_{n=-\infty}^{\infty} \mathrm{e}^{-\pi n^2/t} \qquad (t > 0). \tag{1.10}$$

Proof of Lemma 1.4 For $v \in \mathbb{R}$, $\tau = iy$, $y > 0$ we have the Fourier expansion

$$f(v) := \sum_{n=-\infty}^{\infty} \mathrm{e}^{\pi i\tau(n+v)^2} = \sum_{k=-\infty}^{\infty} c_k \mathrm{e}^{2\pi ikv}, \tag{1.11}$$

since $f(v)$ is periodic with period 1 and $f(v) \in C^1[0, 1]$. Hence with $A = -2\pi ik$, $B = \pi y$, we have for the Fourier coefficients c_k the expression

$$\begin{aligned} c_k &= \int_0^1 \sum_{n=-\infty}^{\infty} \mathrm{e}^{\pi i\tau(n+v)^2 - 2\pi ikv}\, \mathrm{d}v \\ &= \sum_{n=-\infty}^{\infty} \int_0^1 \mathrm{e}^{\pi i\tau(n+v)^2 - 2\pi ik(n+v)}\, \mathrm{d}v \\ &= \int_{-\infty}^{\infty} \mathrm{e}^{-\pi yv^2 - 2\pi ikv}\, \mathrm{d}v = \int_{-\infty}^{\infty} \mathrm{e}^{Av - Bv^2}\, \mathrm{d}v \\ &= \sqrt{\frac{\pi}{B}}\mathrm{e}^{A^2/(4B)} = \frac{1}{\sqrt{y}}\mathrm{e}^{-\pi k^2/y}. \end{aligned} \tag{1.12}$$

The change of order of summation and integration in (1.12) is justified by absolute convergence. Here we used the classical integral

$$\int_{-\infty}^{\infty} \exp(At - Bt^2)\, \mathrm{d}t = \sqrt{\frac{\pi}{B}} \exp\left(\frac{A^2}{4B}\right) \qquad (\mathbb{R}\mathrm{e}\, B > 0). \tag{1.13}$$

Setting $v = 0, i\tau = i^2 y = -t, y = t$ in (1.11) and (1.12), we obtain (1.10) of Lemma 1.4. By analytic continuation it is seen that (1.10) remains valid for $\mathbb{R}\mathrm{e}\, t > 0$. If we define the theta-function as

$$\vartheta(t) = \sum_{n=1}^{\infty} \mathrm{e}^{-\pi n^2 t} \qquad (\mathbb{R}\mathrm{e}\, t > 0), \tag{1.14}$$

then (1.10) yields the transformation formula

$$\vartheta(t) = \frac{1}{2\sqrt{t}}\left(2\vartheta\left(\frac{1}{t}\right) + 1\right) - \frac{1}{2} \qquad (\mathbb{R}\mathrm{e}\, t > 0). \tag{1.15}$$

Proof of Theorem 1.1 We start from

$$\Gamma(\tfrac{1}{2}s) = \int_0^{\infty} \mathrm{e}^{-x} x^{s/2-1}\, \mathrm{d}x \qquad (\sigma > 0),$$

which is just (1.2) with $s/2$ in place of s. If $n \in \mathbb{N}$, we write $\pi n^2 x$ in place of x to obtain

$$\Gamma(\tfrac{1}{2}s) = \pi^{s/2} n^s \int_0^{\infty} \mathrm{e}^{-\pi n^2 x} x^{s/2-1}\, \mathrm{d}x \qquad (\sigma > 0),$$

or

$$n^{-s} = \frac{\pi^{s/2}}{\Gamma(s/2)} \int_0^\infty e^{-\pi n^2 x} x^{s/2-1}\, dx \qquad (\sigma > 0).$$

Summation over n gives, for $\sigma > 1$,

$$\zeta(s) = \sum_{n=1}^\infty n^{-s} = \frac{\pi^{s/2}}{\Gamma(s/2)} \sum_{n=1}^\infty \int_0^\infty e^{-\pi n^2 x} x^{s/2-1}\, dx.$$

Since the series

$$\sum_{n=1}^\infty \int_0^\infty \left| e^{-\pi n^2 x} x^{s/2-1} \right| dx = \sum_{n=1}^\infty \int_0^\infty e^{-\pi n^2 x} x^{\sigma/2-1}\, dx = \sum_{n=1}^\infty \Gamma(\sigma/2) \pi^{-\sigma/2} n^{-\sigma}$$

converges for $\sigma > 1$, we can change the order of summation and integration to obtain

$$\zeta(s) = \frac{\pi^{s/2}}{\Gamma(s/2)} \int_0^\infty \vartheta(x) x^{s/2-1}\, dx. \tag{1.16}$$

In view of (1.15) we may write (1.16) as

$$\begin{aligned}
\pi^{-s/2}\Gamma(s/2)\zeta(s) &= \int_0^1 x^{s/2-1}\vartheta(x)\, dx + \int_1^\infty x^{s/2-1}\vartheta(x)\, dx \\
&= \int_0^1 x^{s/2-1}\Big(x^{-1/2}\vartheta\Big(\frac{1}{x}\Big) + \frac{1}{2}x^{-1/2} - \frac{1}{2}\Big)\, dx \\
&\quad + \int_1^\infty x^{s/2-1}\vartheta(x)\, dx \\
&= \frac{1}{s-1} - \frac{1}{s} + \int_0^1 x^{s/2-3/2}\vartheta(1/x)\, dx + \int_1^\infty x^{s/2-1}\vartheta(x)\, dx \\
&= \frac{1}{s(s-1)} + \int_1^\infty \Big(x^{-s/2-1/2} + x^{s/2-1}\Big)\vartheta(x)\, dx. \qquad (1.17)
\end{aligned}$$

Note first that the last expression in (1.17) remains invariant if s is replaced by $1 - s$. Secondly, the last integral in (1.17) converges uniformly (since $x \ge 1$ in the integrand) in any strip

$$-\infty < a \le \sigma = \mathrm{Re}\, s \le b < +\infty.$$

Consequently the last integral in (1.17) represents an entire function of s. Therefore

$$\pi^{-s/2}\Gamma(s/2)\zeta(s) - \frac{1}{s(s-1)}$$

is an entire function of s. Since $\pi^{s/2}/\Gamma(s/2)$ is an entire function (because $\Gamma(s)$ has no zeros),

$$\zeta(s) - \frac{1}{s(s-1)}\frac{\pi^{s/2}}{2\Gamma(s/2)}$$

is also an entire function. Further, since $s\Gamma(s/2) = 2\Gamma(s/2+1)$, it follows that

$$\zeta(s) - \frac{1}{s-1}\frac{\pi^{s/2}}{2\Gamma(s/2+1)}$$

is an entire function. Since $\sqrt{\pi}/(2\Gamma(3/2)) = 1$, we have that $\zeta(s) - 1/(s-1)$ is an entire function, thus $\zeta(s)$ is regular in $\mathbb{C}$ except for a simple pole at $s = 1$ with residue 1.

This discussion shows that (1.17) provides analytic continuation of $\zeta(s)$ to $\mathbb{C}$, as well as the functional equation (1.5).

Corollary 1.5 *If we define*

$$\eta(s) = \pi^{-s/2}\Gamma(s/2)\zeta(s), \qquad \xi(s) = \tfrac{1}{2}s(s-1)\eta(s), \tag{1.18}$$

then $\xi(s)$ is an entire function of s satisfying the functional equation $\xi(s) = \xi(1-s)$. It is real for $t = 0$ and $\sigma = 1/2$ and $\xi(0) = \xi(1) = 1/2$.

1.3 Properties of Hardy's function

We continue with the discussion of $Z(t)$. Recall that the zeta, sine and the gamma-function take conjugate values at conjugate points. Hence it follows from (1.7) and (1.9) that

$$\overline{\chi(\tfrac{1}{2}+it)} = \chi(\tfrac{1}{2}-it) = \chi^{-1}(\tfrac{1}{2}+it),$$

so that (1.4) gives $Z(t) \in \mathbb{R}$ when $t \in \mathbb{R}$, and $|Z(t)| = |\zeta(\frac{1}{2}+it)|$. Thus the zeros of $\zeta(s)$ on the "critical line" $\mathrm{Re}\, s = 1/2$ are in one-to-one correspondence with the real zeros of $Z(t)$. This property makes $Z(t)$ an invaluable tool in the study of the zeros of the zeta-function on the critical line. If we use (1.5) and (1.6) we have

$$\left(\chi(\tfrac{1}{2}+it)\right)^{-1/2} = \pi^{-it/2}\frac{\Gamma^{1/2}(\frac{1}{4}+\frac{1}{2}it)}{\Gamma^{1/2}(\frac{1}{4}-\frac{1}{2}it)} = \pi^{-it/2}\frac{\Gamma(\frac{1}{4}+\frac{1}{2}it)}{|\Gamma(\frac{1}{4}+\frac{1}{2}it)|} := e^{i\theta(t)},$$

say, where $\theta(t)$ is a smooth function for which

$$\theta(t) = -\frac{1}{2i}\log\chi(\tfrac{1}{2}+it), \qquad \theta'(t) = -\frac{1}{2}\frac{\chi'(\frac{1}{2}+it)}{\chi(\frac{1}{2}+it)}.$$

Note that also

$$\theta(t) = \mathbb{Im}\,\left\{\log\Gamma(\tfrac14 + \tfrac12 it)\right\} - \tfrac12 t\log\pi \;\in \mathbb{R}\mathrm{e} \tag{1.19}$$

if $t \in \mathbb{R}$, thus $\theta(0) = 0$. The function $\theta(t)$ is odd, since in view of $\chi(s)\chi(1-s) = 1$ we have

$$\begin{aligned}\theta(-t) &= -\frac{1}{2i}\log\chi(\tfrac12 - it) = -\frac{1}{2i}\log\frac{1}{\chi(\tfrac12 + it)}\\ &= \frac{1}{2i}\log\chi(\tfrac12 + it) = -\theta(t).\end{aligned}$$

It is also monotonic increasing for $t \ge 7$, which follows from formulas (1.21)-(1.22) below. We may write $Z(t)$ alternatively as

$$Z(t) = \mathrm{e}^{i\theta(t)}\zeta(\tfrac12 + it), \quad \mathrm{e}^{i\theta(t)} := \pi^{-it/2}\frac{\Gamma(\tfrac14 + \tfrac12 it)}{|\Gamma(\tfrac14 + \tfrac12 it)|} \quad (\theta(t) \in \mathbb{R}). \tag{1.20}$$

It is also useful to note that $Z(t)$ is an even function of t, because

$$\begin{aligned}Z(-t) &= \zeta(\tfrac12 - it)\big(\chi(\tfrac12 - it)\big)^{-1/2} = \zeta(\tfrac12 + it)\big(\chi(\tfrac12 - it)\big)^{1/2}\\ &= \zeta(\tfrac12 + it)\big(\chi(\tfrac12 + it)\big)^{-1/2} = Z(t).\end{aligned}$$

We have the explicit representation

$$\theta(t) = \frac{t}{2}\log\frac{t}{2\pi} - \frac{t}{2} - \frac{\pi}{8} + \Delta(t). \tag{1.21}$$

Here (see Lemma 5.1 for a proof; here $\Delta(t)$ is not to be confused with the error term in the Dirichlet divisor problem)

$$\Delta(t) := \frac{t}{4}\log\left(1 + \frac{1}{4t^2}\right) + \frac14\arctan\frac{1}{2t} + \frac{t}{2}\int_0^\infty \frac{\psi(u)}{(u+\tfrac14)^2 + (\tfrac{t}{2})^2}\,\mathrm{d}u \tag{1.22}$$

with

$$\psi(x) = x - [x] - \tfrac12 = -\sum_{n=1}^\infty \frac{\sin(2n\pi x)}{n\pi} \qquad (x \notin \mathbb{Z}).$$

The representation (1.21)-(1.22) follows from *Stirling's formula* (see (2.14)) for the gamma-function in the form

$$\log\Gamma(s) = (s - 1/2)\log s - s + \log\sqrt{2\pi} - \int_0^\infty \frac{\psi(u)}{u+s}\,\mathrm{d}u,$$

which in turn is a consequence of the product formula

$$\frac{1}{\Gamma(s)} = s\exp(\gamma s)\prod_{n=1}^{\infty}\left(1+\frac{s}{n}\right)\mathrm{e}^{-s/n}. \tag{1.23}$$

Note that (1.23) valid for $s \in \mathbb{C}$, and can serve as a definition of $\Gamma(s)$ equivalent to (1.2).

The expression (1.21) is very useful, since it allows one to evaluate explicitly all the derivatives of $\theta(t)$. For $t \to \infty$ it is seen that $\Delta(t)$ admits an asymptotic expansion in terms of negative powers of t, and from (1.19) and Stirling's formula it is found that (B_k is the kth Bernoulli number)

$$\Delta(t) \sim \sum_{n=1}^{\infty}\frac{(2^{2n}-1)|B_{2n}|}{2^{2n}(2n-1)2nt^{2n-1}}. \tag{1.24}$$

The meaning of $\sim$ in (1.24) is that, for an arbitrary integer $N \ge 1$, $\Delta(t)$ equals the sum of the first N terms of the series in (1.24), plus the error term, which is $O_N(t^{-2N-1})$. In general we shall have, for $k \ge 0$ and suitable constants $c_{k,n}$,

$$\Delta^{(k)}(t) \sim \sum_{n=1}^{\infty} c_{k,n}t^{1-2n-k}. \tag{1.25}$$

Thus (1.21) and (1.24) give

$$\theta(t) \sim \frac{t}{2}\log\frac{t}{2\pi} - \frac{t}{2} - \frac{\pi}{8} + \sum_{n=1}^{\infty}\frac{(2^{2n-1}-1)|B_{2n}|}{2^{2n}(2n-1)2nt^{2n-1}}, \tag{1.26}$$

and we also have asymptotic expansions for the derivatives of $\theta(t)$. In particular, we have the approximations

$$\begin{aligned}\theta(t) &= \frac{t}{2}\log\frac{t}{2\pi} - \frac{t}{2} - \frac{\pi}{8} + \frac{1}{48t} + \frac{7}{5760t^3} + O\left(\frac{1}{t^5}\right),\\ \theta'(t) &= \frac{1}{2}\log\frac{t}{2\pi} + O\left(\frac{1}{t^2}\right),\\ \theta''(t) &= \frac{1}{2t} + O\left(\frac{1}{t^3}\right),\end{aligned} \tag{1.27}$$

which are sufficiently sharp for many applications.

1.4 The distribution of zeta-zeros

In what concerns the distribution of zeros of $\zeta(s)$, it is known that $\zeta(s)$ has no zeros in the region

$$\sigma \ge 1 - C(\log t)^{-2/3}(\log\log t)^{-1/3} \qquad (C > 0,\ t \ge t_0 > 0). \tag{1.28}$$

This result, the strongest so-called *zero-free region* for $\zeta(s)$ even today, was obtained by an application of I. M. Vinogradov's method of exponential sums. In a modern form, the crucial bound which implies (1.28) states (see, e.g., [Iv1, chapter 6]) that

$$\sum_{N<n\le N_1\le 2N} n^{it} \ll N \exp\Big(-\frac{C\log^3 N}{\log^2 t}\Big) \qquad (C>0) \tag{1.29}$$

for $N_0 \le N \le \frac{1}{2}t,\ t \ge t_0$. From (1.5) it follows that $\zeta(-2n) = 0$ for $n \in \mathbb{N}$. These zeros are the only real zeros of $\zeta(s)$, and are called the *trivial zeros* of $\zeta(s)$. In 1859, B. Riemann [Rie] calculated a few complex zeros of $\zeta(s)$ and found that they lie on the line $\mathbb{R}\mathrm{e}\, s = \frac{1}{2}$, which is called the *critical line* in the theory of $\zeta(s)$. The first ten pairs of complex zeros (arranged in size according to their absolute value) are (see, e.g., C. B. Haselgrove [Has])

$$\frac{1}{2} \pm i14.134\,725\ldots,\quad \frac{1}{2} \pm i21.022\,039\ldots,\quad \frac{1}{2} \pm i25.010\,857\ldots,$$
$$\frac{1}{2} \pm i30.424\,876\ldots,\quad \frac{1}{2} \pm i32.935\,061\ldots,\quad \frac{1}{2} \pm i37.586\,178\ldots,$$
$$\frac{1}{2} \pm i40.918\,719\ldots,\quad \frac{1}{2} \pm i43.327\,073\ldots,\quad \frac{1}{2} \pm i48.005\,150\ldots,$$
$$\frac{1}{2} \pm i49.773\,832\ldots.$$

The number of complex zeros $\rho = \beta + i\gamma$ of $\zeta(s)$ with $0 < \gamma \le T$ (multiplicities included) is denoted by $N(T)$. The asymptotic formula for $N(T)$ is the famous *Riemann-von Mangoldt formula*. It was enunciated by B. Riemann [Rie] in 1859, but proved by H. von Mangoldt [Man] in 1895. We state it here as follows.

Theorem 1.6 *Let*

$$S(T) := \frac{1}{\pi}\arg\zeta(\tfrac{1}{2} + iT). \tag{1.30}$$

Then

$$N(T) = \frac{T}{2\pi}\log\Big(\frac{T}{2\pi}\Big) - \frac{T}{2\pi} + \frac{7}{8} + S(T) + O\Big(\frac{1}{T}\Big), \tag{1.31}$$

where the O-term is a continuous function of T, and

$$S(T) = O(\log T). \tag{1.32}$$

Here $\arg\zeta(\frac{1}{2} + iT)$ is evaluated by continuous variation starting from $\arg\zeta(2) = 0$ and proceeding along straight lines, first up to $2 + iT$ and then to

$1/2+iT$, assuming that T is not an ordinate of a zeta zero. If T is an ordinate of a zero, then we set $S(T)=S(T+0)$.

Remark 1.7 On the RH (the Riemann hypothesis, that all complex zeros of $\zeta(s)$ have real parts equal to 1/2) one can slightly improve (1.32) and obtain that (see [Tit3])

$$S(T) = O\left(\frac{\log T}{\log\log T}\right). \tag{1.33}$$

Proof of Theorem 1.6 Let $\mathcal{D}$ be the rectangle with vertices $2\pm iT, -1\pm iT$, where T (>3) is not an ordinate of a zero. The function $\xi(s)$, defined by (1.18), has $2N(T)$ zeros in the interior of $\mathcal{D}$, and none on the boundary. Therefore we have

$$N(T) = \frac{1}{4\pi}\mathbb{I}\mathrm{m}\left(\int_{\mathcal{D}}\frac{\xi'(s)}{\xi(s)}\,\mathrm{d}s\right). \tag{1.34}$$

Logarithmic differentiation of (1.18) gives

$$\frac{\xi'(s)}{\xi(s)} = \frac{1}{s}+\frac{1}{s-1}+\frac{\eta'(s)}{\eta(s)},$$

where $\eta(s)$ is also given by (1.18). Observe first that

$$\mathbb{I}\mathrm{m}\left\{\int_{\mathcal{D}}\left(\frac{1}{s}+\frac{1}{s-1}\right)\mathrm{d}s\right\} = 4\pi.$$

Next, note that $\eta(s)=\eta(1-s)$ and $\eta(\sigma\pm it)$ are conjugates, so that

$$\int_{\mathcal{D}}\left(\frac{\eta'(s)}{\eta(s)}\,\mathrm{d}s\right) = 4\mathbb{I}\mathrm{m}\int_{\mathcal{L}}\left(\frac{\eta'(s)}{\eta(s)}\,\mathrm{d}s\right),$$

where $\mathcal{L}$ consists of the segments $[2,\ 2+iT]$ and $[2+iT,\ 1/2+iT]$. Therefore

$$\begin{aligned}\mathbb{I}\mathrm{m}\int_{\mathcal{L}}\left(\frac{\eta'(s)}{\eta(s)}\,\mathrm{d}s\right) &= \mathbb{I}\mathrm{m}\left\{\int_{\mathcal{L}}\left(-\frac{1}{2}\log\pi+\frac{1}{2}\frac{\Gamma'(s/2)}{\Gamma(s/2)}+\frac{\zeta'(s)}{\zeta(s)}\right)\mathrm{d}s\right\}\\ &= -\frac{1}{2}(\log\pi)T+\mathbb{I}\mathrm{m}\left(\int_{\mathcal{L}}\frac{\Gamma'(s/2)}{2\Gamma(s/2)}\,\mathrm{d}s+\int_{\mathcal{L}}\frac{\zeta'(s)}{\zeta(s)}\,\mathrm{d}s\right).\end{aligned}$$

Note that

$$\mathbb{I}\mathrm{m}\left(\int_{\mathcal{L}}\frac{\Gamma'(s/2)}{2\Gamma(s/2)}\,\mathrm{d}s\right) = \mathbb{I}\mathrm{m}\,\log\Gamma(\tfrac{1}{4}+\tfrac{1}{2}iT), \tag{1.35}$$

and using Stirling's formula in the form (2.16) we have

$$\mathbb{I}\mathrm{m}\left(\int_{\mathcal{L}}\frac{\Gamma'(s/2)}{2\Gamma(s/2)}\,\mathrm{d}s\right) = \frac{1}{2}T\log\left(\frac{T}{2}\right)-\frac{T}{2}-\frac{\pi}{8}+O\left(\frac{1}{T}\right),$$

and to prove Theorem 1.6 it remains to show that

$$\mathbb{I}\mathrm{m}\left\{\int_{1/2+iT}^{2+iT}\frac{\zeta'(s)}{\zeta(s)}\,\mathrm{d}s\right\} = O(\log T), \tag{1.36}$$

since the integral over the other segment of $\mathcal{L}$ is clearly bounded. To prove (1.36) we shall need some estimates involving ζ'/ζ. We use the expression

$$\frac{\zeta'(s)}{\zeta(s)} = B - \frac{1}{s-1} + \tfrac{1}{2}\log\pi - \frac{\Gamma'(s/2+1)}{2\Gamma(s/2+1)} + \sum_\rho\left(\frac{1}{s-\rho}+\frac{1}{\rho}\right). \tag{1.37}$$

This formula, where ρ denotes complex (non-trivial) zeros of $\zeta(s)$, and $B = \log 2 + \frac{1}{2}\log\pi - 1 - \frac{1}{2}\gamma$, follows by logarithmic differentiation of the product formula

$$f(s) = \mathrm{e}^{A+Bs}\prod_{n=1}^{\infty}(1 - s/\rho_n)\mathrm{e}^{s/\rho_n} \tag{1.38}$$

for suitable constants A, B if $f(s)$ is an integral function of order 1 with zeros $\rho_1, \rho_2, \ldots$. Taking in (1.38) $f(s) = \xi(s)$, which is an integral function of order 1, we obtain (1.37).

Now we suppose that $t \ge 2$, $1 \le \sigma \le 2$. The gamma-term in (1.37) is $\ll \log t$ and consequently

$$-\,\mathbb{R}\mathrm{e}\,\frac{\zeta'(s)}{\zeta(s)} < C\log t - \sum_\rho \mathbb{R}\mathrm{e}\left(\frac{1}{s-\rho}+\frac{1}{\rho}\right) \qquad (1 \le \sigma \le 2,\ t \ge 2). \tag{1.39}$$

In (1.39) we take $s = 2 + iT$. Since $\frac{\zeta'}{\zeta}(2+iT) \ll 1$, we obtain

$$\sum_\rho \mathbb{R}\mathrm{e}\left(\frac{1}{s-\rho}+\frac{1}{\rho}\right) < C\log T. \tag{1.40}$$

If $\rho = \beta + i\gamma$ is a non-trivial zero of $\zeta(s)$, then

$$\mathbb{R}\mathrm{e}\,\frac{1}{\rho} = \frac{\beta}{\beta^2+\gamma^2} > 0$$

and

$$\mathbb{R}\mathrm{e}\,\frac{1}{s-\rho} = \frac{2-\beta}{(2-\beta)^2+(T-\gamma)^2} \ge \frac{1}{4+(T-\gamma)^2},$$

hence (1.40) gives

$$\sum_{\rho} \frac{1}{1+(T-\gamma)^2} \ll \log T, \tag{1.41}$$

where summation is over all non-trivial zeros ρ of $\zeta(s)$. The bound (1.41) immediately gives

$$N(T+1) - N(T) \ll \sum_{\rho} \frac{1}{1+(T-\gamma)^2} \ll \log T, \tag{1.42}$$

that is, each strip $T < t \le T+1$ contains fewer than $C \log T$ zeros of $\zeta(s)$ for some absolute constant $C > 0$. Again using (1.37) with $s = \sigma + it$, $-1 \le \sigma \le 2$ and $2 + it$, where $t > 2$ is not an ordinate of any ρ, and subtracting, we obtain

$$\frac{\zeta'(s)}{\zeta(s)} = \sum_{\rho} \left(\frac{1}{s-\rho} - \frac{1}{2+it-\rho} \right) + O(\log t). \tag{1.43}$$

In (1.43) for the terms with $|\gamma - t| \ge 1$ we have

$$\begin{aligned} \left|(s-\rho)^{-1} - (2+it-\rho)^{-1}\right| &= (2-\sigma)\left|(s-\rho)(2+it-\rho)\right|^{-1} \\ &\le 3\left|\gamma - t\right|^{-2} \end{aligned}$$

if $-1 \le \sigma \le 2$. Hence by (1.37) the portion of the sum in (1.43) for which $|\gamma - t| \ge 1$ is $\ll \log t$, and for $|\gamma - t| < 1$ we have $|2 + it - \rho| \ge 1$, and the number of such ρ is $\ll \log t$ by (1.42). Thus we obtain

$$\frac{\zeta'(s)}{\zeta(s)} = \sum_{\rho, |\gamma - t| < 1} \frac{1}{s-\rho} + O(\log t) \qquad (-1 \le \sigma \le 2). \tag{1.44}$$

Now the proof of (1.36) easily follows, since by (1.44) (here Δ denotes the variation of the argument)

$$\begin{aligned} \mathbb{Im}\left(\int_{1/2+iT}^{2+iT} \frac{\zeta'(s)}{\zeta(s)} \, \mathrm{d}s \right) &= \mathbb{Im}\left(\int_{1/2+iT}^{2+iT} \sum_{\rho, |\gamma - t| < 1} \frac{1}{s-\rho} \, \mathrm{d}s \right) + O(\log T) \\ &= \sum_{\rho, |\gamma - t| < 1} \Delta \arg(s-\rho) + O(\log T) = O(\log T), \end{aligned}$$

since $|\Delta \arg(s - \rho)| < \pi$ on $[1/2 + iT,\ 2 + iT]$ and (1.42) holds. This completes the proof of Theorem 1.6. Note, however, that by following the preceding proof we obtain, in view of (1.35), the following corollary.

Corollary 1.8

$$N(T) = \frac{1}{\pi}\theta(T) + 1 + S(T). \tag{1.45}$$

Formula (1.45) shows the important connection between the functions $N(T)$, $\theta(T)$ and $S(T)$. Recall that $\theta(T)$ is a very smooth function, so that the jumps of $S(T)$ come at the zeros of $\zeta(s)$, which is of course also evident from its definition (1.30). Since (see (1.19)) $\theta(t)$ plays a fundamental rôle in the theory of $Z(t)$, Corollary 1.8 shows an intrinsic connection between $N(T)$, $S(T)$ and $Z(T)$. This is also evident from the relation

$$\log\zeta(\tfrac{1}{2} + it) = \log|Z(t)| + \pi i S(t).$$

Corollary 1.9 *We have*

$$\sum_{|\gamma|\le T}\frac{1}{|\gamma|} \ll \log^2 T, \quad \sum_{|\gamma|>T}\frac{1}{\gamma^2} \ll \frac{\log T}{T}, \quad \gamma_n \sim \frac{2\pi n}{\log n}\ (n\to\infty). \tag{1.46}$$

In (1.46) we denote by $0 < \gamma_1 \le \gamma_2 \le \cdots$ the consecutive ordinates of non-trivial zeros $\rho = \beta + i\gamma$ of $\zeta(s)$. They should not be confused with the Stieltjes constants, defined before (1.2) as the Laurent coefficients (Pierre Alphonse Laurent, July 18, 1813-September 2, 1854, French mathematician) of $\zeta(s)$ at $s = 1$. Both bounds in (1.46) follow by partial summation from (1.30) and (1.31), as well as the asymptotic formula for γ_n in view of the obvious inequality

$$N(\gamma_n - 1) < n \le N(\gamma_n + 1).$$

Finally a few words about the Riemann hypothesis (RH for short), still unsettled at the time of the writing of this text. In his celebrated work [Rie] B. Riemann conjectured that all complex zeros of the zeta-function lie on the critical line. This statement is called the Riemann hypothesis, and is probably the most famous open problem in Mathematics.

The RH implies (see, e.g., [Iv1], [Tit3]) that

$$\zeta(\tfrac{1}{2} + it) \ll \exp\left(\frac{C\log t}{\log\log t}\right) \qquad (C > 0). \tag{1.47}$$

A slightly weaker bound than (1.47), which in practice can often replace the RH, is the bound

$$\zeta(\tfrac{1}{2} + it) \ll_\varepsilon (|t| + 1)^\varepsilon, \tag{1.48}$$

which is known as the *Lindelöf hypothesis* (LH for short). It is also unproved, and it is not known whether (1.48) implies the RH, although this is not very likely. Namely the LH is equivalent (see, [Tit3, theorem 13.5]), to the statement that, for every $\sigma > 1/2$,

$$N(\sigma, T+1) - N(\sigma, T) = o(\log T) \qquad (T \to \infty),$$

where $N(\sigma, T)$ denotes the number of complex zeros $\rho = \beta + i\gamma$ for which $\beta \ge \sigma$, $|\gamma| \le T$. For the best unconditional bounds for $\zeta(\frac{1}{2} + it)$ and related topics, see the survey paper of M. N. Huxley and the present author [HuIv].

Hardy's original application of $Z(t)$ was to show that $\zeta(s)$ has infinitely many zeros on the critical line $\mathbb{R}\mathrm{e}\, s = 1/2$ (see, e.g., E. C. Titchmarsh [Tit3]). This will be discussed in the next chapter. Later A. Selberg (see [Sel] and [Tit3]) obtained that a positive proportion of zeros of $\zeta(s)$ lies on the critical line. This can be stated as

$$N_0(T) \ \ge\ CN(T) \qquad (C > 0, T \ge T_0), \tag{1.49}$$

where $N_0(T)$ denotes the number of zeros of $Z(t)$ in $(0, T]$, or equivalently the number of complex zeros of $\zeta(s)$ on the critical line $\mathbb{R}\mathrm{e}\, s = 1/2$ whose imaginary parts lie in $(0, T]$. Selberg's bound (1.49) improves on $N_0(T) \ge CT$, which is a result of G. H. Hardy and J. E. Littlewood [HaLi2] of 1921. Selberg's bound is one of the most important results of analytic number theory of all time, as it reveals the true order of magnitude of $N_0(T)$. Later work by various scholars led to explicit values of C in (1.49).

Notes

It was actually the great Swiss mathematician Leonhard Euler (April 15, 1707-September 18, 1783) who first used the zeta-function, albeit only for real values of the variable. Besides (1.1), Euler discovered several other identities from zeta-function theory. His paper [Eul] from 1768 contains the assertion

$$\frac{1 - 2^{n-1} + 3^{n-1} - 4^{n-1} + 5^{n-1} - \cdots}{1 - 2^{-n} + 3^{-n} - 4^{-n} + 5^{-n} - \cdots} = \frac{-1 \times 2 \times 3 \times \cdots (n-1)(2^{n-1} - 1)}{(2^{n-1} - 1)\pi^n} \cos(\tfrac{1}{2}\pi n),$$

which he verified for $n = 1$ and $n = 2k$. E. Landau [Lan] wrote Euler's identity as

$$\frac{\lim\limits_{x\to 1} \sum\limits_{n=1}^{\infty} (-1)^{n+1} n^{s-1} x^{n-1}}{\lim\limits_{x\to 1} \sum\limits_{n=1}^{\infty} (-1)^{n+1} n^{-s} x^{n-1}} = -\frac{\Gamma(s)(2^s - 1)}{(2^{s-1} - 1)\pi^s)} \cos(\tfrac{1}{2}\pi s),$$

proved its validity, and showed its equivalence with the functional equation (1.5).

There are many ways to obtain the analytic continuation of $\zeta(s)$ outside the region $\sigma > 1$. One simple way (see T. Estermann [Est2]) is to write, for $\sigma > 1$,

$$\zeta(s) = \sum_{n=1}^{\infty} n^{-s} = \sum_{n=1}^{\infty} \left(n^{-s} - \int_n^{n+1} u^{-s}\, du \right) + \frac{1}{s-1}$$

and to observe that

$$\left| n^{-s} - \int_n^{n+1} u^{-s}\, du \right| = \left| s \int_n^{n+1} \int_u^n z^{-s-1}\, dz\, du \right| \le |s| n^{-\sigma-1}.$$

Hence the second series above converges absolutely for $\sigma > 0$, and by the principle of analytic continuation we obtain

$$\zeta(s) = \sum_{n=1}^{\infty} \left(n^{-s} - \int_n^{n+1} u^{-s}\, du \right) + \frac{1}{s-1} \qquad (\sigma > 0),$$

showing incidentally that $\zeta(s)$ is regular for $\sigma > 0$, except for a simple pole at $s = 1$ with residue equal to 1.

Analytic continuation of $\zeta(s)$ for $\sigma > 0$ is also given by

$$\zeta(s) = (1 - 2^{1-s})^{-1} \sum_{n=1}^{\infty} (-1)^n n^{-s} \tag{1.50}$$

since the series in (1.50) converges for $\sigma > 0$.

For $x > 1$ one has

$$\begin{aligned} \sum_{n \le x} n^{-s} &= \int_{1-0}^{x} u^{-s}\, d[u] = [x]x^{-s} + s \int_1^x [u] u^{-s-1}\, du \\ &= O(x^{1-\sigma}) + s \int_1^x ([u] - u) u^{-s-1}\, du + \frac{s}{s-1} - \frac{s x^{1-s}}{s-1}. \end{aligned}$$

If $\sigma > 1$ and $x \to \infty$, it follows that

$$\zeta(s) = \frac{s}{s-1} + s \int_1^{\infty} ([u] - u) u^{-s-1}\, du.$$

By using the customary notation $\psi(x) = x - [x] - 1/2$, this relation can be written as

$$\zeta(s) = \frac{1}{s-1} + \frac{1}{2} - s \int_1^{\infty} \psi(u) u^{-s-1}\, du. \tag{1.51}$$

Since $\int_y^{y+1} \psi(u)\, du = 0$ for any real y, integration by parts shows that (1.51) provides the analytic continuation of $\zeta(s)$ to the half-plane $\sigma > -1$, and in particular it shows that $\zeta(0) = -1/2$. It also follows that the Laurent expansion of $\zeta(s)$ at its pole $s = 1$ has the form

$$\zeta(s) = \frac{1}{s-1} + \gamma_0 + \gamma_1(s-1) + \gamma_2(s-1)^2 + \cdots,$$

where the so-called *Stieltjes constants* γ_k are given by

$$\begin{aligned} \gamma_k &= \frac{(-1)^{k+1}}{k!} \int_{1-0}^{\infty} x^{-1} (\log x)^k\, d\psi(x) \\ &= \frac{(-1)^k}{k!} \lim_{N \to \infty} \left(\sum_{m \le N} \frac{\log^k m}{m} - \frac{\log^{k+1} N}{k+1} \right), \end{aligned} \tag{1.52}$$

as already mentioned in the text. The formula (1.52) was proved first by T. J. Stieltjes [Sti] in 1905 (Thomas Joannes Stieltjes, December 29, 1856-December 31, 1894, Dutch mathematician). On successive integrations by parts of the integral in (1.51) one can obtain the analytic continuation of $\zeta(s)$ to $\mathbb{C}$.

A detailed discussion of the γ_ks is given by I. M. Israilov [Isr1], [Isr2]. The first three values, to five decimal places, are

$$\gamma_1 = 0.07281\ldots, \quad \gamma_2 = -0.00485\ldots, \quad \gamma_3 = -0.00034\ldots.$$

Not much is known about the properties of the Stieltjes constants γ_k. It is widely believed that they are all irrational, but this has not been proved even for $\gamma = \gamma_0$ (Euler's constant).

A number n is squarefree if $n = 1$ or $n = p_1 \ldots p_r$, where $p_1, \ldots, p_r$ are different primes. A number n is squarefull if $n = 1$ or $n = p_1^{\alpha_1} \ldots p_r^{\alpha_r}$, where $\alpha_1 \geq 2, \ldots, \alpha_r \geq 2$. Various generating functions involving $\zeta(s)$ are thoroughly discussed in chapter 1 of [Iv1], and various divisor problems are discussed in chapters 13 and 14.

The equivalence of the definitions (1.2) and (1.23) for $\Gamma(s)$ is standard. For example, K. Ramachandra [Ram] in the appendix starts from (1.2) and derives several properties of $\Gamma(s)$, including (1.23) and Stirling's formula (2.14). On the other hand, Karatsuba–Voronin [KaVo] (also Montgomery–Vaughan [MoVa], appendix C) in the appendix start from (1.23) and show that (1.2) holds. These works contain all the facts about $\Gamma(s)$ needed in this text.

There are many proofs in the literature of the fundamental functional equation (1.5) for $\zeta(s)$. For example, E. C. Titchmarsh [Tit3] in chapter 2 of his well-known monograph presents seven different proofs of (1.5).

A quick proof of (1.13) is as follows (see (A.38) of [Iv1]). By the principle of analytic continuation it suffices to prove (1.13) for B real and positive, when the change of variable

$$t = \frac{A}{2B} + \frac{x}{\sqrt{B}}$$

gives

$$\int_{-\infty}^{\infty} \exp(At - Bt^2)\,\mathrm{d}t = B^{-1/2}\exp(A^2/(4B)) \int_{-\infty}^{\infty} \mathrm{e}^{-x^2}\,\mathrm{d}x = (\pi/B)^{1/2}\exp(A^2/(4B)).$$

Recall that the Bernoulli numbers B_k are defined by the series expansion

$$\frac{z}{\mathrm{e}^z - 1} = \sum_{k=0}^{\infty} B_k \frac{z^k}{k!} \qquad (|z| < 2\pi),$$

so that $B_0 = 1$, $B_1 = -1/2$, $B_2 = 1/6$, $B_4 = -1/30$, $B_6 = 1/42$ etc., and $B_{2k+1} = 0$ for $k \geq 1$. A classical formula (see [Iv1], theorem 1.4 for a proof) is that

$$\zeta(2k) = \frac{(-1)^{k+1}(2\pi)^{2k} B_{2k}}{2(2k)!} \qquad (k \in \mathbb{N}),$$

so that in particular

$$\zeta(2) = \sum_{n=1}^{\infty} n^{-2} = \frac{\pi^2}{6}, \quad \zeta(4) = \sum_{n=1}^{\infty} n^{-4} = \frac{\pi^4}{90}.$$

Not much is known about the numbers $\zeta(2n+1)$, $n \in \mathbb{N}$. R. Apéry [Ape] (Roger Apéry, November 14, 1916-December 18, 1994, a Greek-French mathematician) proved in 1978 that $\zeta(3)$ is irrational. This result has incited much subsequent research. For example, T. Rivoal [Riv1] proved that infinitely many of the numbers $\zeta(2n+1)$ are irrational. In [Riv2] he proved that one of the nine numbers $\zeta(2n+1)$ for $2 \leq n \leq 10$ is irrational. Rivoal's method was generalized by V. Zudilin [Zud], who obtained a number of irrationality results for $\zeta(2n+1)$.

Hardy's function $Z(t)$ was named after Godfrey Harold "G. H." Hardy FRS (February 7, 1877-December 1, 1947), one of the greatest mathematicians of his time. He is best known for his numerous achievements in number theory and mathematical analysis, often obtained in joint works with J. E. Littlewood (June 9, 1885-September 6, 1977). Starting in 1914, he was the mentor of the famous Indian mathematician Srinivasa Ramanujan (December 22, 1887-April 26, 1920), a relationship that has become celebrated. He was a lecturer at Cambridge from 1906, which he left in 1919 to take the Savilian Chair of Geometry at Oxford. He returned to Cambridge in 1931, where he was Sadleirian Professor until 1942. Besides writing numerous research articles of highest quality (these may be found in [Har7]), he wrote several well-known books, such as [Har4], [Har5] and [Har6]. His essay from 1940 on the aesthetics of mathematics, *A Mathematician's Apology* [Har3], is often considered as one of the best insights into the mind of a working mathematician written for the layman. His textbook *An Introduction to the Theory of Numbers* [HaWr], written jointly with E. M. Wright (Sir Edward Maitland Wright, February 13, 1906-February 2, 2005), is one of the best introductory texts on number theory ever written.

Stirling's formula for $\Gamma(s)$ exists in many forms. See (2.14) for a sharp version of this result. It is named after the Scottish mathematician James Stirling (May 1692-December 5, 1770).

The functional equation of $\zeta(s)$ in a certain sense characterizes it completely. This was established long ago by H. Hamburger [Ham] (Hans Ludwig Hamburger, August 5, 1889, Berlin-August 14, 1956, a German mathematician), who proved the following result (see also chapter 2 of [Cha]). Let G be an integral function of finite order, P a polynomial, and let the series

$$f(s) = G(s)/P(s) = \sum_{n=1}^{\infty} a_n n^{-s}$$

converge absolutely for $\sigma > 1$. If

$$\pi^{-s/2}\Gamma(s/2)f(s) = \pi^{-(1-s)/2}\Gamma((1-s)/2)g(s),$$

where $g(1-s) = \sum_{n=1}^{\infty} b_n n^{-s}$ converges absolutely for $\sigma < -\alpha < 0$, then

$$f(s) = a_1\zeta(s) = g(s).$$

Ivan Matveevich Vinogradov (September 14, 1891-March 20, 1983) was a leading Soviet mathematician. He served as director of the Steklov Mathematical Institute for 49 years. For his method of exponential sums see [Vin1] and [Vin2]. For the sharpest value of the constant C in (1.29) and related bounds, see the papers of K. Ford [For1], [For2].

An extensive account on the RH is to be found in the monograph of P. Borwein *et al.* [BCRW]. There is much numerical evidence favoring the RH, and for the calculations involving the zeros of $\zeta(s)$ see e.g., the works of A. M. Odlyzko [Odl1], [Odl2], H. te Riele and J. van de Lune [RiLu], and van de Lune *et al* [LRW]. An excellent account of the RH and LH is to be found in the classical book of E. C. Titchmarsh [Tit3]. Edward Charles "Ted" Titchmarsh (June 1, 1899-January 18, 1963 Oxford) was a leading British mathematician. He was a student of G. H. Hardy, and is known for work in analytic number theory, Fourier analysis and other parts of mathematical analysis. He was Savilian Professor of Geometry at the University of Oxford from 1932 to 1963.

A function $f(s)$, regular over $\mathbb{C}$, is called an *integral* (or *entire*) function of finite order if

$$|f(s)| \ll \exp\Big(B|s|^A\Big)$$

for some constants $A, B\ (\geq 0)$ as $|s| \to \infty$. The order of $f(s)$ is the lower bound of A for which the above bound holds. The study of integral functions of finite order was developed at the end of the nineteenth century by J. Hadamard, who showed that these functions can be written as an infinite product containing factors of the form $s - s_0$ corresponding to the zero s_0 of the function in question. The integral function often used in the theory of $\zeta(s)$ is $\xi(s)$, which is an integral function of order one (see (1.18)). To see this, note that from (1.51) we have

$$\log|(1-s)\zeta(s)| \ll \log|s| + 1$$

uniformly for $\sigma \geq \frac{1}{2}$, and by Stirling's formula it follows that

$$\log |\xi(s)| \ll |s|(\log |s| + 1)$$

uniformly for $\sigma \geq \frac{1}{2}$. Since $\xi(1-s) = \xi(s)$, this bounds holds uniformly for $\sigma \leq \frac{1}{2}$, too. Further, by Stirling's formula for real s we have $\log \xi(s) \sim \frac{1}{2} s \log s$ as $s \to \infty$, which shows that the order of $\xi(s)$ is exactly unity.

The $\ll$-constant in (1.42) was made explicit by G. Csordas *et al.* [COSV], who proved that

$$N(T+1) - N(T) \leq \log T \qquad (T \geq 3 \cdot 10^8),$$

which is an inequality that is easy to remember.

For an explicit value of C in (1.47), see K. Soundararajan [Sou4]. He proved that one can take $C = 3/8$. A sharper result, also on the RH, was obtained by V. Chandee and K. Soundararajan [ChSo], namely

$$|\zeta(\tfrac{1}{2} + it)| \ll \exp\left\{ \frac{\log 2}{2} \frac{\log t}{\log\log t} \left(1 + O\left(\frac{\log\log\log t}{\log\log t}\right)\right)\right\}.$$

On the other hand, it is known that there are arbitrarily large values of t for which one has unconditionally

$$|\zeta(\tfrac{1}{2} + it)| > \exp\left\{(1 + o(1))\sqrt{\frac{\log t}{\log\log t}}\right\} \qquad (t \to \infty), \tag{1.53}$$

as shown by K. Soundararajan [Sou3].

The function $S(T)$ is relatively small, and one can make the bounds in (1.32) and (1.33) explicit. Namely it was proved by T. S. Trudgian [Tru3] that, for $T > \mathrm{e}$,

$$|S(T)| \leq 1.998 + 0.17 \log T.$$

This is an unconditional result. On the RH, K. Ramachandra and A. Sankaranarayanan [RaSa2] showed that

$$|S(T)| \leq 1.2 \frac{\log T}{\log\log T} \qquad (T > T_0).$$

This was improved, again on the RH, by D. A. Goldston and S. M. Gonek [GoGo]. They proved that, for $0 < h \leq \sqrt{t}$,

$$|S(t+h) - S(t)| \leq (\tfrac{1}{2} + o(1)) \frac{\log T}{\log\log T} \qquad (T \to \infty),$$

and deduced, via $\int_0^T S(t)\,\mathrm{d}t \ll \log T$ (this is an unconditional result of J. E. Littlewood [Lit]), that under the RH,

$$|S(T)| \leq \left(\tfrac{1}{2} + o(1)\right) \frac{\log T}{\log\log T} \qquad (T \to \infty), \tag{1.54}$$

and a bound analogous to (1.54) for $m(\frac{1}{2} + i\gamma)$, the multiplicity of the zero $\frac{1}{2} + i\gamma$ ($\gamma > 0$), also under the RH. Related results are to be found in the works of M. A. Korolev [Kor1], [Kor2] and Karatsuba-Korolev [KaKo]. For the moments of $S(t)$ one has the classical unconditional result of A. Selberg [Sel] (Atle Selberg, June 14, 1917-August 6, 2007, Norwegian mathematician known for his work in analytic number theory, and in the theory of automorphic forms, in particular bringing them into relation with spectral theory) that, for fixed $k \in \mathbb{N}$,

$$\int_0^T |S(t)|^{2k}\,\mathrm{d}t = \frac{(2k)!}{k!(2\pi)^{2k}} T(\log\log T)^k + O(T(\log\log T)^{k-1/2}). \tag{1.55}$$

This suggests that $S(t)/\sqrt{\log\log t}$ resembles a Gaussian random variable with mean 0 and variance $2\pi^2$. A. Ghosh [Gho] established this in 1983.

Several omega results involving the functions $S(T)$ and $S_1(T) = \int_0^T S(t)\,dt$ are proved by K.-M. Tsang [Tsa1]. As is well-known (see Chapter IX of [Tit3]), these functions are closely related to the distribution of the imaginary parts of the zeros of $\zeta(s)$. A. Selberg [Sel] has proved that unconditionally one has

$$S(T) = \Omega_\pm\left((\log T)^{1/3}(\log\log T)^{-7/3}\right),$$

$$S_1(T) = \Omega_+\left((\log T)^{1/2}(\log\log T)^{-4}\right), \quad S_1(T) = \Omega_-\left((\log T)^{1/3}(\log\log T)^{-10/3}\right).$$

Tsang refined Selberg's arguments, based on the evaluation of high moments of $S(T)$ (see (11.31)), and proved that

$$S(T) = \Omega_\pm\left((\log T/\log\log T)^{1/3}\right), \tag{1.56}$$

$$S_1(T) = \Omega_+\left((\log T)^{1/2}(\log\log T)^{-9/4}\right), \quad S_1(T) = \Omega_-\left((\log T)^{1/3}(\log\log T)^{-4/3}\right),$$

and if one assumes the Riemann hypothesis, then

$$S_1(T) = \Omega_\pm\left((\log T)^{1/2}(\log\log T)^{-3/2}\right).$$

It may be conjectured that both $S(T)$ and $S_1(T)$ are of the order $(\log T)^{1/2+o(1)}$ as $T \to \infty$, although it is known, for example, only that $S(T) = O(\log T)$ (and $O(\log T/\log\log T)$ if the RH holds), so that there is still a considerable gap between O- and Ω-results.

K.-M. Tsang [Tsa2] has shown that unconditionally

$$\left(\sup_{T<t\le 2T}\log|\zeta(\tfrac12+it)|\right)\left(\sup_{T<t\le 2T}\pm S(t)\right) \gg \frac{\log T}{\log\log T},$$

which means that the result holds once with $+S(t)$ and once with $-S(t)$. He also proved that

$$S_1(t) = \int_0^t S(u)\,du = \Omega_+\left((\log t)^{1/2}(\log\log t)^{-3/2}\right).$$

These results supplement that of H. L. Montgomery [Mon2]; they are good when $\sigma > 1/2$ is fixed. Montgomery showed that, for $1/2 < \sigma < 1$ fixed, one has

$$\log|\zeta(\sigma+it)| = \Omega_+\left(\log^{1-\sigma} t(\log\log t)^{-\sigma}\right), \tag{1.57}$$

and under the RH that

$$\log|\zeta(\tfrac12+it)| = \Omega\left\{\frac{1}{20}\sqrt{\frac{\log t}{\log\log t}}\right\}.$$

K. Ramachandra and A. Sankaranarayanan [RaSa1] used Montgomery's method, made optimal use of the parameter α in his proof and obtained a result which is the perhaps the limit of Montgomery's method, containing explicit evaluation of the constant implied by the Ω-symbol in (1.57). As to the true order of $\zeta(\frac12+it)$, it is a deep open question, and one can only make guesses. For example, D. W. Farmer *et al.* [FGH] conjecture that, as $T \to \infty$,

$$\max_{t\in[0,T]}|Z(t)| = \max_{t\in[0,T]}|\zeta(\tfrac12+it)| = \exp\left((1+o(1))\sqrt{\frac12\log T\log\log T}\right)$$

and

$$\limsup_{t\to\infty} \frac{S(t)}{\sqrt{\log t \log\log t}} = \frac{1}{\pi\sqrt{2}}.$$

Their arguments are based on methods involving random matrix theory (see, for example, the books of G. W. Anderson *et al.* [AGZ] and M.L. Mehta [Meh] for an account on random matrices). In analytic number theory, the distribution of zeros of the Riemann zeta-function (and other L-functions) is often modeled by the distribution of eigenvalues of certain random matrices (see J. Keating [Kea]). The connection was first discovered by Hugh Montgomery (see [Mon1]) and Freeman J. Dyson. It is connected to the *Hilbert-Pólya conjecture* (David Hilbert, January 23, 1862-February 14, 1943, a German mathematician, and George Pólya, December 13, 1887-September 7, 1985, a Hungarian mathematician) that the imaginary parts of the zeros of $\zeta(s)$ correspond to the eigenvalues of an unbounded self-adjoint operator.

For a thorough discussion on upper bounds for $N(\sigma, T)$, see [Iv1, chapter 11].

Selberg's method for detecting zeros on the critical line is expounded in many texts, such as [Tit3], [Iv1] and [KaVo]. After Selberg's pioneering work, it was later used and refined by several mathematicians, most notably by N. Levinson [Lev] (Norman Levinson, August 11, 1912-October 10, 1975, American mathematician, made major contributions in the study of Fourier transforms, complex analysis, non-linear differential equations, number theory, and signal processing). Levinson showed that more than a third of zeros lie on the critical line (i.e. $C = 1/3$ in (1.49)). A simplification of Levinson's method is to be found in the work [CoGh2] of J. B. Conrey and A. Ghosh, and in the recent work [You] of M. P. Young. The research was carried on by other mathematicians and, for example, J. B. Conrey [Con2], showed that at least two fifth of the zeros of $\zeta(s)$ are on the critical line. In [CGG3], Conrey *et al.* investigated the occurrence of simple zeros on the critical line. The starting point is the obvious observation that ρ is a simple zero of $\zeta(s)$ if and only if $\zeta'(\rho) \neq 0$, and thus by Cauchy's inequality

$$\left| \sum_{0<\gamma\le T} B(\rho)\zeta'(\rho) \right|^2 \le N_s(T) \sum_{0<\gamma\le T} \left|B(\rho)\zeta'(\rho)\right|^2,$$

where $N_s(T)$ is the number of simple zeros $\rho = \beta + i\gamma$ of $\zeta(s)$ with ordinates in $(0, T]$, and $B(s)$ is any analytic function. To minimize the loss in applying Cauchy's inequality we want to choose B so that $B\zeta'$ is close to constant, and therefore B is chosen as a suitable *Dirichlet series mollifer* for $\zeta'(s)$. The choice of the mollifier is incidentally the most complicated task in this problem. Assuming the RH, they show that $N_s(T)/N(T) \ge 19/27 = 0.703\,703\ldots$ as $T \to \infty$.

The latest development concerning Selberg's method involve the works of Bui *et al.* [BCY], and by S. Feng [Fen]. The former proved that at least 41.05% of the zeros of $\zeta(s)$ are on the critical line (i.e. one can take $C = 0.4105$ in (1.49)) and at least 40.58% of the zeros of $\zeta(s)$ are simple and on the critical line. At about the same time Feng independently claimed that at least 41.73% of the zeros of $\zeta(s)$ are on the critical line and at least 40.75% of them are simple and on the critical line. Bui *et al.* found some mistakes in Feng's work and in the second version of his paper in the ArXiv (http://arxiv.org/abs/1003.0059), he removed the result on simple zeros, and now claimed only that $C = 0.4128$. Anyway, as it stands at the moment, Feng seems to be the record holder for the proportion of zeros, while Bui *et al.* are the ones for simple zeros.

So far it is not known whether the simplicity of zeros ($\zeta'(\rho) \neq 0$ if $\zeta(\rho) = 0$) and the RH imply one another – it could happen, as far as we know, that both statements are true, both are false, or that one is true and one is false!

2

The zeros on the critical line

2.1 The infinity of zeros on the critical line

The distribution of zeros of $\zeta(s)$ on the critical line $\mathbb{R}\mathrm{e}\, s = 1/2$ is one of the central topics in zeta-function theory. The aim of this section is to prove that $\zeta(s)$ has an infinity of zeros on the critical line. This assertion is equivalent to showing that $Z(t)$ has arbitrarily large zeros in absolute value and, since $Z(t)$ is even, we may consider only positive zeros. Suppose on the contrary that, for $T \geq T_0$, the function $Z(t)$ does not change sign. Then

$$\int_T^{2T} |Z(t)|\,\mathrm{d}t \;=\; \left|\int_T^{2T} Z(t)\,\mathrm{d}t\right|. \tag{2.1}$$

But it is not difficult to show that

$$\int_T^{2T} |Z(t)|\,\mathrm{d}t \;=\; \int_T^{2T} |\zeta(\tfrac{1}{2}+it)|\,\mathrm{d}t \;\gg\; T. \tag{2.2}$$

On the other hand, to bound the integral on the right-hand side of (2.1) we can use the approximate functional equation

$$Z(t) = 2\sum_{n\leq\sqrt{t/(2\pi)}} n^{-1/2}\cos\left(t\log\frac{\sqrt{t/(2\pi)}}{n} - \frac{t}{2} - \frac{\pi}{8}\right) + O\left(t^{-1/4}\right). \tag{2.3}$$

This is a weakened form of the so-called Riemann-Siegel formula, and for its proof see Chapter 4 (also [Iv1] or [Tit3]).

If the expression in (2.3) is integrated over $[T,\ 2T]$ the error term produces $O(T^{3/4})$. The integral of the sum in (2.3) is

$$2\mathbb{R}\mathrm{e}\left\{\sum_{n\le\sqrt{T/\pi}} n^{-1/2}\int_{\max(T,2\pi n^2)}^{2T} \exp\Big(iF(t,n)\Big)\,\mathrm{d}t\right\}$$

with

$$F(t,n) \ :=\ t\log\frac{\sqrt{t/(2\pi)}}{n}-\frac{t}{2}-\frac{\pi}{8},$$

so that

$$F'(t,n)=\frac{\mathrm{d}F(t,n)}{\mathrm{d}t}=\log\frac{\sqrt{t/(2\pi)}}{n},\qquad F''(t,n)=\frac{1}{2t}.$$

If we now apply the second derivative test (see Lemma 2.3 below), we infer that our integral is

$$\ll \sum_{n\le\sqrt{T/\pi}} n^{-1/2}\cdot T^{1/2} \ \ll\ T^{3/4}.$$

It follows that

$$\int_T^{2T} Z(t)\,\mathrm{d}t \ =\ O(T^{3/4}). \tag{2.4}$$

Thus from (2.1), (2.2) and (2.4) we obtain

$$T \ \ll\ \int_T^{2T} |Z(t)|\,\mathrm{d}t \ \ll\ T^{3/4},$$

which is a contradiction. In fact, the argument actually shows that there is always a zeros of $Z(t)$ in $[T, T+CT^{3/4}]$ for suitable $C>0$. Therefore we have proved the following.

Theorem 2.1 *There are* $\gg T^{1/4}$ *zeros of* $\zeta(s)$ *of the form* $\frac{1}{2}+it$, *where* $t\in[0,T]$.

We conclude this section by giving as lemmas the first and the second derivative tests. These are standard and well-known results, and therefore the proofs are not given. In the next section we shall, however, also formulate and prove a general result (easily generalized to much more general Dirichlet series), which implies the bound in (2.2).

Lemma 2.2 *Let* $F(x)$ *be a real, differentiable function such that* $F'(x)$ *is monotonic and* $|F'(x)|\ge m\,(>0)$ *for* $a\le x\le b$. *If* $G(x)$ *is a positive, monotonic* *function for $a \le x \le b$ such that $|G(x)| \le G$, then*

$$\int_a^b G(x)\,\mathrm{e}^{iF(x)}\,\mathrm{d}x \ \le\ 4Gm^{-1}. \tag{2.5}$$

Lemma 2.3 *Let $F(x)$ be a real, twice-differentiable function for $a \le x \le b$ such that $|F''(x)| \ge m\,(>0)$. If $G(x)$ is as in Lemma 2.2, then*

$$\int_a^b G(x)\,\mathrm{e}^{iF(x)}\,\mathrm{d}x \ \le\ 8Gm^{-1/2}. \tag{2.6}$$

2.2 A lower bound for the mean values

There are many mean value results for the lower bound of powers of $\zeta(s)$, some of which are easily generalized to more general L-functions. We now present one such result, which is of independent interest. This is the following.

Theorem 2.4 *If $k \ge$ is a fixed integer, $\sigma \ge \frac{1}{2}$ is fixed,*

$$12\log\log T \le H \le T, \qquad T \ge T_0 > 0,$$

then uniformly in σ

$$\int_{T-H}^{T+H} |\zeta(\sigma+it)|^k\,\mathrm{d}t \ \gg\ H. \tag{2.7}$$

Proof of Theorem 2.4 Let $\sigma_1 = \sigma+2$, $s_1 = \sigma_1 + it$, $T - \frac{1}{2}H \le t \le T + \frac{1}{2}H$. Then $\zeta(s_1) \gg 1$ and therefore

$$\int_{T-\frac{1}{2}H}^{T+\frac{1}{2}H} |\zeta(\sigma_1+it)|^k\,\mathrm{d}t \ \gg\ H. \tag{2.8}$$

Now let $\mathcal{E}$ be the rectangle with vertices $\sigma + iT \pm iH, \sigma_2 + iT \pm iH$ ($\sigma_2 = \sigma + 3$) and let X be a parameter that satisfies

$$T^{-c} \le X \le T^{c} \tag{2.9}$$

for some constant $c > 0$. The residue theorem gives then

$$\frac{1}{\mathrm{e}}\zeta^k(s_1) = \frac{1}{2\pi i}\int_{\mathcal{E}} \frac{\zeta^k(w)}{w-s_1}\exp\left(-\cos\Big(\frac{w-s_1}{3}\Big)\right) X^{s_1-w}\,\mathrm{d}w.$$

On $\mathcal{E}$ we have $|\mathbb{R}\mathrm{e}\,\frac{1}{3}(w-s_1)| \le 1$, and on its horizontal sides

$$\left|\mathbb{I}\mathrm{m}\left(\frac{w-s_1}{3}\right)\right| \ge \frac{1}{3}\cdot\frac{H}{2} \ge 2\log\log T.$$

Note that for $w = u + iv$ $(u, v \in \mathbb{R})$ we have

$$|\exp(-\cos w)| = \left|\exp\left(-\tfrac{1}{2}(e^{iw} + e^{-iw})\right)\right|$$
$$= \left|\exp\left(-\tfrac{1}{2}(e^{iu}e^{-v} + e^{-iu}e^{v})\right)\right| = \exp(-\cos u \cdot \cosh v).$$

The above function $\exp(-\cos w)$, which may justly be called *Ramachandra's kernel function*, sets the limit to the lower bound for H (a multiple of $\log\log T$) in Theorem 2.4. Actually, according to W. K. Hayman, there does not exist a regular function of a similar type which decreases faster than a double-order exponential. Therefore Ramachandra's kernel function is optimal. This does not preclude that the lower bound in Theorem 2.4 for H holds for some function smaller than a multiple of $\log\log T$, but such a result (if true) certainly cannot be obtained by a variant of the above method.

To continue with the proof of the theorem, note that, if w lies on the horizontal sides of $\mathcal{E}$, we have

$$\left|\exp\left(-\cos\left(\frac{w - s_1}{3}\right)\right)\right|$$
$$\le \exp\left(-\tfrac{1}{2}\cos 1 \exp(2\log\log T)\right) = \exp\left(-\tfrac{1}{2}\cos 1(\log T)^2\right).$$

Therefore the condition (2.9) ensures that, for suitable $C, c_1 > 0$,

$$\zeta^k(\sigma_1 + it) \ll X^2 \int_{T-H}^{T+H} |\zeta(\sigma + iv)|^k \exp\left(-c_1 e^{|v-t|/3}\right) dv$$
$$+ X^{-1} \int_{T-H}^{T+H} \exp\left(-c_1 e^{|v-t|/3}\right) dv + e^{-C\log^2 T}.$$

Integrating this estimate over $t \in [T - \frac{1}{2}H, T + \frac{1}{2}H]$ and using (2.8) we obtain

$$H \ll X^2 \int_{T-H}^{T+H} |\zeta(\sigma + iv)|^k dv \left(\int_{T-\frac{1}{2}H}^{T+\frac{1}{2}H} \exp\left(-c_1 e^{|v-t|/3}\right) dt\right)$$
$$+ X^{-1} \int_{T-H}^{T+H} dv \left(\int_{T-\frac{1}{2}H}^{T+\frac{1}{2}H} \exp\left(-c_1 e^{|v-t|/3}\right) dt\right)$$
$$\ll X^2 \int_{T-H}^{T+H} |\zeta(\sigma + iv)|^k dv + X^{-1}H. \tag{2.10}$$

Now let

$$I := \int_{T-H}^{T+H} |\zeta(\sigma + iv)|^k dv,$$

and choose $X = H^{\varepsilon}$. Then (2.10) gives $I \gg H^{1-2\varepsilon}$, showing that I cannot be too small. Then we choose $X = H^{1/3}I^{-1/3}$, so that (since $\zeta(\frac{1}{2}+it) \ll |t|^{1/6}$) trivially

$$T^{-k/18} \ll X \ll H \ll T,$$

and (2.9) is satisfied. With this choice of X (2.10) reduces to $H \ll H^{2/3}I^{1/3}$, and (2.7) follows. Slightly sharper results than (2.7), involving powers of $\log\log T$, are known. They are extensively discussed, for example, by K. Ramachandra in [Ram].

Remark 2.5 If one wants only to prove the lower bound in (2.2), this can be done very simply. Observe that

$$\int_T^{2T} |\zeta(\tfrac{1}{2}+it)|\,\mathrm{d}t \geq \left|\int_T^{2T} \zeta(\tfrac{1}{2}+it)\,\mathrm{d}t\right| = \left|\int_{1/2+iT}^{1/2+2iT} \zeta(s)\,\mathrm{d}s\right|.$$

But on using the trivial, uniform bound

$$\zeta(\sigma+it) \ll |t|^{1/4} \qquad (\tfrac{1}{2} \leq \sigma \leq 2,\ |t| \geq 2),$$

we see that the last integral equals

$$\begin{aligned}
&\left(\int_{1/2+iT}^{2+iT} + \int_{2+iT}^{2+2iT} + \int_{2+2iT}^{1/2+2iT}\right)\zeta(s)\,\mathrm{d}s \\
&= \left[s - \sum_{n=2}^{\infty}\frac{1}{n^s\log n}\right]_{s=2+iT}^{2+2iT} + \int_{1/2}^{2} O(t^{1/4})\,\mathrm{d}\sigma \\
&= iT + O(T^{1/4}).
\end{aligned}$$

This yields the lower bound in (2.2).

2.3 Lehmer's phenomenon

Our second aim is to discuss the so-called *Lehmer phenomenon*, which involves Hardy's function and its zeros. Namely $Z(t)$, defined by (1.4), has a negative local maximum $-0.526\,25\ldots$ at $t = 2.475\,75\ldots$. This is the only known occurrence of a negative local maximum, while no positive local minimum is known. *The Lehmer phenomenon* is the fact that the graph of $Z(t)$ sometimes barely crosses the t-axis. This means that the absolute value of the maximum or minimum of $Z(t)$ between its two consecutive zeros is small. For instance, Lehmer's phenomenon shows the delicacy of the RH, and the possibility that a counterexample to the RH may be found numerically. For should it happen that, for $t \ge t_0$, $Z(t)$ attains a negative local maximum or a positive local minimum, then the RH would be disproved. This assertion is a consequence of the following

Theorem 2.6 *If the RH is true, then the graph of $Z'(t)/Z(t)$ is strictly monotonically decreasing between the zeros of $Z(t)$ for $t \ge t_0$.*

Namely, suppose that $Z(t)$ has a negative local maximum or a positive local minimum between its two different consecutive zeros γ_n and γ_{n+1}. Then $Z'(t)$ would have at least two distinct zeros x_1 and x_2 (say $x_1 < x_2$) in (γ_n, γ_{n+1}), and hence so would $Z'(t)/Z(t)$. But the monotonicity property gives

$$\frac{Z'(x_1)}{Z(x_1)} > \frac{Z'(x_2)}{Z(x_2)},$$

which is a contradiction, since $Z'(x_1) = Z'(x_2) = 0$. In other words, under the RH the zeros of $Z(t)$ and $Z'(t)$ are interlacing, that is, for $t \ge t_0$ there is precisely one zero of $Z'(t)$ between two consecutive zeros of $Z(t)$.

Proof of Theorem 2.6 Let us consider the function $\xi(s)$, defined by (1.18). From the product representation

$$\xi(s) = \mathrm{e}^{A+Bs} \prod_\rho \left(1 - \frac{s}{\rho}\right) \mathrm{e}^{s/\rho}$$

one obtains unconditionally, by logarithmic differentiation,

$$\frac{\xi'(s)}{\xi(s)} = B + \sum_\rho \Big(\frac{1}{s-\rho} + \frac{1}{\rho}\Big) \tag{2.11}$$

with

$$B = \log 2 + \frac{1}{2}\log\pi - 1 - \frac{1}{2}C_0.$$

Here $\rho = \beta + i\gamma$ denotes complex zeros of $\zeta(s)$ and $C_0 = -\Gamma'(1)$ is Euler's constant (to distinguish it from $\gamma = \operatorname{Im}\rho$). In Chapter 1 it was shown (see (1.20)) that

$$Z(t) = \chi^{-1/2}(\tfrac{1}{2}+it)\zeta(\tfrac{1}{2}+it) = \frac{\pi^{-it/2}\,\Gamma(\frac{1}{4}+\frac{1}{2}it)\zeta(\frac{1}{2}+it)}{|\Gamma(\frac{1}{4}+\frac{1}{2}it)|},$$

so that in view of (1.20) we may write

$$\xi(\tfrac{1}{2}+it) = -f(t)Z(t), \quad f(t) := \tfrac{1}{2}\pi^{-1/4}(t^2+\tfrac{1}{4})|\Gamma(\tfrac{1}{4}+\tfrac{1}{2}it)|.$$

Consequently logarithmic differentiation gives

$$\frac{Z'(t)}{Z(t)} = -\frac{f'(t)}{f(t)} + i\,\frac{\xi'(\frac{1}{2}+it)}{\xi(\frac{1}{2}+it)}. \tag{2.12}$$

Assume now that the RH is true. Then by using (2.11) with $\rho = \frac{1}{2} + i\gamma$, $s = \frac{1}{2} + it$ we obtain, if $t \neq \gamma$,

$$\left(\frac{i\xi'(\frac{1}{2}+it)}{\xi(\frac{1}{2}+it)}\right)' = -\sum_{\gamma} \frac{1}{(t-\gamma)^2} < -C(\log\log t)^2 \quad (C > 0)$$

for $t \geq t_0$, since on the RH, we have the bound

$$\gamma_{n+1} - \gamma_n \ll \frac{1}{\log\log\gamma_n} \tag{2.13}$$

for the gap between consecutive zeros on the critical line. The bound (2.13) follows from (1.31) and (1.33). Namely, for $H > 0$, $C_1 > 0$ and $T \geq T_0$ we have

$$\begin{aligned} N(T+H) - N(T) &= \left[\frac{t}{2\pi}\log\left(\frac{t}{2\pi}\right) - \frac{t}{2\pi} + \frac{7}{8} + S(t) + O\left(\frac{1}{t}\right)\right]_{t=T}^{T+H} \\ &= \int_T^{T+H} \frac{1}{2\pi}\log\left(\frac{t}{2\pi}\right)\mathrm{d}t + O\left(\frac{\log T}{\log\log T}\right) \\ &\geq \frac{H}{2\pi}\log\left(\frac{T}{2\pi}\right) - C_1\frac{\log T}{\log\log T} > 0, \end{aligned}$$

if $H = C/\log\log T$ with $C > 2\pi C_1$. Thus with this H the interval $[T,\ T+H]$ always contains a γ_n and (2.13) follows.

On the other hand, we shall use the classical *Stirling formula* for the gamma-function. The full form of this assertion states that

$$\begin{aligned} \log\Gamma(s+b) &= (s + b - \tfrac{1}{2})\log s - s + \tfrac{1}{2}\log 2\pi \\ &\quad + \sum_{j=1}^{K} \frac{(-1)^j B_{j+1}(b)}{j(j+1)s^j} + O_\delta\left(\frac{1}{|s|^{K+1}}\right), \end{aligned} \tag{2.14}$$

which is valid for b a constant, any fixed integer $K \geq 1$, $|\arg s| \leq \pi - \delta$ for $\delta > 0$, where the points $s = 0$ and the neighborhoods of the poles of $\Gamma(s+b)$ are excluded, and the $B_j(b)$s are *Bernoulli polynomials*; for this see, for example,

[EMOT]. As a corollary one has, uniformly for $0 \le \sigma \le 1$,

$$\Gamma(s) = \sqrt{2\pi}\,|t|^{\sigma-1/2} \exp\left\{-\tfrac{1}{2}\pi|t| + i\left(t\log|t| - t + \frac{\pi t}{2|t|}\left(\sigma - \frac{1}{2}\right)\right)\right\}$$
$$\times \left(1 + \frac{i}{2t}(\sigma - \sigma^2 - \tfrac{1}{6}) + O(|t|^{-2})\right). \tag{2.15}$$

One also has

$$\frac{\Gamma'(s)}{\Gamma(s)} = \log s - \frac{1}{2s} + O\Big(\frac{1}{|s|^2}\Big), \tag{2.16}$$

where $|\arg s| \le \pi - \delta$, $|s| \ge \delta$ and $\delta > 0$ is fixed, and of course both (2.15) and (2.16) are special cases of a full asymptotic expansion. From Stirling's formula and the symmetric functional equation (1.5) for $\zeta(s)$ one obtains, as first approximations,

$$\chi(s) = \left(\frac{2\pi}{t}\right)^{\sigma+it-1/2} e^{i(t+\pi/4)}\left(1 + O\left(\frac{1}{t}\right)\right),$$
$$\frac{\chi'(s)}{\chi(s)} = -\log\Big(\frac{t}{2\pi}\Big) + O\left(\frac{1}{t}\right). \tag{2.17}$$

To complete the proof, note that since $\log|z| = \Re e \log z$, it is readily found that

$$\frac{d}{dt}\left(\frac{f'(t)}{f(t)}\right) \ll \frac{1}{t},$$

so that from (2.12) it follows that $(Z'(t)/Z(t))' < 0$ if $t \ge t_0$, which implies Theorem 2.6. Actually the value of t_0 may be easily effectively determined and seen not to exceed 1000. Since $Z(t)$ has no positive local minimum or negative local maximum for $3 \le t \le 1000$, it follows that the RH is false if we find (numerically) the occurrence of a single negative local maximum (besides the one at $t = 2.475\,75\ldots$) or a positive local minimum of $Z(t)$. It seems appropriate to quote in concluding H. M. Edwards [Edw], who says that Lehmer's phenomenon "must give pause to even the most convinced believer of the Riemann hypothesis."

2.4 Gaps between consecutive zeros on the critical line

Let $0 < \gamma_1 \le \cdots \le \gamma_n \le \cdots$ denote the zeros of $Z(t)$ for $t > 0$, or equivalently let γ_n denote the imaginary part of the nth zero of $\zeta(s)$ of the form $s = \frac{1}{2} + it, t > 0$. Although all known zeros of $Z(t)$ are simple, the existence of multiple zeros has never been ruled out. Under the RH we have the bound (2.13), and we seek an unconditional bound for $\gamma_{n+1} - \gamma_n$, the gap between consecutive zeros on the critical line. To obtain such a result, we shall contrast the behavior of the integrals

$$I_1 := \int_{T-U}^{T+U} Z(u)\exp\Big(-(T-u)^2U^{-2}L\Big)\,\mathrm{d}u \tag{2.18}$$

and

$$I_2 := \int_{T-U}^{T+U} |Z(u)|\exp\Big(-(T-u)^2U^{-2}L\Big)\,\mathrm{d}u, \tag{2.19}$$

where

$$T^\varepsilon \ll U \ll T^{1/4}, \quad L = (\log T)^{1+\varepsilon}. \tag{2.20}$$

Suppose that $Z(t)$ has no zeros in $[T-U,\ T+U]$. Then

$$|I_1| \;=\; I_2, \tag{2.21}$$

and we shall show that (2.21) is impossible with a suitable choice of U. From (2.19) we have first

$$I_2 \gg \int_{T-UL^{-1/2}}^{T+UL^{-1/2}} |Z(u)|\,\mathrm{d}u \gg UL^{-1/2} \tag{2.22}$$

by Theorem 2.4. Next, by using (2.3) and making the change of variable $u = T + v$ in (2.18) we obtain

$$\begin{aligned} I_1 &= \int_{-U}^{U} Z(T+v)\exp\Big(-v^2U^{-2}L\Big)\,\mathrm{d}v \\ &= 2\sum_{n\le Q(T)} n^{-1/2}\mathbb{R}\mathrm{e}\left\{\int_{-U}^{U}\exp\Big(if(T+v)-v^2U^{-2}L\Big)\,\mathrm{d}v\right\} + O(UT^{-1/4}) \end{aligned}$$

with

$$f(x) = f(x,n) := x\log\left(\frac{n}{Q}\right) + \frac{x}{2} + \frac{\pi}{8}, \qquad Q = Q(x) = (x/(2\pi))^{1/2},$$

so that

$$f'(x) = \log\left(\frac{n}{Q}\right), \qquad f''(x) = -\frac{1}{2x}.$$

Note that here we used the trivial estimate ($Q = Q(T)$)

$$\int_{-U}^{U}\sum_{Q<n\le((T+v)/(2\pi))^{1/2}} n^{-1/2}\exp(\cdots)\,\mathrm{d}v \ll UT^{-1/4}.$$

Next, by Taylor's formula we have

$$f(T+v) = f(T) + v\log\left(\frac{n}{Q}\right) + O(v^2T^{-1}) \qquad (|v| \le U),$$

and consequently

$$I_1 = 2\sum_{n\le Q} n^{-1/2}\mathbb{R}\mathrm{e}\left\{\int_{-U}^{U} \exp\Big(if(T) + iv\log(n/Q) - v^2U^{-2}L\Big)\,\mathrm{d}v\right\}$$
$$+ O(UT^{-1/4}) + O(U^3T^{-3/4}).$$

However, trivial estimation and (1.13) with $A = i\log(n/Q)$, $B = LU^{-2}$ give

$$\int_{-U}^{U} \exp\Big(if(T) + iv\log(n/Q) - v^2U^{-2}L\Big)\,\mathrm{d}v$$
$$= \exp(if(T))\int_{-\infty}^{\infty} \exp\Big(iv\log(n/Q) - v^2U^{-2}L\Big)\,\mathrm{d}v + O(1/T)$$
$$= \sqrt{\frac{\pi}{L}}U\exp(if(T))\exp\left(-\frac{U^2\log^2(n/Q)}{4L}\right) + O(1/T).$$

This yields, since $U \ll T^{1/4}$,

$$I_1 = 2\sqrt{\frac{\pi}{L}}U\mathbb{R}\mathrm{e}\left\{\sum_{n\le Q} n^{-1/2}\exp\left(if(T) - \frac{U^2\log^2(n/Q)}{4L}\right)\right\} + O(UT^{-1/4}). \tag{2.23}$$

Let

$$P = [Q] = \left[\sqrt{\frac{T}{2\pi}}\right], \qquad n = P - m.$$

Consider first those n in (2.23) for which

$$m > QU^{-1}L^{1+\varepsilon} = (2\pi)^{-1/2}T^{1/2}U^{-1}(\log T)^{(1+\varepsilon)^2}.$$

But we have

$$U^2L^{-1}\log^2(n/Q) = U^2L^{-1}\left\{\log(1-(m+O(1))Q^{-1})\right\}^2 \ge \tfrac{1}{2}L^{1+\varepsilon}.$$

This means that the second term in the exponential in (2.23) makes the contribution of these n to I_1 negligible, that is, they are in total certainly $O(UT^{-1/4})$. For the remaining n we obtain by partial summation

$$I_1 \ll UT^{-1/4} + UT^{-1/4}L^{-1/2}\sum_{M}\max_{M'}\Bigg|\sum_{M<m\le M'\le 2M}\exp\Big(iT\log(P-m)\Big)\Bigg|. \tag{2.24}$$

The maximum is taken over M' satisfying $M < M' \le 2M$, and $\sum_M$ denotes summation over $O(\log T)$ values $M = 2^{-j}QU^{-1}L^{1+\varepsilon}$, $j = 1, 2, \ldots$, so that the exponential sum in (2.24) is "short" in the sense that $M = o(P)$ as $T \to \infty$. Therefore the problem reduces to the estimation of the exponential sum

$$S = S(M, M', T) := \sum_{M<m\le M'\le 2M} \exp\bigl(iT\log(P-m)\bigr), \quad P = \left[\sqrt{\frac{T}{2\pi}}\right], \tag{2.25}$$

where $M \ll T^{1/2}U^{-1}L^{1+\varepsilon}$. We shall take advantage of the special structure of the exponential sum in (2.25), as m stays not close to P. Namely from the definition of P we have $T = 2\pi(P+\theta)^2$ for some $0 \le \theta < 1$, and thus

$$\begin{aligned} T\log(P-m) - T\log P &= -T\sum_{m=1}^{\infty}(m/P)^j j^{-1} \\ &= -2\pi Pm - 2\pi(2P\theta+\theta^2)mP^{-1} - \pi m^2 \\ &\quad -2\pi(2P\theta+\theta^2)m^2(2P^2)^{-1} \\ &\quad -T\Bigl\{m^3/(3P^3) + m^4/(4P^4) + \cdots\Bigr\}. \end{aligned}$$

Taking into account that $\exp(2\pi ir) = 1$ for $r \in \mathbb{Z}$, and considering separately even and odd m (to get rid of πm^2) we obtain

$$|S| \le |S'| + |S''|,$$

where S' comes from even m and equals

$$S' := \sum_{M_1<m\le M_1'\le 2M_1} \exp(2\pi i F(m)) \qquad (M \ll M_1 \ll M), \tag{2.26}$$

with

$$F(x) := c_1x + c_2x^2 + \frac{T}{2\pi}\Bigl\{(2x)^3/(3P^3) + (2x)^4/(4P^4) + \cdots\Bigr\}$$

and

$$c_1 = 2(2\theta P + \theta^2)P^{-1} = O(1), \qquad c_2 = c_1P^{-1} = O(P^{-1}).$$

The expression for S'' (coming from odd m) is similar to S', and thus it will be sufficient to estimate only S' in (2.26). For the relevant range when $M \gg T^{1/4}$ and $M_1 \le x \le 2M_1$ we have $|F'(x)| \gg 1$ and

$$\begin{aligned} |F^{(k)}(x)| &\asymp M^{3-k}T^{-1/2} \qquad (k = 1, 2, 3), \\ |F^{(k)}(x)| &\asymp T^{1-k/2} \ll M^{3-k}T^{-1/2} \qquad (k > 3), \end{aligned}$$

where the $\asymp$ constants depend only on k. The first result follows on applying (6.13) of Lemma 6.6 with $\lambda_2 = MT^{-1/2}$, on noting that the values of M for which $M \le T^\varepsilon$ make a negligible contribution. The remaining values of M make a contribution to I_1 which is

$$\begin{aligned}&\ll UT^{-1/4}L\sum_{M>T^\varepsilon}\left(M^{3/2}T^{-1/4}+T^{1/4}M^{-1/2}\right)\\&\ll UT^{-1/4}L(T^{3/4}U^{-3/2}T^{-1/4}+T^{-1/4}T^{-\varepsilon/2}).\end{aligned}$$

Then from (2.21) and (2.22) we get a contradiction if $U = CT^{1/6}L^{3/2}$. This implies that

$$\gamma_{n+1}-\gamma_n \ll_\varepsilon \gamma_n^{1/6}(\log\gamma_n)^{3/2+\varepsilon}, \tag{2.27}$$

but by more sophisticated techniques better bounds than (2.27) can be obtained. Namely we may estimate S' and S'' by the theory of *exponent pairs* (see [Iv1, chapter 2] or [GrKo]) as

$$S' \ll A^\kappa M^\lambda, \tag{2.28}$$

where (κ, λ) is an exponent pair and

$$A = \max_{M_1\le M\le 2M_1}|F'(x)| \ll M^2T^{-1/2}. \tag{2.29}$$

Thus for $M \gg T^{1/4}$ we use (2.28) and (2.29), while for $M \ll T^{1/4}$ we use lemma 2.6 of [Iv1], taking

$$a = M_1,\quad b = M_1',\quad f(x) = T\log(P-x),\quad \lambda_3 = T^{-1/2}$$

to estimate S'. Then we obtain

$$S \ll M^{2\kappa+\lambda}T^{-\kappa/2} + T^{5/24}. \tag{2.30}$$

Inserting this bound in (2.24) and summing over various M we obtain

$$I_1 \ll_\varepsilon UT^{-1/4} + UT^{-1/24}\log T + UT^{(\kappa+\lambda)/2-1/4}U^{-(2\kappa+\lambda)}(\log T)^{2\kappa+\lambda-1/2+\varepsilon}. \tag{2.31}$$

Then again from (2.21) and (2.22) we get

$$UL^{-1/2} \ll_\varepsilon UT^{-1/24}\log T + UT^{(\kappa+\lambda)/2-1/4}U^{-(2\kappa+\lambda)}(\log T)^{2\kappa+\lambda-1/2+\varepsilon},$$

which is a contradiction if for a suitable $\delta = \delta(\varepsilon)$ (which tends to 0 as $\varepsilon \to 0$)

$$U = T^{(2\kappa+2\lambda-1)/(8\kappa+4\lambda)}(\log T)^{1+\delta}. \tag{2.32}$$

Now we use lemma 2.8 and lemma 2.9 of [Iv1] to deduce that, if (κ, λ) is an exponent pair, then

$$\left(\frac{\lambda}{2\kappa+2}, \frac{2\kappa+1}{\kappa+2}\right)$$

is also an exponent pair. Replacing (κ, λ) in (2.32) by this last exponent pair we obtain the condition for U in the form

$$U = T^{(\kappa+\lambda)/(4\kappa+4\lambda+2)}(\log T)^{1+\delta}. \tag{2.33}$$

Thus for the choice of U in (2.33) the equality $|I_1| = I_2$ is impossible, which means that we have proved the following.

Theorem 2.7 *If (κ, λ) is an exponent pair, then we have*

$$\gamma_{n+1} - \gamma_n \ll_\varepsilon \gamma_n^{(\kappa+\lambda)/(4\kappa+4\lambda+2)}(\log \gamma_n)^{1+\varepsilon}. \tag{2.34}$$

Corollary 2.8 *Taking R. A. Rankin's exponent pair (see [Ran])*

$$(\kappa, \lambda) = (\tfrac{1}{2}\alpha + \varepsilon, \tfrac{1}{2} + \tfrac{1}{2}\alpha + \varepsilon), \qquad \alpha = 0.329\,0213\,56\ldots$$

one obtains

$$\gamma_{n+1} - \gamma_n \ll \gamma_n^{\beta}, \quad \beta = 0.155\,945\,83\ldots + \varepsilon. \tag{2.35}$$

Theorem 2.7 deals with the size of individual gaps between the γ_ns. One can also address this problem statistically, namely consider the number of gaps of size at least V (>0) for $\gamma_n \le T$. To this end let us define

$$R = R(T, V) = \sum_{0<\gamma_n\le T, \gamma_{n+1}-\gamma_n\ge V} 1.$$

Then we have the following.

Theorem 2.9 *We have*

$$R \ll TV^{-2}\min(\log T, V^{-1}\log^5 T). \tag{2.36}$$

Proof of Theorem 2.9 Let

$$500\log T \le V \le T^{1/6},\ L = 10\sqrt{\log T},\ G = VL^{-1},\ \tfrac{1}{2}T \le t \le T, \tag{2.37}$$

and

$$I_1(t) := \int_{t-\frac{1}{4}V}^{t+\frac{1}{4}V} |Z(u)| \exp\big(-(t-u)^2G^{-2}\big)\,\mathrm{d}u,$$

$$I_2(t) := \int_{t-\frac{1}{4}V}^{t+\frac{1}{4}V} Z(u) \exp\big(-(t-u)^2G^{-2}\big)\,\mathrm{d}u. \tag{2.38}$$

The upper bound for V comes from the bound in Theorem 2.7. The lower bound for V may be assumed, since if it does not hold, then the trivial bound $R \ll T/V$ improves on (2.36). The starting point of the proof is the fact that $I_1(t) = |I_2(t)|$ if $Z(u)$ has no zeros for $t - \frac{1}{4}V \le u \le t + \frac{1}{4}V$. We now need the following.

Lemma 2.10 *Let the above hypotheses hold and let $Q(t) = \sqrt{t/(2\pi)}$, $X = T^{1/2}V^{-1}L^2$. Then*

$$I_1(t) \gg G \tag{2.39}$$

and

$$I_2(t) \ll \int_{-\frac{1}{4}V}^{\frac{1}{4}V} \Bigg| \sum_{Q(t+u)-X \le n \le Q(t+u)} n^{-1/2-it-iu} \Bigg| \mathrm{e}^{-u^2G^{-2}}\, \mathrm{d}u + 1. \tag{2.40}$$

Proof of Lemma 2.10 This is similar to the reasoning employed in the proof of Theorem 2.7. We have

$$I_1(t) \gg \int_{t-V/L}^{t+V/L} |Z(u)|\, \mathrm{d}u = \int_{t-G}^{t+G} |\zeta(\tfrac{1}{2} + it)|\, \mathrm{d}u$$
$$\gg G(\log G)^{1/4} \gg G(\log V)^{1/4}.$$

This bound is even slightly stronger than what is asserted by (2.39), and would improve the bounds in (2.36) by the factors $(\log V)^{-1/2}$ and $(\log V)^{-1}$, respectively, in (2.36). Here we used a bound of K. Ramachandra's [Ram] (with $k = 1$) for the integral of $|\zeta(\frac{1}{2} + it)|$ in short intervals. To obtain (2.40), note that for $|u| \le \frac{1}{4}V \ll T^{1/6}$ we have, by (2.3),

$$Z(t+u) = 2\Re\mathrm{e}\left\{ \sum_{n \le Q(t+u)} n^{-1/2} \exp\left(i(t+u)\log\frac{Q(t+u)}{n} - i\frac{(t+u)}{2} - \frac{\pi i}{8}\right)\right\} + O(T^{-1/4}).$$

We simplify the expression in the exponential by Taylor's formula, noting that

$$(t+u)\log Q(t+u) - \tfrac{1}{2}u = (t+u)Q(t) + O(u^2T^{-1}).$$

Hence

$$I_2(t) = \int_{-\frac{1}{4}V}^{\frac{1}{4}V} Z(t+u)\mathrm{e}^{-u^2G^{-2}}\, \mathrm{d}u = 2\Re\mathrm{e}\, I_2' + O(GT^{-1/4}), \tag{2.41}$$

where

$$I_2' := \int_{-\frac{1}{4}V}^{\frac{1}{4}V} \sum_{n\le Q(t+u)} n^{-1/2} \exp\left(i(t+u)\log\frac{Q(t)}{n} - i\frac{t}{2} - \frac{\pi i}{8} - u^2G^{-2}\right) du.$$

Further, observing that $Q(t+u) - Q(t) \ll 1$ for $|u| \le \frac{1}{4}V$, we have

$$I_2' \ll \int_{-\frac{1}{4}V}^{\frac{1}{4}V} \Big| \sum_{Q(t+u)-X\le n\le Q(t+u)} n^{-1/2-it-iu} \Big| e^{-u^2G^{-2}}\, du$$
$$+ \sum_{n<Q(t)-X} n^{-1/2} \Big| \int_{-\frac{1}{4}V}^{\frac{1}{4}V} \exp\Big(iu\log\frac{Q(t)}{n} - u^2G^{-2}\Big) du \Big| + 1, \quad (2.42)$$

and

$$X \le \frac{1}{5}T^{1/2} \le \frac{2\sqrt{\pi}}{5} Q(t).$$

Using (1.13) and the fact that, for $n < Q(t) - X$,

$$\frac{1}{4}G^2\left(\log\frac{Q(t)}{n}\right)^2 \ge \frac{1}{4}G^2\left(\frac{X}{Q(t)}\right)^2 \ge \frac{\pi L^2}{4} > 50\log T$$

we obtain

$$\begin{aligned}
&\int_{-\frac{1}{4}V}^{\frac{1}{4}V} \exp\Big(iu\log\frac{Q(t)}{n} - u^2G^{-2}\Big) du \\
&\quad = \int_{-\infty}^{\infty} \exp\Big(iu\log\frac{Q(t)}{n} - u^2G^{-2}\Big) du + O(T^{-6}) \\
&\quad = \sqrt{\pi}G \exp\left(\frac{1}{4}G^2\Big(\log\frac{Q(t)}{n}\Big)^2\right) + O(T^{-6}) \\
&\quad \ll T^{-6}.
\end{aligned}$$

Therefore the sum in (2.42) is $\ll 1$, and (2.40) follows from (2.41) and (2.42).

We pass now to the proof of the first bound in (2.36), namely

$$R \ll TV^{-2}\log T.$$

Clearly it suffices to consider only those gaps $\gamma_{n+1} - \gamma_n$ which lie in $[\frac{1}{2}T, T]$. Suppose that τ and $\tau + U$ are two consecutive zeros of $Z(t)$ such that $[\tau, \tau+U] \subseteq [\frac{1}{2}T, T]$, $U \ge V$ and (2.37) holds. Then if $t \in [\tau + \frac{1}{4}V, \tau + \frac{3}{4}V]$, we have $I_1(t) = |I_2(t)|$ in the notation of (2.38). Consequently Lemma 2.10 gives

$$1 \ll G^{-1} \int_{t-\frac{1}{4}V}^{t+\frac{1}{4}V} \Big| \sum_{Q(u)-X\le n\le Q(u)} n^{-1/2-iu} \Big| e^{-(t-u)^2G^{-2}}\, du.$$

Hence for any fixed $k \in \mathbb{N}$ we obtain by Hölder's inequality

$$1 \ll G^{-1} \int_{t-\frac{1}{4}V}^{t+\frac{1}{4}V} \Big| \sum_{Q(u)-X \le n \le Q(u)} n^{-1/2-iu} \Big|^{2k} \mathrm{e}^{-(t-u)^2 G^{-2}} \, \mathrm{d}u, \tag{2.43}$$

since the change of variable $u = t + Gx$ gives

$$\int_{t-\frac{1}{4}V}^{t+\frac{1}{4}V} \mathrm{e}^{-(t-u)^2 G^{-2}} \, \mathrm{d}u = G \int_{-V/(4G)}^{V/(4G)} \mathrm{e}^{-x^2} \, \mathrm{d}x \le G \int_{-\infty}^{\infty} \mathrm{e}^{-x^2} \, \mathrm{d}x = \sqrt{\pi} G. \tag{2.44}$$

By hypothesis, the set of numbers $t \in [\frac{1}{2}T + \frac{1}{4}V, T - \frac{1}{4}V]$ for which (2.43) holds has measure $\gg RV$. We denote this set by E, integrate (2.43) over E, and use (2.44). We obtain then

$$\begin{aligned} RV &\ll G^{-1} \int_{t \in E} \int_{t-\frac{1}{4}V}^{t+\frac{1}{4}V} \Big| \sum_{Q(u)-X \le n \le Q(u)} n^{-1/2-iu} \Big|^{2k} \mathrm{e}^{-(t-u)^2 G^{-2}} \, \mathrm{d}u \, \mathrm{d}t \\ &\le G^{-1} \int_{T/2}^{T} \Big| \sum \Big|^{2k} \left(\int_{u-\frac{1}{4}V}^{u+\frac{1}{4}V} \mathrm{e}^{-(t-u)^2 G^{-2}} \, \mathrm{d}t \right) \mathrm{d}u \ll \int_{T/2}^{T} \Big| \sum \Big|^{2k} \mathrm{d}u, \end{aligned}$$

hence

$$R \ll V^{-1} \int_{T/2}^{T} \Big| \sum_{Q(t)-X \le n \le Q(t)} n^{-1/2-it} \Big|^{2k} \mathrm{d}t \qquad (k \in \mathbb{N}). \tag{2.45}$$

This is the key estimate for the proof of Theorem 2.9. In (2.45) we take first $k = 1$. Developing the square we obtain

$$\begin{aligned} R &\ll V^{-1} \int_{T/2}^{T} \sum_{Q(t)-X \le n \le Q(t)} n^{-1} \, \mathrm{d}t \\ &\quad + V^{-1} \int_{T/2}^{T} \sum_{Q(t)-X \le m,n \le Q(t); m \ne n} (mn)^{-1/2} (m/n)^{it} \, \mathrm{d}t \\ &= \sum\nolimits_1 + \sum\nolimits_2, \end{aligned} \tag{2.46}$$

say. Trivially we have, since $X = T^{1/2} V^{-1} L^2$,

$$\sum\nolimits_1 \ll V^{-1} T^{1/2} \max_{t \in [T/2, T]} \Big(\sum_{Q(t)-X \le n \le Q(t)} 1 \Big) \ll TV^{-2} L^2 \ll TV^{-2} \log T.$$

To estimate $\sum_2$ we change the order of summation and integration. For a given pair m, n such that $|m - n| \le X$ and $m \asymp T^{1/2}$ the variable of integration t

runs over a certain subinterval of $[T/2, T]$. Hence integration gives

$$\begin{aligned}\sum\nolimits_2 &\ll V^{-1}T^{-1/2} \sum_{m \asymp T^{1/2}, 0<m-n\le X} \left(\log \frac{m}{n}\right)^{-1} \\ &\ll V^{-1} \sum_{m \asymp T^{1/2}} \sum_{0<m-n\le X} \frac{1}{m-n} \ll T^{1/2}V^{-1}\log T,\end{aligned}$$

and our bound follows from (2.46) and the bounds for $\sum_1$ and $\sum_2$.

To prove the other bound in (2.36), namely

$$R \ll TV^{-3}\log^5 T, \tag{2.47}$$

we need another lemma, which is elementary, but somewhat involved.

Lemma 2.11 *Let $0 < \Delta < 1$ and $M \ge M_0$ be given, and suppose that natural numbers m_1, n_1, m_2, n_2 satisfy*

$$\begin{aligned}&\frac{1}{100}M^{1/2} \le m_1 \le M^{1/2},\ |n_1 - m_1| \le M^{1/2}\Delta,\\ &|n_2 - m_1| \le M^{1/2}\Delta,\ |m_2 - m_1| \le M^{1/2}\Delta.\end{aligned}$$

If $N(M, \Delta; r)$ denotes the number of quadruples (m_1, n_1, m_2, n_2) such that $m_1n_1 - m_2n_2 = r$ for a given $r \in \mathbb{N}$, then for any fixed $\varepsilon > 0$ we have uniformly in r

$$N(M, \Delta; r) \ll_\varepsilon M\Delta^2\log^2 M + M^{1/2+\varepsilon}. \tag{2.48}$$

Moreover, if $K(M, \Delta)$ denotes the number of quadruples (m_1, n_1, m_2, n_2) such that $m_1n_1 - m_2n_2 = 0$ with

$$|M^{1/2} - m_j| \le M^{1/2}\Delta, \quad |M^{1/2} - n_j| \le M^{1/2}\Delta \qquad (j = 1, 2),$$

then

$$K(M, \Delta) \ll M\Delta^2\log^2 M. \tag{2.49}$$

Proof of Lemma 2.11 Consider first (2.48) and set

$$n_1 = m_1 + \lambda, \quad m_2 = m_1 + \mu, \quad n_2 = m_1 - \mu - \nu$$

for some integers λ, μ, ν. Then there exists a one-to-one correspondence between (m_1, n_1, m_2, n_2) and (m_1, λ, μ, ν), and moreover $|\lambda| \le M^{1/2}\Delta$, $|\mu| \le M^{1/2}\Delta$, $|\nu| \le 2M^{1/2}\Delta$. Hence the equation $m_1n_1 - m_2n_2 = r$ becomes

$$\mu^2 + \mu\nu + m_1(\lambda + \nu) = r, \tag{2.50}$$

whence $1 \le r \le 4M\Delta$. If $\lambda + \nu = 0$ in (2.50), then $\mu(\mu + \nu) = r$ has $\ll d(r) \ll M^\varepsilon$ pairs μ, ν as solutions. Since $\lambda = -\nu$ and $m_1 \le M^{1/2}$, this means that (2.50) has $\ll_\varepsilon M^{1/2+\varepsilon}$ solutions if $\lambda + \nu = 0$ and $r \ge 1$. For a given pair λ, ν such that $|\lambda + \nu| \ge 1$ it follows that $\mu^2 + \mu\nu \equiv r \bmod(|\lambda + \nu|)$, and this quadratic congruence in μ has $d(|\lambda + \nu|)$ solutions $\bmod(|\lambda + \nu|)$. Since $|\lambda + \nu| \le 3M^{1/2}\Delta$ and m_1 is uniquely determined from (2.50) once μ, λ and ν are known, it follows on setting $f(k) = \sum_{k=\lambda+\nu} 1$ that

$$\begin{aligned} N(M, \Delta; r) &\ll_\varepsilon \sum_{\nu,\lambda,|\lambda+\nu|\ge 1} d(|\lambda + \nu|) \frac{\mathcal{M}^{1/2}\Delta}{|\lambda + \nu|} + M^{1/2+\varepsilon} \\ &\ll_\varepsilon M^{1/2}\Delta \sum_{1\le k\le 3\mathcal{M}^{1/2}\Delta} d(k) f(k) k^{-1} + M^{1/2+\varepsilon} \\ &\ll_\varepsilon M\Delta^2 \log^2 M + M^{1/2+\varepsilon}, \end{aligned}$$

since obviously $f(k) \ll M^{1/2}\Delta$. This proves (2.48).

The proof of (2.49) is analogous to the proof of (2.48), since we have (2.50) with $r = 0$. If $|\lambda + \nu| \ge 1$ in (2.50), then as in the previous case we obtain the bound $M\Delta^2 \log^2 M$. If $\lambda + \nu = 0$, then $\mu = 0$ or $\mu + \nu = 0$. Since now there are $\ll M^{1/2}\Delta$ choices for m_1, in either case there are $\ll M\Delta^2$ solutions of (2.50) with $r = 0$, hence (2.49) follows.

With the aid of Lemma 2.11 the proof of (2.47) is not difficult. We use (2.45) with $k = 2$. Then we obtain

$$\begin{aligned} R &\ll V^{-1} \int_{T/2}^{T} \sum_{Q(t)-X\le m_1,n_1,m_2,n_2\le Q(t)} (m_1 n_1 m_2 n_2)^{-1} \left(\frac{m_1 n_1}{m_2 n_2}\right)^{it} \mathrm{d}t \\ &\ll V^{-1}(S_1 + S_2), \end{aligned} \tag{2.51}$$

say, where in S_1 we have $m_1 n_1 = m_2 n_2$, and in S_2 we have $m_1 n_1 \ne m_2 n_2$. To estimate S_1 we use (2.49) of Lemma 2.11 with

$$M = t/(2\pi), \quad \tfrac{1}{2}T \le t \le T, \quad 0 < \Delta = \sqrt{\frac{2\pi T}{t}} V^{-1} L^2 < \tfrac{2}{5}\sqrt{\pi} < 1.$$

We obtain

$$\begin{aligned} S_1 &\ll T^{-1} \int_{T/2}^{T} \Bigg(\sum_{Q(t)-X\le m_1,n_1,m_2,n_2\le Q(t); m_1 n_1 = m_2 n_2} 1 \Bigg) \mathrm{d}t \\ &\ll \sup_{t\in[T/2,T]} K\left(\frac{t}{2\pi}, \Delta\right) \ll T V^{-2} L^4 \log^2 T \ll T V^{-2} \log^4 T. \end{aligned} \tag{2.52}$$

As in the analogous proof of the first part of the theorem, in S_2 we change the order of summation and integration. We have

$$S_2 = \sideset{}{^*}\sum_{\sqrt{\frac{T}{4\pi}}-X\le m_1,n_1,m_2,n_2\le\sqrt{\frac{T}{2\pi}};m_1n_1\ne m_2n_2} (m_1n_1m_2n_2)^{-1/2}\int_A^B \left(\frac{m_1n_1}{m_2n_2}\right)^{it} \mathrm{d}t,$$

where $\sum^*$ denotes summation over those quadruples (m_1, n_1, m_2, n_2) such that $A < B$, where

$$A := \max\Bigl(\tfrac12 T,\ 2\pi m_1^2,\ 2\pi m_2^2,\ 2\pi n_1^2,\ 2\pi n_2^2\Bigr),$$

$$B := \min\Bigl(T,\ 2\pi(m_1+X)^2,\ 2\pi(n_1+X)^2,\ 2\pi(m_2+X)^2,\ 2\pi(n_2+X)^2\Bigr).$$

In particular, it follows that

$$|n_1 - m_1| \le X, \quad |n_2 - m_1| \le X, \quad |m_2 - m_1| \le X,$$

and $m_1n_1 - m_2n_2 \ll TV^{-1}L^2$. Thus writing

$$m_1n_1 - m_2n_2 = r \quad (\text{for } m_1n_1 > m_2n_2),$$

we shall use (2.48) of Lemma 2.11 with $M = t/(2\pi)$, $\Delta = \sqrt{2\pi}V^{-1}L^2$ to estimate S_2. Noting that

$$\left|\int_A^B \left(\frac{m_1n_1}{m_2n_2}\right)^{it} \mathrm{d}t\right| \le 2\left(\log\left|\frac{m_1n_1}{m_2n_2}\right|\right)^{-1} \ll \frac{m_2n_2}{r} + 1 \ll \frac{T}{r},$$

we obtain

$$\begin{aligned}S_2 &\ll \sum_{1\le r\ll TV^{-1}L^2} \frac1r \sideset{}{^*}\sum_{\sqrt{\frac{T}{4\pi}}-X\le m_1,n_1,m_2,n_2\le\sqrt{\frac{T}{2\pi}};m_1n_1-m_2n_2=r} 1\\ &\ll \sum_{1\le r\ll TV^{-1}L^2} \frac1r N\Bigl(\frac{T}{2\pi},\Delta;r\Bigr)\\ &\ll (TV^{-2}L^4\log^2 T + T^{1/2+\varepsilon})\sum_{1\le r\le T}\frac1r \ll TV^{-2}\log^5 T. \qquad (2.53)\end{aligned}$$

The estimate (2.47) follows from (2.51)-(2.53). This completes the proof of Theorem 2.9. It is possible, of course, to use (2.45) with an integer $k \ge 3$, but it does not seem that this would lead to improvements over (2.36).

From Theorem 2.9 (cf. (2.47)) one obtains the following.

Corollary 2.12 *Unconditionally we have*

$$\sum_{0<\gamma_n\le T} (\gamma_{n+1}-\gamma_n)^3 \ll T\log^6 T. \qquad (2.54)$$

Namely the sum on the left-hand side of (2.54) is, by (2.47) and $\zeta(\frac{1}{2}+it) \ll t^{1/6}$,

$$\begin{aligned}
&\sum_{V=T^{1/6}2^{-j},j=1}^{O(\log T)} \sum_{\gamma_n\le T, V<\gamma_{n+1}-\gamma_n\le 2V} (\gamma_{n+1}-\gamma_n)^3 \\
&\le \sum_{V=T^{1/6}2^{-j},j=1}^{O(\log T)} 8V^3 \sum_{\gamma_n\le T, V<\gamma_{n+1}-\gamma_n\le 2V} 1 \\
&\ll \sum_{V=T^{1/6}2^{-j},j=1}^{O(\log T)} V^3\cdot TV^{-3}\log^5 T \ll T\log^6 T.
\end{aligned}$$

In concluding, let us mention a problem involving distribution of gaps between the consecutive zeros of $Z(t)$. Let γ_n^+ be the point in $[\gamma_n, \gamma_{n+1}]$ where $|Z(t)|$ is maximal, so that

$$|Z(\gamma_n^+)| = \max_{\gamma_n\le t\le\gamma_{n+1}} |Z(t)|, \tag{2.55}$$

where $0 < \gamma_1 \le \gamma_2 \le \ldots$ are the zeros of $Z(t)$. The point γ_n^+ is, in view of Theorem 2.6, unique if the RH holds. The information on γ_n^+ sheds light on the distribution of values of $Z(t)$. One of the questions one may ask is as follows.

Problem 2.13 Is it true that (unconditionally) we always have

$$\gamma_{n+1} - \gamma_n \ll |Z(\gamma_n^+)|? \tag{2.56}$$

From the known bounds this appears to be true, as we know (see the discussion in the Notes below) that we actually have (cf. (2.35)) $\beta \le 0.154\,7397\ldots + \varepsilon$, which is smaller than the best exponent $c = 32/205 = 0.156\,09\ldots$ such that $Z(\gamma_n^+) \ll_\varepsilon \gamma_n^{c+\varepsilon}$, which is due to M. N. Huxley [Hux2].

Closely related to Problem 2.13 is

Problem 2.14 Which is larger on the average, $|Z(\gamma_n^+)|$ or $\gamma_{n+1} - \gamma_n$? In other words, the problem is to compare (unconditionally) the order of the functions

$$G(T) := \sum_{0<\gamma_n\le T, |Z(\gamma_n^+)|\le\gamma_{n+1}-\gamma_n} 1, \quad H(T) := \sum_{0<\gamma_n\le T, |Z(\gamma_n^+)|\ge\gamma_{n+1}-\gamma_n} 1.$$

Problem 2.14 addresses the distribution of the gaps and $|Z(\gamma_n^+)|$ statistically. Of course, if (2.56) holds, then Problem 2.14 is trivial.

Notes

The formula (2.3) is a so-called *approximate functional equation* (henceforth AFE for short) for $Z(t)$. The AFEs are a very useful tool in analytic number theory, and their purpose is to express an L-function in the region where its Dirichlet series (Johann Peter Gustav Lejeune Dirichlet, February 13, 1805-May 5, 1859, German mathematician credited with the modern formal definition of a function) does not converge by a finite sum of *Dirichlet polynomials* (finite sums of the form $\sum f(n)n^{-s}$) plus a (good) error term. To get a better error term one often introduces *smoothing functions* in sums involving Dirichlet polynomials. We shall deal with the AFEs for $\zeta(s)$ and its powers in more detail later in Chapter 4. The AFEs for $Z^k(t)$ follow then from those for $\zeta^k(s)$ and the asymptotic formula (2.17) for $\chi(s)$. Strictly speaking, they are not AFEs, as $Z^k(t)$ is not an L-function, but the terminology nevertheless makes sense.

G. H. Hardy [Har1], [Har2] was the first to prove in 1914 that there are infinitely many zeros of $\zeta(s)$ which satisfy $\mathbb{R}\mathrm{e}\, s = \frac{1}{2}$. The function $Z(t)$ (denoted originally by $X(t)$) was introduced for the first time in the classical paper of G. H. Hardy and J. E. Littlewood [HaLi1], where many results on $\zeta(s)$ and the distribution of primes are obtained. It was inspired by Hardy's earlier work on the zeros of $\zeta(s)$, so that standard terminology used today, namely *Hardy's function* for $Z(t) = \zeta(\frac{1}{2} + it)\chi^{-1/2}(\frac{1}{2} + it)$, is appropriate.

For a proof of (2.2), see, for example, [Iv1, chapter 9]. Theorem 9.6 in that reference says: If $k \geq 1$ is a fixed integer, $\sigma \geq \frac{1}{2}$ is fixed, $\log^{1+\varepsilon} T \leq Y \leq T$, $T \geq T_0$, then uniformly in σ

$$\int_{T-Y}^{T+Y} |\zeta(\sigma + it)|^k \,\mathrm{d}t \gg Y.$$

Theorem 2.4 in the text is of a similar nature, but it has as the lower bound for Y not $\log^{1+\varepsilon} T$, but $12 \log\log T$. The result is due to Balasubramanian-Ramachandra (see [Ram] for this and many other related results), and also [Iv4, chapter 1] for a proof). Kanakanahalli Ramachandra, August 18, 1933-January 17, 2011, was an Indian mathematician, who was working in analytic number theory.

The remark of W. K. Hayman that Ramachandra's kernel is essentially optimal, was made in a letter to the author in July 1990.

In the proof of Theorem 2.4 the bound

$$\zeta(\tfrac{1}{2} + it) \ll |t|^c \tag{2.57}$$

with $c = 1/6$ is mentioned. From (2.3) by trivial estimation one obtains the value $c = 1/4$, which may be considered as "trivial." The estimation of the lowest bound of c in (2.57) is one of the most important problems of zeta-function theory. The value $c = \varepsilon$ is the so-called Lindelöf hypothesis (LH). For a history of the values of c for which (2.57) holds see, for example, the notes to chapter 7 of [Iv1]. Currently the best value is $c = 32/205 + \varepsilon = 0.15609\cdots + \varepsilon$, due to M. N. Huxley [Hux2]. It is obtained by an elaboration of the *Bombieri-Iwaniec method* for the estimation of exponential sums (see [BoIw1], [BoIw2], [Hux1], [Hux2] and [HuIv]). The limit of the method appears to be the value $c = 0.15 + \varepsilon$, which shows how far the contemporary methods are from obtaining the LH that $c = \varepsilon$.

D. H. Lehmer (Derrick Henry "Dick" Lehmer, February 23, 1905-May 22, 1991, American mathematician) in his works [Leh1] and [Leh2] made significant contributions to the theory of the zeta-function. The reader is referred to A. M. Odlyzko [Odl3] for a thorough discussion relating to Lehmer's phenomenon. For instance, A. M. Odlyzko *op. cit.* found (many more examples occur in subsequent computations) 1976 values of n such that $|Z(\frac{1}{2}\gamma_n + \frac{1}{2}\gamma_{n+1})| < 0.0005$ in the block that he investigated. Several extreme examples are also given by van de Lune *et al.* in [LRW].

The proof of Theorem 2.6 follows the one given in H. M. Edwards [Edw]. In what concerns possible improvements of the bound (2.13) for $\gamma_{n+1} - \gamma_n$, note that we have shown that

$$N(T+H) - N(T) \geq \frac{H}{2\pi}\log\left(\frac{T}{2\pi}\right) + S(T+H) - S(T) + O\left(\frac{1}{T}\right) \quad (0 < H \leq T).$$

Suppose that $S(T) \ll \log^a T$ for some $1/3 \leq a \leq 1$ (see (1.56)). Then the above formula gives $N(T+H) - N(T) > 0$ for $H = C\log^{a-1} T$ and sufficiently large $C > 0$. This implies the estimate

$$\gamma_{n+1} - \gamma_n \ \ll \ (\log \gamma_n)^{a-1}.$$

The *Bernoulli polynomials* are defined as

$$B_n(x) \ = \ \sum_{k=0}^{n} \binom{n}{k} B_k x^{n-k},$$

where the B_ks are *Bernoulli numbers* (see the Notes of Chapter 1). They are named after Jacob Bernoulli (also known as James or Jacques) (December 27, 1654-August 16, 1705), who was one of the many prominent mathematicians in the Bernoulli family. Accordingly the first three Bernoulli polynomials are

$$B_0(x) = 1, \quad B_1(x) = x - \frac{1}{2}, \quad B_2(x) = x^2 - x + \frac{1}{6}, \quad B_3(x) = x^3 - \frac{3}{2}x^2 + \frac{x}{2}.$$

The Fourier series of the Bernoulli polynomials is

$$B_n(x) \ = \ -\frac{n!}{(2\pi i)^n} \sum_{k=-\infty, k\neq 0}^{\infty} \frac{e^{2\pi i k x}}{k^n},$$

and the series converges absolutely for $n \geq 2$. Jean Baptiste Joseph Fourier (March 21, 1768-May 16, 1830) was a great French mathematician and physicist, best known for initiating the investigation of Fourier series and their applications to problems of heat transfer and vibrations. The Fourier transform and Fourier's Law are also named in his honour.

Several analogs of the Riemann hypothesis have already been proved. The proof of the Riemann hypothesis for varieties over finite fields by P. Deligne [Deli] in 1974 is perhaps the single strongest theoretical reason in favor of the RH. This result provides some evidence for the more general conjecture that all zeta-functions associated with automorphic forms satisfy a form of the Riemann hypothesis, which includes the classical RH as a special case. Similarly the *Selberg zeta-function* $\mathcal{Z}(s)$ (see [Sel]) satisfies the analog of the RH, and is in some ways similar to the Riemann zeta-function, having a functional equation and an infinite product expansion analogous to the Euler product expansion. Actually, in the case of the full modular group, it is $\mathcal{Z}(s)/((2s-1)\zeta(2s)\Gamma(s)$ that satisfies the RH. The Euler product

$$\mathcal{Z}(s) = \prod_{\{P_0\}} \prod_{k=0}^{\infty} (1 - N(P_0)^{-s-k}) \qquad (\mathbb{R}\mathrm{e}\, s > 1)$$

is not a product over the rational primes, but over norms of certain conjugacy classes of groups. For the definition of the norm and other relevant facts on $\mathcal{Z}(s)$, the reader is referred, for example, to chapter 10 of H. Iwaniec [Iwa1]. Note that $\mathcal{Z}(s)$ does not possess a Dirichlet series representation. Also $\mathcal{Z}(s)$ is an entire function of order 2, while $(s-1)\zeta(s)$ is an entire function of order 1. Furthermore, it has a zero at $s = 1$. For these reasons $\mathcal{Z}(s)$ cannot be compared too closely to $\zeta(s)$.

One should also mention the *Davenport-Heilbronn zeta-function* (Harold Davenport FRS, October 30, 1907-June 9, 1969; Hans Arnold Heilbronn, October 8, 1908-April 28, 1975). This is a zeta-function (Dirichlet series) which satisfies a functional equation similar to the classical functional equations (1.5)-(1.7) for $\zeta(s)$. It has other analogies with $\zeta(s)$, like having infinitely many zeros on the critical line $\sigma = 1/2$, but for this zeta-function the analog of the RH does not hold. This function was introduced by H. Davenport and H. Heilbronn [DaHe] as

$$f(s) = 5^{-s}\left\{\zeta\left(s, \frac{1}{5}\right) + \tan\theta\,\zeta\left(s, \frac{2}{5}\right) - \tan\theta\,\zeta\left(s, \frac{3}{5}\right) - \zeta\left(s, \frac{4}{5}\right)\right\},$$

where $\theta = \arctan(\sqrt{10 - 2\sqrt{5}} - 2)/(\sqrt{5} - 1)$ and, for $\mathbb{R}\mathrm{e}\, s > 1$,

$$\zeta(s, a) = \sum_{n=0}^{\infty}(n + a)^{-s} \qquad (0 < a \le 1)$$

is the familiar *Hurwitz zeta-function* (named after the German mathematician Adolf Hurwitz, March 26, 1859-November 9, 1919), defined for $\mathbb{R}\mathrm{e}\, s \le 1$ by analytic continuation. With the above choice of θ (see [DaHe], [KaVo] or [Tit3]) it can be shown that $f(s)$ satisfies the functional equation

$$f(s) = X(s)f(1 - s), \qquad X(s) = \frac{2\Gamma(1 - s)\cos(\frac{\pi s}{2})}{5^{s-\frac{1}{2}}(2\pi)^{1-s}},$$

whose analogy with the functional equations (1.5)-(1.7) for $\zeta(s)$ is evident. Let $1/2 < \sigma_1 < \sigma_2 < 1$. Then it can be shown (see chapter 6 of [KaVo]) that $f(s)$ has infinitely many zeros in the strip $\sigma_1 < \sigma = \mathbb{R}\mathrm{e}\, s < \sigma_2$, and it also has (see chapter 10 of [Tit3]) an infinity of zeros in the half-plane $\sigma > 1$, while from the product representation in (1.1) it follows that $\zeta(s) \neq 0$ for $\sigma > 1$, so that in the half-plane $\sigma > 1$ the behavior of zeros of $\zeta(s)$ and $f(s)$ is different. Actually the number of zeros of $f(s)$ for which $\sigma > 1$ and $0 < t = \mathbb{I}\mathrm{m}\, s \le T$ is $\gg T$, and similarly each rectangle $0 < t \le T$, $1/2 < \sigma_1 < \sigma \le \sigma_2 \le 1$ contains at least $c(\sigma_1, \sigma_2)T$ zeros of $f(s)$. R. Spira [Spi] found that $0.808\,517 + 85.699\,348i$ (the values are approximate) is a zero of $f(s)$ lying in the critical strip $0 < \sigma < 1$, but not on the critical line $\sigma = 1/2$. On the other hand, A. A. Karatsuba [Kar2] (Anatolii Alexeevitch Karatsuba, January 31, 1937-September 28, 2008, Russian mathematician) proved that the number of zeros $\frac{1}{2} + i\gamma$ of $f(s)$ for which $0 < \gamma \le T$ is at least $T(\log T)^{1/2-\varepsilon}$ for any given $\varepsilon > 0$ and $T \ge T_0(\varepsilon)$. This bound is weaker than A. Selberg's classical result [Sel] that there are $\gg T\,\log T$ zeros $\frac{1}{2} + i\gamma$ of $\zeta(s)$ for which $0 < \gamma \le T$. From the Riemann-von Mangoldt formula (1.30)-(1.31) it follows that, up to the value of the $\ll$-constant, Selberg's result on $\zeta(s)$ is best possible. There are certainly $\ll T\,\log T$ zeros $\frac{1}{2} + i\gamma$ of $f(s)$ for which $0 < \gamma \le T$ and it may be that almost all of them lie on the critical line $\sigma = 1/2$, although this has not been proved yet. The Davenport-Heilbronn zeta-function is not the only example of a zeta-function that exhibits the phenomena described above, and many so-called *Epstein zeta-functions*, named after the German mathematician Paul Epstein, July 24, 1871-August 11, 1939 (see, e.g., p. 42 of [Iv1]), also have complex zeros off their respective critical lines. The reader is referred to the paper of E. Bombieri and D. Hejhal [BoHe] for some interesting results, and also to the paper of E. Bombieri and A. Ghosh [BoGh].

The numerical verification that many zeros (i.e. up to height T for some explicit, large T) of an L-function lie on the critical line seems at first sight to be strong evidence for the corresponding RH. However, analytic number theory has had many conjectures supported by large amounts of numerical evidence that turn out to be false. For example, see the discussion related to the so-called *Skewes number* for a notorious example (see, e.g., [Iv18]). The problem is that the behavior of a function is often influenced by very slowly increasing functions such as $\log\log T$, that tend to infinity, but do it so slowly that this cannot be detected by computation. Such functions occur in the theory of the zeta-function controlling the behavior of its zeros; for example the function $S(T)$ above has average size around $(\log\log T)^{1/2}$ (follows, e.g., from (1.55)). As $S(T)$ jumps by at least two at any counterexample to the Riemann hypothesis, one might expect any counterexamples to the RH to start appearing only when $S(T)$ becomes large. It is never much more than three as far as it has been calculated, but $S(T)$ is known to be unbounded (see (1.56)), suggesting that calculations may not have yet reached the region of typical behavior of the zeta-function.

Among the recent texts on the RH the reader should see the papers of E. Bombieri [Bom1] and J. B. Conrey [Con4]. They, as well as many other works, are included in the monograph [BCRW] of P. Borwein *et al.* In particular, this work also contains the present author's paper [Iv18]. It points out some doubts about the truth of the RH, including a discussion of the Lehmer phenomenon (section 3). The paper of P. Sarnak [Sar2] builds on [Bom2] and deals with the grand Riemann hypothesis (that the non-trivial zeros of all automorphic L-functions lie on the critical line 1/2).

It is well known that on the Riemann hypothesis the zeros of $Z(t)$ and those of $Z'(t)$ are interlacing; this is actually a corollary of Theorem 2.4. R. J. Anderson [And] proves a similar result for $Z'(t)$: if the RH is true then there is a $t_0 > 0$ such that, for $t > t_0$, the function $Z''(t)$ has exactly one zero between consecutive zeros of $Z'(t)$. In order to prove this result, Anderson introduces an analytic function whose zeros on the critical line are connected with those of $Z'(t)$. This is the function

$$\eta(s) := \zeta(s) - 2\frac{\chi(s)}{\chi'(s)}\zeta'(s),$$

which is not to be confused with (1.18). Thus

$$-i\frac{Z'(t)}{\theta'(t)} = \mathrm{e}^{i\theta(t)}\eta(\tfrac{1}{2} + it).$$

The last relation shows that the zeros of $Z'(t)$ and $\eta(\frac{1}{2} + it)$ coincide and have the same multiplicity.

Lemma 2.6 of [Iv1] states the following: let $k \ge 2$ be a fixed integer and let $f(x) \in C^k[a, b]$. If $b \ge a + 1$, $K = 2^{k-1}$ and for $a \le x \le b$

$$0 < \lambda_k \le f^{(k)}(x) \le A\lambda_k \quad (\text{or } \lambda_k \le -f^{(k)}(x) \le A\lambda_k,\ A > 1),$$

then

$$\sum_{a<n\le b} \mathrm{e}(f(n)) \ll A^{2/K}(b-a)\lambda_k^{1/(2K-2)} + (b-a)^{1-2/K}\lambda_k^{-1/(2K-2)}. \tag{2.58}$$

This bound follows by induction, the case $k = 2$ being Lemma 6.6. It is useful when $f^{(k-1)}(x)$ vanishes in $[a, b]$.

The definition of an exponent pair (κ, λ) is as follows. Suppose that, corresponding to every $s > 0$, there exist two numbers r and c depending only on s ($r \ge 4$ and $0 < c < \frac{1}{4}$) such that

$$\sum_{a<n\le b} \mathrm{e}(f(n)) \ll z^\kappa a^\lambda \tag{2.59}$$

holds with the $\ll$-constant depending only on s and u, whenever $u > 1$, $1 \le a < b < au$, $y > 0$, $z = ya^{-s}$ (> 1), $f(n)$ is any real function with the first r derivatives in the interval $a \le n \le b$, $a, b \in \mathbb{Z}$, and, for $a \le n \le b$, $0 \le j \le r - 1$,

$$\left|f^{(j+1)}(n) - (-1)^j ys(s+1)\cdots(s+j-1)n^{-s-j}\right| < cys(s+1)\cdots(s+j-1)n^{-s-j}. \tag{2.60}$$

Then (κ, λ) is an exponent pair.

Note that z is effectively $f'(a)$, and in practice it often suffices to replace the cumbersome condition (2.60) by the less stringent

$$AB^{1-r} \ll |f^{(r)}(x)| \ll AB^{1-r} \qquad (B \le x \le 2B,\ r \in \mathbb{N}),$$

with the $\ll$-constant depending on r alone, and $A > 1/2$ is assumed. Then, similarly to (2.58), we have

$$\sum_{B<n\le B+h} \mathrm{e}(f(n)) \ll A^{\kappa} B^{\lambda} \qquad (B \ge 1,\ 1 < h \le B).$$

Some common exponent pairs are $(0,1)$, $(\frac12,\frac12)$, $(\frac16,\frac46)$, $(\frac27,\frac47)$, and the exponent pair $(\frac12\alpha+\varepsilon, \frac12+\frac12\alpha+\varepsilon)$, used in (2.35). This exponent pair with $\alpha = 0.155\,945\,83\ldots$ is due to R. A. Rankin [Ran] (Robert Alexander Rankin, October 27, 1915-January 27, 2001, a Scottish mathematician who worked in analytic number theory). He showed that it is optimal ((κ,λ) with $\kappa+\lambda$ minimal) in terms of classic processes (so-called A, B-processes and convexity) for obtaining exponent pairs. Namely if (κ,λ), (κ_1,λ_1) are exponent pair, then so are the pairs

$$A(\kappa,\lambda) = \left(\frac{\kappa}{2\kappa+2}, \frac12 + \frac{\lambda}{2\kappa+2}\right),$$
$$B(\kappa,\lambda) = \left(\lambda - \tfrac12, \kappa + \tfrac12\right),$$
$$C(t)(\kappa,\lambda)(\kappa_1,\lambda_1) = \left(t\kappa + (1-t)\kappa_1, t\lambda + (1-t)\lambda_1\right)$$

for any real t satisfying $0 \le t \le 1$. There is a conjecture that $(\varepsilon, \frac12+\varepsilon)$ is an exponent pair. This is very strong, since by (2.1) it implies the LH. If this exponent pair is used in conjunction with (2.32), we obtain that $\gamma_{n+1} - \gamma_n \ll_\varepsilon \gamma_n^\varepsilon$ holds.

More recent research brought new exponent pairs, which are being based on the so-called *Bombieri-Iwaniec method* (see their joint works [BoIw1], [BoIw2], and also the work [HuIv] for a short account, and M. N. Huxley's monograph [Hux1] for more details). M. N. Huxley [Hux2] showed that, for $\theta > 32/205$, $(\theta, \theta+1/2)$ is an exponent pair. This shows that if

$$\rho = \inf\left(\kappa+\lambda\ :\ (\kappa,\lambda) \text{ is an exponent pair}\right),$$

then $\rho \le 333/410 = 0.812\,1951\ldots$, which improves on Rankin's value $\rho \le 0.829\,0213\,56\ldots$. This will yield a small improvement of (2.35), namely $\beta = 0.154\,7397\ldots + \varepsilon$. Note that this value is better than the value $c = 32/205 = 0.156\,09\ldots$, the exponent of M. N. Huxley [Hux2] for the order of $\zeta(\frac12+it)$. See also the discussion in [Iv2].

Theorem 2.9 is from the joint work [IvJu] of M. Jutila and the present author. It also contains a corresponding result for gaps between consecutive zeros on the critical line of zeta-functions of holomorphic cusp forms. Clearly the method of proof can be used to deal with functions from the general class $\mathcal{S}$ of L-functions, which will be considered in the next chapter.

To obtain (2.43) one applies Hölder's inequality in the form

$$\int_a^b |f(u)g(u)|\,\mathrm{d}u \le \left(\int_a^b |f(u)|^p\,\mathrm{d}u\right)^{1/p} \left(\int_a^b |g(u)|^q\,\mathrm{d}u\right)^{1/q},$$

where $p, q > 0$, $1/p + 1/q = 1$, and $f(x)$, $g(x)$ are continuous on $[a,b]$. The bound (2.43) follows then with $p = 2k$, $q = 2k/(2k-1)$,

$$f(u) = \sum n^{-1/2-iu}\left\{\exp(-(t-u)^2G^{-2})\right\}^{1/2k}, \quad g(u) = \left\{\exp(-(t-u)^2G^{-2})\right\}^{(2k-1)/(2k)}.$$

In [CoGh1] J. B. Conrey and A. Ghosh assume the RH. Let γ_n^+ be as in (2.55). The result of their paper is then that

$$\frac{1}{N(T)} \sum_{0<\gamma_n\le T} |Z(\gamma_n^+)|^2 \sim (\mathrm{e}^2 - 5)\log T \qquad (T\to\infty). \tag{2.61}$$

Thus, on average, $|Z(\gamma_n^+)|^2$ is larger than the mean value of $|Z(t)|^2 = |\zeta(\frac{1}{2}+it)|^2$ by a factor of $(e^2-5) = 1.1945\ldots$. It would be interesting to find an unconditional estimate for the sum in (2.63). Some results related to this problem will be discussed at the end of Chapter 11. In [Con1], J. B. Conrey considered sums of $|Z(\gamma_n^+)|^4$, and recently M. B. Milinovich [Mil] proved (all on the RH) that

$$(\log T)^{k^2-\varepsilon} \ll_{k,\varepsilon} \frac{1}{N(T)} \sum_{0<\gamma_n\le T} |Z(\gamma_n^+)|^{2k} \ll_{k,\varepsilon} (\log T)^{k^2+\varepsilon}$$

for $k \in \mathbb{N}$.

R. R. Hall in his works [Hal1]-[Hal7] made significant contributions to problems concerning the gaps $\gamma_{n+1}-\gamma_n$ and γ_n^+. In [Hal3] he gives upper bounds, when $k = 1, 2$, for

$$M^{(k)}(T,\theta) := \sum_{\gamma_n\le T,\gamma_{n+1}-\gamma_n\le\frac{2\pi\theta}{\log T}} (\gamma_n^+)^k,$$

where θ (>0) is a given number. Bounds of the form

$$M^{(k)}(T,\theta) \le H_k(\theta)(1+o(1))T\log^{k^2+1}T \quad (k=1,2;\ T\to\infty)$$

are sought, where $H_k(\theta)$ is to be a continuous, strictly increasing, bounded function for $\theta > 0$ such that $H_k(0) = 0$. This interesting problem of the distribution of zeros, connected with H. L. Montgomery's pair correlation conjecture, was already investigated by Hall in [Hal1]. The paper [Hal3] contains sharper results, namely it is shown that

$$H_1(\theta) = \frac{\pi^3\theta^3}{480} \qquad \left(0<\theta\le\frac{5\sqrt{2}}{\pi\sqrt{3}} = 1.299\,4946\ldots\right),$$

$$H_2(\theta) = \frac{\pi\theta^3}{840} \qquad \left(0<\theta\le\frac{\sqrt{35}}{\pi\sqrt{3}} = 1.087\,2352\ldots\right).$$

Moreover, the values $H_1(\infty) = 0.232\,002\,60\ldots$ and $H_2(\infty) = 0.129\,099\,44\ldots/\pi^3$ are obtained. The key ingredient in the proof, lemma 1, is a result of intrinsic interest. It gives an upper bound inequality for $\max_{a\le x\le b}|y(x)|$ (a and b being zeros of $y(x)\in C^2[a,b]$), in terms of mean square integrals over $[a,b]$ of y, y' and y''. The limitations of the method are clearly explained by the author. In [Hal6] Hall proves that, unconditionally,

$$\Lambda := \limsup_{n\to\infty}\frac{\gamma_{n+1}-\gamma_n}{(2\pi/\log\gamma_n)} \ge 2.630\,637\ldots,$$

and that

$$\limsup_{n\to\infty}\frac{v_{n+1}-v_n}{(\pi/\log\gamma_n)} \ge 2.160\,246\ldots,$$

where $\{v_n\}_{n\ge1}$ is the sequence of the positive roots of $Z(t)Z'(t) = 0$. Several authors have given lower bounds for Λ, assuming the RH. These are, for example, H. L. Montgomery and A. M. Odlyzko [MOd], J. Mueller [Mue], J. B. Conrey *et al.* [CGG1]. The same authors proved in [CGG2] that $\Lambda > 2.68$, if the generalized Riemann hypothesis is assumed. Namely let

$$L(s,\chi) := \sum_{n=1}^{\infty}\chi(n)n^{-s} \qquad (\mathbb{R}\mathrm{e}\, s > 1)$$

be the L-function associated with the Dirichlet character $\chi(n)$ to a given modulus. The generalized Riemann hypothesis asserts that for every Dirichlet character χ and every complex number s with $L(s,\chi) = 0$: if $0 < \mathbb{R}\mathrm{e}\, s < 1$, then $\mathbb{R}\mathrm{e}\, s = 1/2$.

The latest result on Λ, assuming the RH, is due to J. Bregberg [Bre] who proves $\Lambda \geq 2.766$. His work contains a detailed discussion of previous results on Λ. On the grand Riemann hypothesis the latest result is $\Lambda > 3.033$, due to H. Bui [Bui].

Hall's paper [Hal6] contains several other results of interest, for example, the asymptotic formula

$$\int_0^T Z(t)Z(t+a)\,\mathrm{d}t = \frac{\sin\alpha/2}{\alpha/2}T\log T + \Big(2\gamma - 1 - \log(2\pi)\Big)T\cos\alpha/2 + O\left(\frac{\alpha T}{\log T} + T^{1/2}\log T\right), \tag{2.62}$$

where $a = \alpha/\log T$, γ is Euler's constant, and the O-term is uniform for $a \ll 1$. The formula contains a proof of Hardy's result that $Z(t)$ has infinitely many sign changes; one has only to fix α ($= 3\pi$ say), so that the sine is negative. A result similar to (2.62) is Theorem 6.7. Hall (*op. cit.*) also proves that

$$\int_0^T |Z(t)Z(t+a)|^2\,\mathrm{d}t = \frac{12}{\pi^2}\left(\frac{\cos\alpha - 1 + \alpha^2/2}{\alpha^4}\right)T\log^4 T + O(T\log^3 T),$$

$$\int_0^T \left|\frac{d}{\mathrm{d}t}Z(t)Z(t+a)\right|^2\mathrm{d}t = \frac{24}{\pi^2}\left(\frac{1 - \cos\alpha - \alpha^2/2 + \alpha^4/24}{\alpha^6}\right)T\log^6 T + O(T\log^5 T).$$

Again, both of these formulas are uniform for $a \ll 1$.

H. L. Montgomery [Mon2] studied the distribution of the differences $\gamma - \gamma'$, where γ, γ' denote imaginary parts of zeta-zeros on the critical line. Let

$$F(x,T) := \sum_{0<\gamma\leq T, 0<\gamma'\leq T} x^{i(\gamma-\gamma')}w(\gamma-\gamma'), \qquad w(u) := \frac{4}{4+u^2}.$$

Assuming the RH, he proved that

$$F(x,T) \sim \frac{T}{2\pi}\log x + \frac{T}{2\pi x^2}(\log T)^2 \qquad (T \to \infty)$$

for $1 \leq x \leq T$ (actually he proved this for $1 \leq x \leq o(T)$ and the full range was established by D. A. Goldston [Gol1]). Montgomery conjectured that, as $T \to \infty$,

$$F(x,T) \sim \frac{T}{2\pi}\log T \qquad (T \leq x \leq T^M)$$

for $M > 1$ fixed. This is known as the *strong pair correlation conjecture*. From this one derives the *(weak) pair correlation conjecture* that, under the RH, for fixed $\alpha < \beta$,

$$\sum_{0<\gamma,\gamma'\leq T, \alpha/L\leq\gamma-\gamma'\leq\beta/L} 1 = \Big(1+o(1)\Big)\left\{\int_\alpha^\beta \left(1 - \left(\frac{\sin\pi u}{\pi u}\right)^2\right)du + \delta(\alpha,\beta)\right\}TL, \tag{2.63}$$

where $L = \log(T/(2\pi))$, $T \to \infty$, γ, γ' denote imaginary parts of zeta-zeros on the critical line, and

$$\delta(\alpha,\beta) = \begin{cases} 1 & \text{if } 0 \in [\alpha,\beta], \\ 0 & \text{if } 0 \notin [\alpha,\beta]. \end{cases}$$

It follows from the RH and (2.63) that almost all zeta-zeros on the critical line are simple, that is,

$$\sum_{\gamma\leq T, \rho \text{ simple}} 1 = \Big(1+o(1)\Big)N(T) = \Big(1+o(1)\Big)\frac{T}{2\pi}\log T \qquad (T \to \infty).$$

Assuming only the RH, he proved that

$$\sum_{\gamma \le T, \rho \text{ simple}} 1 \ge \left(\frac{2}{3} + o(1)\right) \frac{T}{2\pi} \log T \qquad (T \to \infty).$$

There is an extensive literature on the pair correlation conjecture, see, for example, the papers of D. A. Goldston [Gol2], [Gol3] and Goldston-Montgomery [GoMo].

3

The Selberg class of L-functions

3.1 The axioms of Selberg's class

The Riemann zeta-function $\zeta(s) = \sum_{n=1}^{\infty} n^{-s}$ ($\mathbb{R}\mathrm{e}\, s > 1$) is the simplest and most important example of a Dirichlet series (or L-function) of the form

$$F(s) = \sum_{n=1}^{\infty} a_F(n) n^{-s} \tag{3.1}$$

with an arithmetic significance. One feels that the essential properties of $\zeta(s)$, namely the Euler product (1.1) and the functional equation (1.5) should, in some form, hold true for a general class of L-functions given by (3.1). The best known such generalization is the *Selberg class* $\mathcal{S}$, defined axiomatically by A. Selberg [Sel] as follows.

(a) The series in (3.1) converges absolutely for $\sigma = \mathbb{R}\mathrm{e}\, s > 1$ with $a_F(1) = 1$, and the coefficients $a_F(n)$ satisfy the bound $a_F(n) \ll_\varepsilon n^\varepsilon$ for every $\varepsilon > 0$.

(b) There is a natural number m_F such that $(s-1)^{m_F} F(s)$, which is of a finite order, admits analytic continuation that is regular in $\mathbb{C}$.

(c) There is a function

$$\Phi_F(s) := Q_F^s \Gamma_F(s) F(s) \qquad (Q_F > 0) \tag{3.2}$$

with

$$\Gamma_F(s) := \prod_{j=1}^{r_F} \Gamma(\lambda_j(F)s + \mu_j(F)) \qquad (\lambda_j(F) > 0,\ \mathbb{R}\mathrm{e}\, \mu_j(F) \ge 0).$$

Then the functional equation

$$\Phi_F(s) = \omega_F \overline{\Phi}_F(1-s) \qquad (|\omega_F| = 1) \tag{3.3}$$

holds for $s \in \mathbb{C}$, where for any function $f(s)$ we define $\bar{f}(s) = \overline{f(\bar{s})}$.

(d) We have

$$\log F(s) = \sum_{n=2}^{\infty} b_F(n)\Lambda(n)/(n^s \log n),$$

where $b_F(n) \ll n^\theta$ for some $0 \le \theta < \frac{1}{2}$, and $\Lambda(n)$ is the von Mangoldt function.

The quantity

$$d_F := 2\sum_{j=1}^{r_F} \lambda_j(F) \tag{3.4}$$

plays an important rôle in the theory of $\mathcal{S}$. It is called *the degree of* F, and it is conjectured that it is always a non-negative integer. A brief discussion on d_F will be given in Section 3.3.

From $a_F(n) \ll_\varepsilon n^\varepsilon$ it follows that $b_F(p^k) \ll_{k,\varepsilon} p^\varepsilon$, and for convenience one sets $b_F(n) \equiv 0$ if n is not a prime power. Furthermore the coefficients $a_F(n)$ are multiplicative functions of n (see [CoGh3]). Moreover, the Euler product

$$F(s) = \prod_p F_p(s), \quad F_p(s) = \sum_{m=0}^{\infty} a_F(p^m)p^{-ms}$$

is absolutely convergent for $\sigma > 1$ and, for every prime p, $F_p(s)$ is absolutely convergent for $\sigma > 0$. The factors $F_p(s)$ are called *the Euler factors* of $F(s)$. Note that usually $(F_p(s))^{-1}$ is a polynomial in p^{-s}, of degree independent of p for all but finitely many primes p. We also remark that

$$b_F(p) = a_F(p), \quad b_F(p^\alpha) = a_F(p^\alpha) - \frac{1}{\alpha}\sum_{j=1}^{\alpha} j a_F(p^{\alpha-j}) b_F(p^j) \quad (\alpha > 1).$$

Note that $F_p(s) \ne 0$ for $\sigma > \theta$ and every p, since $b_F(n) \ll n^\theta$. Therefore we also have the zero-free region $F(s) \ne 0$ for $\sigma > 1$. Like in the theory of $\zeta(s)$, it is an important problem to find a better zero-free region for $F(s)$.

3.2 The analogs of Hardy's and Lindelöf's function for $\mathcal{S}$

We can write the functional equation (3.3) for $F(s)$ as

$$F(s) = H(s)\bar{F}(1-s), \quad H(s) := \omega_F Q_F^{1-2s} \prod_{j=1}^{r_F} \frac{\Gamma(\lambda_j(1-s)+\overline{\mu}_j(F))}{\Gamma(\lambda_j s + \mu_j(F))}. \tag{3.5}$$

For $t \geq 1$ and uniformly in bounded σ, it follows by (2.15) that

$$H(\sigma+it) = (\lambda_F Q_F^2 t^{d_F})^{1/2-\sigma-it} \exp\big(itd_F + \tfrac{1}{4}i\pi(\mu_F - d_F)\big) \times \left(\omega_F + O\Big(\frac{1}{t}\Big)\right)$$

with

$$\mu_F := 2\sum_{j=1}^{r_F}(1-2\mu_j), \quad \lambda_F := \prod_{j=1}^{r_F} \lambda_j^{2\lambda_j}.$$

This gives the functional equation for $F(s)$ in the form

$$F(s) = H(s)\bar{F}(1-s), \quad H(s) \asymp |t|^{d_F(1/2-\sigma)} Q_F^{1-2\sigma} \qquad (|t| \to \infty), \tag{3.6}$$

where $a \asymp b$ means that $b \ll |a| \ll b$.

If the coefficients $a_F(n)$ in (3.1) are real, then $\bar{F}(1-s) = F(1-s)$, so that the functional equation (3.6) takes the form

$$F(s) = H(s)F(1-s), \quad H(s)H(1-s) = 1,$$

which is in complete analogy with the functional equation for $\zeta(s)$ in the form

$$\zeta(s) = \chi(s)\zeta(1-s), \quad \chi(s)\chi(1-s) = 1.$$

Then the analog of Hardy's function for $F(s)$ may be defined as

$$Z_F(t) := F(\tfrac{1}{2}+it)\big(H(\tfrac{1}{2}+it)\big)^{-1/2} \quad (t \in \mathbb{R}). \tag{3.7}$$

Moreover, if ω_F and the $\mu_j(F)$s in (3.5) are real, then we have $|Z_F(t)| = |F(\frac{1}{2}+it)|$ and

$$\overline{Z_F(t)} = F(\tfrac{1}{2}-it)\big(H(\tfrac{1}{2}-it)\big)^{1/2} = F(\tfrac{1}{2}+it)\big(H(\tfrac{1}{2}+it)\big)^{-1/2} = Z_F(t),$$

so that $Z_F(t)$ is real-valued and even, just like Hardy's function $Z(t)$ itself.

A general problem concerning the Dirichlet series $F(s)$ is to analyze the growth of $|F(\sigma+it)|$ when σ is fixed. One can define *the Lindelöf function* for $F \in \mathcal{S}$ as

$$\mu_F(\sigma) := \limsup_{t\to\infty} \frac{\log|F(\sigma+it)|}{\log t} \qquad (\sigma \in \mathbb{R}). \tag{3.8}$$

The function $\mu_F(\sigma)$, which is non-negative, convex downwards and monotonic non-increasing, provides important information about the growth of $F(s)$. It is clear that $\mu_F(\sigma) = 0$ for $\sigma > 1$, since the series defining $F(s)$ converges absolutely in this region. The assertion that $\mu_F(1/2) = 0$ (or equivalently, that $\mu_F(\sigma) = 0$ for $\sigma \ge 1/2$) is known as *the Lindelöf hypothesis* for $\mathcal{S}$, in analogy with the classical Lindelöf hypothesis for $\zeta(s)$ that $\zeta(\frac{1}{2}+it) \ll_\varepsilon |t|^\varepsilon$. We have

$$\mu_F(\sigma) \le \begin{cases} 0 & \text{if } \sigma > 1, \\ \frac{1}{2}d_F(1-\sigma) & \text{if } 0 \le \sigma \le 1, \\ (\frac{1}{2}-\sigma)d_F & \text{if } \sigma < 0. \end{cases} \tag{3.9}$$

The first bound in (3.9) is trivial, and the third follows from (3.6) and the first. The second follows from the first, third and convexity (the *Phragmén-Lindelöf principle*). Of special interest are the bounds for $F(s)$ on the critical line $\mathbb{R}\mathrm{e}\, s = 1/2$. From (3.9) one has

$$F(\tfrac{1}{2}+it) \ll_\varepsilon |t|^{d_F/4+\varepsilon},$$

which is known as *the convexity bound.* Improving convexity bounds is one of the major problems of L-functions in general. For $\zeta(s)$ it is known that $\mu(1/2) \le 32/205 = 0.156\,09\ldots$, which is the best known value due to M. N. Huxley [Hux2]. See also [HuIv] and (4.30).

3.3 The degree d_F and the invariants of $\mathcal{S}$

So far it is known (see Kaczorowski-Perelli [KaPe1], [KaPe2]) that there are no elements in $\mathcal{S}$ if $0 < d_F < 1$ and (see [KaPe5]) $1 < d_F < 2$. Clearly $\zeta(s)$ and the shifted Dirichlet L-functions $L(s+i\theta,\chi)\,(\theta \in \mathbb{R})$ associated to primitive characters χ belong to $\mathcal{S}$ with $d_F = 1$, and these are essentially the functions in $\mathcal{S}$ with $d_F = 1$ (a simpler proof of the classification of functions in $\mathcal{S}$ with $d_F = 1$ is given by K. Soundararajan [Sou2]). Trivially the squares of functions in $\mathcal{S}$ with $d_F = 1$ satisfy $d_F = 2$, and other well-known examples with $d_F = 2$ are the zeta-functions of holomorphic and non-holomorphic cusp forms. However, the classification of all functions in $\mathcal{S}$ with $d_F = 2$ is still not complete, and represents a difficult problem. Other well-known functions of $\mathcal{S}$ besides $\zeta(s)$ (and its integral powers) are $\zeta_K(s)$, the Dedekind zeta-function of an algebraic field of degree K, the Rankin-Selberg convolution

$$L_{f\times\bar{g}}(s)=\sum_{n=1}^{\infty}a(n)\overline{b(n)}n^{-s}$$

of any two normalized holomorphic newforms (where $a(n)$, $b(n)$ are the Fourier coefficients of the modular forms $f(z)$, $g(z)$, respectively), the Hecke L-functions $L_K(s,\chi)$ with primitive Hecke characters χ (Erich Hecke, September 20, 1887-February 13, 1947, a German mathematician), and the normalized L-functions $L(s,f)$ associated with newforms $f(z)$ for the Hecke groups $\Gamma_0(q)$. Moreover, the Artin L-functions $L(s,K/k,\chi)$ belong to $\mathcal{S}$ (Emil Artin, 1898-1962, an Austrian-American mathematician of Armenian descent) if the *Artin conjecture* holds true (see, e.g., J. Steuding [Ste], chapter 13.7), while the cuspidal $GL(n)$ automorphic L-functions belong to $\mathcal{S}$ provided that the *Ramanujan conjecture* (that $a_F(n)\ll_\varepsilon n^\varepsilon$ in axiom (a)) holds true.

The first to prove that if $d_F>0$, then $d_F\geq 1$ was H.-E. Richert [Ric], followed by S. Bochner [Boc]. In short, Bochner's argument runs as follows. Let $F(s)=\sum_{n=1}^{\infty}f(n)n^{-s}$. From the Mellin inversion formula for $\Gamma(s)$ (see (A.7) of [Iv1]) one has, for $x>0$,

$$\sum_{n=1}^{\infty}f(n)\mathrm{e}^{-nx}=\frac{1}{2\pi i}\int_{2-i\infty}^{2+i\infty}F(s)x^{-s}\Gamma(s)\,\mathrm{d}s.$$

Because of the Phragmén-Lindelöf principle and (3.6), $F(s)$ has polynomial growth in $|t|=\mathbb{I}\mathrm{m}\,s$ in any vertical strip. Hence if we move the line of integration to the left, then the residue theorem gives (the poles are possibly at $s=1$ and simple poles at non-positive integers)

$$\sum_{n=1}^{\infty}f(n)\mathrm{e}^{-nx}=\frac{p(\log x)}{x}+\sum_{n=0}^{\infty}\frac{(-1)^nF(-n)}{n!}x^n \tag{3.10}$$

with a certain polynomial p. Suppose that $0<d_F<1$. By (3.5), Stirling's formula in the form

$$\log\Gamma(z)=(z-\tfrac{1}{2})-z+\tfrac{1}{2}\log 2\pi+O\Big(\frac{1}{|z|}\Big)\qquad(|z|\to\infty),$$

and the well-known relation $\Gamma(z)\Gamma(1-z)=\pi/\sin(\pi z)$ we find that

$$H(-n)=\exp\left(2\sum_{j=1}^{r_F}\lambda_j n\log n+O_F(n)\right)=\exp\Big(d_F n\log n+O_F(n)\Big),$$

where $H(s)$ is defined by (3.5). By using the standard approximation

$$n! = \sqrt{2\pi n}\Big(\frac{n}{\mathrm{e}}\Big)^n\Big(1 + O\Big(\frac{1}{n}\Big)\Big)$$

and the fact that $0 < d_F < 1$, it follows that the sum on the right-hand side of (3.10) converges for all $x \in \mathbb{C}$. Hence the left-hand side is analytic in $\mathbb{C}\backslash\{x \le 0 : x \in \mathbb{R}\}$. Note that the left-hand side is also $2\pi i$ periodic. Therefore

$$\varphi(z) := \sum_{n=1}^{\infty} f(n)\mathrm{e}^{-nz}$$

is an entire function, and we have $(z = x + iy)$

$$f(n)\mathrm{e}^{-nx} = \int_0^{2\pi} \varphi(x + iy)\mathrm{e}^{-iny}\,\mathrm{d}y \ll n^{-2},$$

on integrating twice by parts (or differentiating twice over x). If we then taken $x = 1/n$ we obtain that $f(n) \ll n^{-2}$. This implies that the Dirichlet series defining $F(s)$ converges for $\mathrm{Re}\, s > -1$. However, this is impossible, since we can again relate $F(-\frac{1}{2} + it)$ to $F(\frac{3}{2} - it)$ by (3.6) to obtain that $F(-\frac{1}{2} + it)$ cannot be bounded.

The function $\Gamma_F(s)$ in axiom (c) is called the γ-factor, and its factors $\Gamma\big(\lambda_j(F)s + \mu_j(F)\big)$ are called the Γ-factors. The form of the γ-factor of a given $F(s) \in \mathcal{S}$ is clearly not unique. This is due to *Legendre's duplication formula* (see (1.8), Adrien-Marie Legendre, September 18, 1752-January 10, 1833, French mathematician)

$$\Gamma(s)\Gamma(s + \tfrac{1}{2}) = 2^{1-2s}\sqrt{\pi}\,\Gamma(2s),$$

so that, for example, the functional equation for $\zeta(s)$ can be also written as

$$\begin{aligned}&\Big(\frac{\pi}{2}\Big)^{-s/2}\Gamma\Big(\frac{s}{4}\Big)\Gamma\Big(\frac{s}{4} + \frac{1}{4}\Big)\zeta(s)\\ &= \Big(\frac{\pi}{2}\Big)^{-(1-s)/2}\Gamma\Big(\frac{1-s}{4}\Big)\Gamma\Big(\frac{1-s}{4} + \frac{1}{4}\Big)\zeta(1-s).\end{aligned}$$

Other formulas that can be also used for the transformation of the functional equation are $s\Gamma(s) = \Gamma(s + 1)$, the functional equation for the gamma-function, and the *Legendre-Gauss multiplication formula* (Johann Carl Friedrich Gauss, April 30, 1777-February 23, 1855, famous German mathematician)

$$\Gamma(s) = m^{s-1/2}(2\pi)^{(1-m)/2}\prod_{k=0}^{m-1}\Gamma\Big(\frac{s+k}{m}\Big).$$

In other words, writing

$$\boldsymbol{\lambda} = (\lambda_1, \ldots, \lambda_r), \qquad \boldsymbol{\mu} = (\mu_1, \ldots, \mu_r),$$

the data $(Q, \boldsymbol{\lambda}, \boldsymbol{\mu}, \omega)$ are not uniquely defined by $F(s)$. This gives rise to the notion of an *invariant*, i.e. an expression defined in terms of the data of $F(s)$ which is uniquely defined by $F(s)$ itself. For example, d_F (see (3.4)) is an invariant of $F(s)$ ($\in \mathcal{S}$) (see, e.g., A. Perellli and J. Kaczorowski [KaPe1]-[KaPe5], A. Perelli [Per] and J. Kaczorowski [Kac] for a survey and results on $\mathcal{S}$). Besides d_F, some other invariants in $\mathcal{S}$ are the conductor q_F and the shift θ_F. They are defined as

$$q_F = (2\pi)^{d_F} Q^2 \prod_{j=1}^{r_F} \lambda_j^{2\lambda_j}, \qquad \theta_F = 2\mathbb{I}\mathrm{m}\Big\{\sum_{j=1}^{r_F} (\mu_j - 1/2)\Big\}.$$

A function $F \in \mathcal{S}$ is *primitive* if $F = F_1 F_2$ ($F_1, F_2 \in \mathcal{S}$) implies $F_1 = 1$ or $F_2 = 1$. Every $F \in \mathcal{S}$ can be factored into a product of primitive factors. For a given $k \in \mathbb{N}$ there always exists a function $F(s) \in \mathcal{S}$ which is primitive and satisfies $d_F = k$.

In his seminal paper [Sel] A. Selberg conjectured that, for every $F \in \mathcal{S}$, there exists an integer n_F such that

$$\sum_{p \le x} |a_F(p)|^2 p^{-1} = n_F \log\log x + O(1),$$

while if $F(s)$, $F_1(s)$ are distinct, primitive elements of $\mathcal{S}$, then $n_F = 1$ and

$$\sum_{p \le x} a_F(p) a_{F_1}(p) p^{-1} = O(1),$$

which is called the *orthonormality property*. He also conjectured the analog of the classical RH for $\mathcal{S}$, namely that all non-trivial zeros of $F(s) \in \mathcal{S}$ satisfy $\mathbb{R}\mathrm{e}\, s = 1/2$ (see next section for precise definition). Similarly, as in the case of $\zeta(s)$, the truth of the general RH for $\mathcal{S}$ is a deep, open problem. Selberg's conjectures imply unique factorization into primitive functions in $\mathcal{S}$. Namely suppose that

$$F = F_1^{d_1} \cdots F_r^{d_r} = F_1^{e_1} \cdots F_r^{e_r}$$

with primitive functions $F_1 \cdots F_r$ and d_i, e_i non-negative integers. Suppose that, for example, $d_1 \neq e_1$. We expand both products into Dirichlet series and equate coefficients of $a_F(p)$. It follows that

$$\sum_{i=1}^{r} d_i a_{F_i}(p) = \sum_{i=1}^{r} e_i a_{F_i}(p). \tag{3.11}$$

Upon multiplying both sides of (3.11) by $\overline{a_{F_1}(p)}/p$ and summing over $p \le x$ we obtain

$$d_1 \sum_{p\le x} \frac{|a_{F_1}(p)|^2}{p} + \sum_{i=2}^{r} d_i \sum_{p\le x} \frac{a_{F_i}(p)\overline{a_{F_1}(p)}}{p}$$
$$= e_1 \sum_{p\le x} \frac{|a_{F_1}(p)|^2}{p} + \sum_{i=2}^{r} e_i \sum_{p\le x} \frac{a_{F_i}(p)\overline{a_{F_1}(p)}}{p}.$$

From this relation by Selberg's conjectures it follows that

$$d_1 \log\log x + O(1) = e_1 \log\log x + O(1),$$

implying $d_1 = e_1$, which is impossible if $d_1 \neq e_1$. It is conjectured that $\mathcal{S}$ coincides with the family of automorphic L-functions, but this is a very difficult open problem.

3.4 The zeros of functions in $\mathcal{S}$

The zero-free half-plane $\sigma > 1$ and the functional equation give rise to the familiar notions of critical strip and critical line, i.e. the strip $0 < \sigma < 1$ and the line $\sigma = 1/2$, respectively. There are zeros of $F(s)$ located at the poles of

$$\prod_{j=1}^{r_F} \Gamma(\lambda_j(F)s + \mu_j(F)),$$

i.e. at

$$\rho = -(k + \mu_j(F))/\lambda_j(F), \quad k = 0, 1, 2, \dots, \quad j = 1, \dots, r_F.$$

These zeros, in analogy with $\zeta(s)$, are called *the trivial zeros* of $F(s)$, and have multiplicity $m(\rho)$ equal to the number of pairs (k, j) with $-(k + \mu_j(F))/\lambda_j(F) = \rho$. Clearly, such zeros lie in the half-plane $\sigma \le 0$. The case $\rho = 0$, if present, requires special attention in view of the possible pole of $F(s)$ at $s = 1$. The other (complex) zeros, located inside the critical strip are called *the non-trivial zeros*. We cannot a priori exclude the possibility that $F(s)$ has a trivial and a non-trivial zero at the same point.

Let $N_F(T)$ denote the number of zeros $\rho_F = \beta_F + i\gamma_F$ with $0 < \gamma_F \le T$ and let

$$S_F(T) := \frac{1}{\pi} \arg F(\tfrac{1}{2} + iT).$$

This is the analog of the zero-counting function $N(T)$ for $\zeta(s)$, introduced in Chapter 1, and arg $F(\frac{1}{2}+iT)$ is defined analogously to arg $\zeta(\frac{1}{2}+iT)$. If T is an ordinate of a zero of $F(s)$, then we set $S_F(T)=S_F(T+0)$.

Similarly to (1.31)-(1.32) one obtains, by using the argument principle, that for certain constants c_F, c'_F, we have

$$\begin{aligned} N_F(T) &= \frac{d_F}{2\pi}T\log T + c_F T + S_F(T) + c'_F + O\left(\frac{1}{T}\right) \\ &= \frac{d_F}{2\pi}T\log T + c_F T + O(\log T). \end{aligned} \tag{3.12}$$

Observe that (3.12) determines the degree d_F uniquely, showing in fact that it is an invariant in $\mathcal{S}$.

We note in concluding that, instead of $\mathcal{S}$, one often works with the more general *extended Selberg class* $\mathcal{S}^\sharp$. The class $\mathcal{S}^\sharp$ satisfies the same axioms (a), (b) and (c) as $\mathcal{S}$, but (d) does not have to hold and $a_F(n)\ll_\varepsilon n^\varepsilon$ also does not have to hold. Typical elements of $\mathcal{S}^\sharp$ are the $\mathbb{R}$-linear combinations of functions in $\mathcal{S}$ satisfying the same functional equation; however, $\mathcal{S}^\sharp$ is much more general and contains also functions of different nature. If $\mathcal{S}^\sharp_d$ denotes the subclass of $\mathcal{S}^\sharp$ consisting of the functions with a given degree d, then it is known that $\mathcal{S}^\sharp_d$ is empty for $0<d<1$ and $1<d<2$.

Notes

There are several ways to make such a generalization of $\zeta(s)$ to include general functions of the form (3.1). For example, see the works of H.-E. Richert [Ric] (Hans-Egon Richert, 1924-1993, German mathematician, known for his work on multiplicative number theory and sieve methods), K. Chandrasekharan and R. Narasimhan [ChNa], I. Piatetski-Shapiro [Pia] and E. Carletti *et al.* [CMoP], for example. See also the survey paper [Sar1] of P. Sarnak on L-functions. However, the axioms of the Selberg class $\mathcal{S}$ appear to be more satisfactory than any other set of axioms. For a comprehensive survey of $\mathcal{S}$, see the papers of Kaczorowski-Perelli [KaPe1] and J. Kaczorowski [Kac]. The latter contains, in particular, definitions and basic notions involved with various L-functions mentioned in Section 3.3.

The axioms defining $\mathcal{S}$ may be motivated in various ways. First, the assumption that $F(s)$ has a pole at $s=1$ is not essential, since if $F(s)$ has a pole at $s=a$ $(\neq 1)$, then the shifted L-function $F(s+a-1)$ has a pole at $s=1$. The significance of the point $s=1$ is that it seems to be the only possible pole for an automorphic L-function. Moreover, such a pole seems to be related to the simple pole of $\zeta(s)$, in the sense that $F(s)\zeta^{-k}(s)$ is an entire function for a suitable $k\in\mathbb{N}$. The axioms $a_F(n)\ll_\varepsilon n^\varepsilon$ (the Ramanujan conjecture for $\mathcal{S}$) and (d) are more of *arithmetic nature*, while the other axioms are more of *analytic nature*. The Ramanujan conjecture $a_F(n)\ll_\varepsilon n^\varepsilon$ appears to be crucial in connection with the RH. Namely, if this axiom is violated, but all other axioms of $\mathcal{S}$ hold, then the RH may not hold, as the following example (due to J. Kaczorowki) shows. Let $\chi(n)$ be a primitive Dirichlet character with $\chi(-1)=-1$ and define $G(s)=L(2s-1/2,\chi)$, which is absolutely convergent for $\sigma>3/4$. It satisfies the functional equation (3.2) with $\lambda=1$, $\mu=1/4$.

It also possesses an Euler product allowing the choice $\theta = 1/4$. If we take $0 < \delta < 1/4$ and set

$$F(s) := G(s-\delta)G(s+\delta),$$

then $F(s)$ satisfies all the axioms of $\mathcal{S}$ but the Ramanujan conjecture, and for a suitable choice of δ it cannot have zeros on the line $\mathbb{R}\mathrm{e}\, s = 1/2$.

See also the paper of J. B. Conrey and A. Ghosh [CoGh3] for another proof that if $d_F > 0$, then $d_F \geq 1$.

The restriction $\mathbb{R}\mathrm{e}\,\mu_j(F) \geq 0$ in the functional equation for $F(s)$ comes from the theory of *Maass waveforms* (see, e.g., H. Iwaniec [Iwa1], [Iwa2], [Iwa3] and Y. Motohashi [Mot5]). If one assumes the existence of an arithmetic subgroup of $\mathrm{SL}_2(\mathbb{R})$ together with a Maass cusp form (Hans Maass, June 17, 1911-April 15, 1992, German mathematician who introduced Maass wave forms, Koecher-Maass series and Maass-Selberg relations), which corresponds to the exceptional eigenvalue, and if we further suppose that all the local roots are sufficiently small, then the L-function associated with the Maass cusp form has a functional equation where the corresponding μ_js have negative real parts, but such an L-function does not satisfy the Riemann hypothesis.

As stated in the text, the first to prove Bochner's theorem was H.-E. Richert [Ric], one year before Bochner (Salomon Bochner, August 20, 1899-May 2, 1982), to answer exactly the following question of A. Selberg: is it true that the Riemann zeta-function has minimal degree? Richert's proof is based on *Perron's inversion formula* (Oskar Perron, May 7, 1880-February 22, 1975, German mathematician). For a proof see, for example, the appendix of [Iv1]. Bochner's theorem was rediscovered (independently) by G. Molteni [Mol], who gives yet another proof of the same result.

One way to see the significance of the condition $0 \leq \theta < 1/2$ in axiom (d) of $\mathcal{S}$ is as follows. For $\theta = 1/2$ there are examples violating the Riemann hypothesis, such as the function

$$F(s) := (1-2^{1-s})\zeta(s) = \sum_{n=1}^{\infty}(-1)^{n-1}n^{-s} \qquad (\mathbb{R}\mathrm{e}\, s > 0).$$

Then $F(s)$ has zeros $s = 1 + 2k\pi i/\log 2$ ($k \in \mathbb{Z}$) off the critical line, but does not have a functional equation analogous to (3.2).

The classical Lindelöf hypothesis for $\zeta(s)$ was enunciated by E. Lindelöf [Lin] in 1908 (Ernst Leonard Lindelöf, March 7, 1870-June 4, 1946, noted Finnish topologist). He expressed the belief that $\zeta(s)$ is even bounded for $\sigma \geq \frac{1}{2} + \varepsilon$, which is false. Namely, for $1/2 < \sigma < 1$ fixed one has

$$\log|\zeta(\sigma+it)| = \Omega_+(\log^{1-\sigma} T(\log\log t)^{-\sigma}),$$

which is a result of H. L. Montgomery [Mon1]. However, Lindelöf's conjecture that $\zeta(\frac{1}{2}+it) \ll_\varepsilon |t|^\varepsilon$ is widely believed to be true. It follows from the Riemann hypothesis (see, e.g., [Iv1], chapter 1), and is strong enough so that often it can be used to replace the Riemann hypothesis in practice.

The general Dirichlet series are of the form $F(s) = \sum_{n=1}^{\infty} f(n)\mathrm{e}^{-\lambda_n s}$, where $\{\lambda_n\}$ is an increasing sequence of reals tending to infinity, while if $\lambda_n \equiv \log n$ it reduces to the ordinary Dirichlet series $F(s) = \sum_{n=1}^{\infty} f(n)n^{-s}$. For an account of general Dirichlet series the reader is referred to the classical treatise of G. H. Hardy and M. Riesz [HaRi]. This work contains, in particular, the proofs of all properties of the Lindelöf function $\mu(\sigma)$ mentioned in the text.

The well-known *maximum modulus principle* for regular functions asserts that if $f(s)$ is regular in a domain and if $|f(s)| \leq C$ on the boundary of the domain, then the same inequality holds for all points of the domain. The Phragmén-Lindelöf principle (Lars Edvard Phragmén, September 2, 1863-March 13, 1937, Swedish mathematician) mentioned in the text, is a generalization of the maximum modulus principle for an unbounded domain. Most often it is used for a strip $\sigma_1 \leq \sigma \leq \sigma_2$ in the s-plane and in this case it may be formulated as follows.

Let $f(s)$ be regular in the region

$$\mathcal{D} := \left\{ s = \sigma + it,\ \sigma_1 \le \sigma \le \sigma_2, -\infty < t < \infty \right\},$$

where one has

$$|f(s)| < A \exp(\mathrm{e}^{C|t|}) \qquad \left(A > 0,\ 0 < C < \frac{\pi}{\sigma_2 - \sigma_1} \right).$$

If $|f(s)| \le B$ holds for $\sigma = \sigma_1$ and $\sigma = \sigma_2$ with some absolute constant $B > 0$, then this inequality holds for all $s \in \mathcal{D}$.

In 1975, S. M. Voronin [Vor] (Sergei Mikhailovich Voronin, March 11, 1946-October 18, 1997, Russian mathematician) proved an important result, called the *universality theorem* for $\zeta(s)$. It states the following. Let $0 < r < 1/4$ and suppose that $f(s)$ is a non-vanishing, continuous function for $|s| \le r$ which is analytic for $|s| < r$. Then for any $\varepsilon > 0$ there exists a real number τ (> 0) such that

$$\max_{|s| \le r} \left| \zeta(s + \tfrac{3}{4} + i\tau) - f(s) \right| < \varepsilon. \tag{3.13}$$

Moreover, the set of τ satisfying (3.13) has positive lower density. Hence any such $f(s)$ can be uniformly approximated by purely imaginary shifts of the zeta-function in the critical strip. A good account is to be found in the monograph of A. A. Karatsuba and Voronin himself [KaVo]. Several authors have generalized the universality theorem to various functions of $\mathcal{S}$. We mention here the work by H. Nagoshi and J. Steuding [NaSt]. Let

$$F(s) = \sum_{n=1}^{\infty} a_F(n) n^{-s} \in \mathcal{S}$$

with d_F the degree of $F(s)$ and $\sigma_F = \max(1/2, 1 - 1/d_F)$. Assume that

$$\lim_{x \to \infty} \frac{\log x}{x} \sum_{p \le x} |a_F(p)|^2 = \kappa\ (>0).$$

Let K be a compact set in the strip $D := \{ s \in \mathbb{C} : \sigma_F < \sigma < 1 \}$ with connected complement, and let $g(s)$ be a non-vanishing, continuous function on K which is analytic in the interior of K. Then for any $\varepsilon > 0$

$$\liminf_{T \to \infty} \frac{1}{T} \mu \left\{ \tau \in [0, T] : \max_{s \in K} |F(s + i\tau) - g(s)| < \varepsilon \right\} > 0,$$

where $\mu(\cdot)$ denotes the Lebesgue measure.

There are several standard applications of universality. For example, any universal L-function $F(s)$ is functionally independent and, in particular, it does not satisfy any algebraic differential equation. For this and other consequences of universality see, for example, A. Laurinčikas [Lau5] and J. Steuding [Ste].

In his seminal paper (see [Sel]), A. Selberg states without proof the following result on the value distribution of $F(s) \in \mathcal{S}$. For $\frac{1}{2} \le \sigma \le \frac{1}{2} + (\log T)^{-\delta}$ $(\delta > 0)$ the function

$$\kappa_F(\sigma, t) := \frac{\log F(\sigma + it)}{\sqrt{\pi \sum_{p<t} |a_F(p)|^2 p^{-2\sigma}}}$$

has a normal Gaussian distribution in $\mathbb{C}$. Also, $\mathbb{R}\mathrm{e}\, \kappa_F(\sigma, t)$ and $\mathbb{I}\mathrm{m}\, \kappa_F(\sigma, t)$ have a normal Gaussian distribution on the real line. More precisely, if $\chi_{a,b}(t)$ is the characteristic function of the interval

(a, b), then

$$\int_0^T \chi_{a,b}\Big(\mathbb{R}\mathrm{e}\,\kappa_F(\sigma, t)\Big)\,\mathrm{d}t = T\int_a^b \mathrm{e}^{-\pi u^2}\,\mathrm{d}u + O\left(T\frac{(\log\log\log T)^2}{\sqrt{\log T}}\right),$$

$$\int_0^T \chi_{a,b}\Big(\mathbb{I}\mathrm{m}\,\kappa_F(\sigma, t)\Big)\,\mathrm{d}t = T\int_a^b \mathrm{e}^{-\pi u^2}\,\mathrm{d}u + O\left(T\frac{\log\log\log T}{\sqrt{\log T}}\right).$$

In particular, these results hold for $\zeta(s)$, and they are discussed in the Notes to Chapter 11, and may be proved by the same methods.

The results that $\mathcal{S}_d^\sharp$ is empty for $0 < d < 1$ and $1 < d < 2$ are due to G. Molteni [Mol], Kaczorowski-Perelli [KaPe3], [KaPe4], while the result for $1 < d < 2$ is to be found in their recent work [KaPe5].

4

The approximate functional equations for $\zeta^k(s)$

4.1 A simple AFE for $\zeta(s)$

We recall the AFE (2.3), which we write again as

$$Z(t) = 2 \sum_{n \le \sqrt{t/(2\pi)}} n^{-1/2} \cos\left(t \log \frac{\sqrt{t/(2\pi)}}{n} - \frac{t}{2} - \frac{\pi}{8}\right) + O\left(t^{-1/4}\right). \tag{4.1}$$

The purpose of this section is to provide a more thorough discussion of the AFEs for $\zeta^k(s)$, $k \in \mathbb{N}$ and fixed k. Most of the results can be generalized to more general L-functions, notably to those belonging to the Selberg class $\mathcal{S}$. However, the accent here is to acquaint the reader with the methods used in obtaining AFEs, and not to provide the results in any form of generality. If one wants to obtain AFEs for $Z^k(t)$, one only has to use the AFEs for $(\zeta(\frac{1}{2} + it))^k$ and the asymptotic formula for $\chi(\frac{1}{2} + it)$ (see the defining relation (1.4) for $Z(t)$). Recall that the first approximation to $\chi(s)$ is (see (2.17))

$$\chi(s) = \left(\frac{2\pi}{t}\right)^{\sigma + it - 1/2} \mathrm{e}^{i(t+\pi/4)} \cdot \left(1 + O\left(\frac{1}{t}\right)\right) \qquad (t \ge t_0 > 0), \tag{4.2}$$

and that the O-term in (4.2) admits an asymptotic expansion in negative powers of t.

The first result which we present is an AFE for $\zeta(s)$ which is of a rather simple nature (see, e.g., [Iv1, chapter 1]). This is

Theorem 4.1 *For* $0 < \sigma_0 \le \sigma \le 2$, $x \ge |t|/\pi$, $s = \sigma + it$,

$$\zeta(s) = \sum_{n \le x} n^{-s} + \frac{x^{1-s}}{s-1} + O(x^{-\sigma}). \tag{4.3}$$

Proof of Theorem 4.1 We have, for $\mathbb{R}\mathrm{e}\, s > 1$ and $N \geq 2$,

$$\sum_{n>N} n^{-s} = \int_N^\infty \tau^{-s}\,\mathrm{d}[\tau] = -N^{1-s} + s\int_N^\infty [\tau]\tau^{-s-1}\,\mathrm{d}\tau$$
$$= \frac{N^{1-s}}{s-1} - \tfrac{1}{2}N^{-s} - s\int_N^\infty \psi(\tau)\tau^{-s-1}\,\mathrm{d}\tau,$$

where $\psi(x) = x - [x] - \frac{1}{2}$. Therefore

$$\zeta(s) = \sum_{n\leq N} n^{-s} + \sum_{n>N} n^{-s}$$
$$= \sum_{n\leq N} n^{-s} + \frac{N^{1-s}}{s-1} - \tfrac{1}{2}N^{-s} - s\int_N^\infty \psi(\tau)\tau^{-s-1}\,\mathrm{d}\tau, \tag{4.4}$$

and by analytic continuation (4.4) is valid for $\sigma > 0$, the last summand being $\ll (1+|t|)N^{-\sigma}$. If $u \geq x$ (≥ 1), we set

$$A(u) := \sum_{x<n\leq u} n^{-it},$$

and apply the following elementary, standard lemma (see [Iv1, chapter 1]) from the theory of exponential sums ($\mathrm{e}(x) := \exp(2\pi i x)$).

Lemma 4.2 *Let $f(x)$ be a real-valued function on the interval $[a, b]$, and let $f'(x)$ be continuous and monotonic on $[a, b]$ and $|f'(x)| \leq \delta < 1$. Then*

$$\sum_{a<n\leq b} \mathrm{e}(f(n)) = \int_a^b \mathrm{e}(f(x))\mathrm{d}x + O\Big((1-\delta)^{-1}\Big).$$

We apply Lemma 4.2 with

$$f(x) = \frac{1}{2\pi}|t|\log x, \quad \delta = \frac{1}{2},$$

provided that $x \geq |t|/\pi$. This gives

$$A(u) = \int_x^u y^{-it}\,\mathrm{d}y + O(1) = \frac{u^{1-it} - x^{1-it}}{1-it} + O(1).$$

For $x \leq N$ partial summation gives

$$\sum_{x<n\leq N} n^{-s} = \sigma\int_x^N u^{-\sigma-1}A(u)\,\mathrm{d}u + A(N)N^{-\sigma}$$
$$= \sigma\int_x^N \frac{u^{-s} - u^{-\sigma-1}x^{1-it}}{1-it}\,\mathrm{d}u + O(x^{-\sigma} + xN^{-\sigma}) + \frac{N^{1-\sigma-it}}{1-it}$$
$$= \frac{N^{1-s}}{1-s} - \frac{x^{1-s}}{1-s} + O(x^{-\sigma} + xN^{-\sigma}).$$

Substituting this expression in (4.4) we finally have, for $\sigma_0 \le \sigma \le 2$,

$$\zeta(s) = \sum_{n \le x} n^{-s} + \frac{x^{1-s}}{1-s} + O(x^{-\sigma}) + O(xN^{-\sigma} + |t|N^{-\sigma}).$$

If we let $N \to \infty$ we obtain (4.3). The basic idea of the proof is to estimate the tails in the series for $\zeta(s)$ by Lemma 4.2. This, however, is particular to the sums of n^{-s}, when the corresponding integral of x^{-s} can be easily evaluated. Thus Theorem 4.1 cannot be easily generalized to other L-functions. Note also that the functional equation for $\zeta(s)$ was not used in the proof, which is one way to show the elementary nature of Theorem 4.1.

4.2 The Riemann-Siegel formula

Theorem 4.3 *Let* $0 \le \sigma \le 1$; $x, y, t > C > 0$; $2\pi xy = t$. *Then uniformly in* σ

$$\zeta(s) = \sum_{n \le x} n^{-s} + \chi(s) \sum_{n \le y} n^{s-1} + O(x^{-\sigma}) + O(t^{1/2-\sigma} y^{\sigma-1}). \tag{4.5}$$

Theorem 4.3 is a classical result of G. H. Hardy and J. E. Littlewood [HaLi3]. A corresponding formula also holds (since $\zeta(\bar{s}) = \overline{\zeta(s)}$) if $t < 0$, with t replaced by $|t|$ in the error term and $2\pi xy = t$. From (4.2) and (4.5) one obtains, on taking $\sigma = 1/2$, $x = y = \sqrt{t/(2\pi)}$, the AFE (4.1) for $Z(t)$. Hardy and Littlewood also proved (*op. cit.*) the analog of (4.5) for $\zeta^2(s)$. This is as follows.

Theorem 4.4 *Let* $0 < \sigma < 1$; $x, y, t > C > 0$; $4\pi^2 xy = t^2$. *Then uniformly in* σ

$$\zeta^2(s) = \sum_{n \le x} d(n) n^{-s} + \chi^2(s) \sum_{n \le y} d(n) n^{s-1} + O(x^{1/2-\sigma} \log t). \tag{4.6}$$

Remark 4.5 From (4.6) with $x = y = t/(2\pi)$, $\sigma = \frac{1}{2}$ and (4.2) we obtain

$$Z^2(t) = 2 \sum_{n \le t/(2\pi)} d(n) n^{-1/2} \cos\left(t \log \frac{t}{2\pi} - t - \frac{\pi}{4}\right) + O(\log t). \tag{4.7}$$

Remark 4.6 The error terms both in (4.5) and (4.6) are best possible (see [Iv1, chapter 4]) in the sense that there are values of x, y, t for which the corresponding error term(s) are of the exact order of magnitude. Theorem 4.3 is a special case of the full *Riemann-Siegel formula*. The result is due to C. L. Siegel [Sie] (Carl Ludwig Siegel, December 31, 1896-April 4, 1981, German mathematician) who was inspired by looking at Riemann's notes on the zeta-function in the Göttingen University library. This is a remarkable achievement and represents one of the deepest results of zeta-function theory. It states the following.

If $0 \le \sigma \le 1$, $m = [\sqrt{t/(2\pi)}]$, $N < At$, where $A > 0$ is sufficiently small, then

$$\begin{aligned}\zeta(s) &= \sum_{n\le m} n^{-s} + \chi(s)\sum_{n\le m} n^{s-1}\\ &\quad+(-1)^{m-1}\mathrm{e}\Big(\frac{1-s}{4}\Big)(2\pi t)^{(s-1)/2}\exp\left(-\frac{it}{2}-\frac{i\pi}{8}\right)\Gamma(1-s)\\ &\quad\times\left\{S_N + O((ANt^{-1})^{N/6})\right\} + O(\mathrm{e}^{-At}),\end{aligned} \tag{4.8}$$

where

$$S_N := \sum_{n=0}^{N-1}\sum_{\nu\le n/2}\frac{n!\,i^{\nu-n}}{\nu!(n-2\nu)!2^n}(2/\pi)^{(n-2\nu)/2}a_m\Psi^{(n-2\nu)}\left(\frac{\eta}{\pi}-2m\right). \tag{4.9}$$

Here the coefficients a_n are given by the recurrence relation

$$(n+1)t^{1/2}a_{n+1} = (\sigma-n-1)a_n + ia_{n-2}$$
$$a_{-2} = a_{-1} = 0, \quad a_0 = 1, \quad (n = 0, 1, 2, \ldots).$$

We have

$$a_n \ll ((5\mathrm{e}/(2n)t^{-1/2})^{n/3}, \ \eta = \sqrt{2\pi t}, \ \Psi(z) = \left(\cos\pi(\tfrac{1}{2}z^2 - z - \tfrac{1}{8})\right)/\cos\pi z.$$

Remark 4.7 Taking $N = 3$ in the Riemann-Siegel formula and simplifying the resulting expression we obtain

$$\begin{aligned}Z(t) &= 2\sum_{n\le m} n^{-1/2}\cos(\theta(t) - \log n)\\ &\quad+(-1)^{m-1}\Big(\frac{2\pi}{t}\Big)^{1/4}\frac{\cos\{t-(2m+1)\sqrt{2\pi t}-\pi/8\}}{\cos\sqrt{2\pi t}} + O(t^{-3/4}).\end{aligned} \tag{4.10}$$

The asymptotic formula (4.10) clearly sharpens (4.1).

Remark 4.8 Taking $t = 2\pi k^2$, $k \in \mathbb{N}$, it is seen that the second main term in (4.10) is $\gg t^{-1/4}$, showing that the error term in (4.5) of Theorem 4.3 is the best possible.

Remark 4.9 We shall prove only Theorem 4.3, as the proofs of (4.6) and the Riemann-Siegel formula (4.8)-(4.9) are more involved (a detailed analysis is given, for example, by H. M. Edwards [Edw]). However, we remark that its proof actually yields, for complex s not equal to the poles of the gamma-factors (this is equation (56) of C. L. Siegel [Sie]),

$$\pi^{-s/2}\Gamma\Big(\frac{s}{2}\Big)\zeta(s)$$
$$= \pi^{-s/2}\Gamma\Big(\frac{s}{2}\Big)\int\limits_{0\swarrow 1} \frac{e^{i\pi x^2}x^{-s}}{e^{i\pi x}-e^{-i\pi x}}\,dx + \pi^{(s-1)/2}\Gamma\Big(\frac{1-s}{2}\Big)\int\limits_{0\searrow 1} \frac{e^{-i\pi x^2}x^{s-1}}{e^{i\pi x}-e^{-i\pi x}}\,dx. \tag{4.11}$$

Here $0 \swarrow 1$ (resp. $0 \searrow 1$) denotes a straight line that starts from infinity in the upper complex half-plane, has slope equal to 1 (resp. to -1), and cuts the real axis between 0 and 1. Setting in (4.11) $s = \frac{1}{2} + it$ we obtain

$$Z(t) \;=\; 2\,\mathbb{R}\mathrm{e}\left(e^{-i\theta(t)}\int\limits_{0\searrow 1} \frac{e^{-i\pi z^2}z^{-1/2+it}}{e^{i\pi z}-e^{-i\pi z}}\,dz\right).$$

As $\mathbb{R}\mathrm{e}\,(iw) = -\mathbb{I}\mathrm{m}\ w$, this can be conveniently written as

$$Z(t) \;=\; \mathbb{I}\mathrm{m}\left(e^{-i\theta(t)}\int\limits_{0\searrow 1} e^{-i\pi z^2}z^{-1/2+it}\,\frac{dz}{\sin(\pi z)}\right). \tag{4.12}$$

To obtain (4.8) from (4.12) one moves the line of integration to the right, keeping it parallel, and cutting the x-axis between m and $m+1$, with $m = [\sqrt{t/(2\pi)}]$ as in (4.8). The sine function in (4.12) has simple poles at integer points. Thus we have to pick the contribution from the residues at $z = 1, 2, \ldots, m$. This will furnish the main term in (4.10). The remaining integral is evaluated by the *saddle point technique* for the evaluation of exponential integrals (see, e.g., [Iv1, chapter 2]). This analysis is delicate and involves a lot of technical details, which are given in [Sie], [Edw] and [Gab].

Proof of Theorem 4.3 It is possible to extend the range for σ in Theorem 4.3 to any strip $-\sigma_0 < \sigma < \sigma_0$ by some minor changes in the argument. This is, however, not important, since it is the critical strip $0 \le \sigma \le 1$ that is the most interesting range for σ. For $\sigma > 0$ we have

$$\int_0^\infty x^{s-1}e^{-nx}\,dx = n^{-s}\int_0^\infty y^{s-1}e^{-y}\,dy = n^{-s}\Gamma(s). \tag{4.13}$$

Hence for $\sigma > 1$ and $m \geq 2$ we obtain from (4.13)

$$\begin{aligned}\zeta(s) &= \sum_{n\leq m} n^{-s} + \sum_{n>m} n^{-s} \\ &= \sum_{n\leq m} n^{-s} + \frac{1}{\Gamma(s)} \int_0^\infty x^{s-1} \left(\sum_{n<m} \mathrm{e}^{-nx} \right) \mathrm{d}x \\ &= \sum_{n\leq m} n^{-s} + \frac{1}{\Gamma(s)} \int_0^\infty \frac{x^{s-1}\mathrm{e}^{-mx}}{\mathrm{e}^x - 1} \mathrm{d}x,\end{aligned}$$

where the inversion of the order of summation and integration is justified by absolute convergence. Now consider the integral

$$I(s) = I(s;m) := \int_C \frac{z^{s-1}\mathrm{e}^{-mz}}{\mathrm{e}^z - 1} \mathrm{d}z,$$

where C is the contour that starts at infinity on the positive real axis, encircles the origin once in the positive direction (excluding the points $\pm 2\pi i, \pm 4\pi i, \ldots$), and returns to infinity. On $I(s)$ we have

$$z^{s-1} = \exp\Big((s-1)\log z\Big),$$

where the logarithm is real at the beginning of the contour, and its imaginary part varies from 0 to 2π around the contour. For $0 < \varepsilon < 1$ we may write

$$\begin{aligned}I(s) &= -\int_\varepsilon^\infty \frac{x^{s-1}\mathrm{e}^{-mx}}{\mathrm{e}^x - 1} \mathrm{d}x + \int_{|z|=\varepsilon} \frac{z^{s-1}\mathrm{e}^{-mz}}{\mathrm{e}^z - 1} \mathrm{d}z + \int_\varepsilon^\infty \frac{(\mathrm{e}^{2\pi i}x)^{s-1}\mathrm{e}^{-mx}}{\mathrm{e}^x - 1} \mathrm{d}x \\ &= (\mathrm{e}^{2\pi i s} - 1) \int_0^\infty \frac{x^{s-1}\mathrm{e}^{-mx}}{\mathrm{e}^x - 1} \mathrm{d}x,\end{aligned}$$

on letting $\varepsilon \to 0$. The loop integral $I(s)$ is uniformly convergent in any finite region in the s-plane, and thus defines an integral function of s. Hence by analytic continuation we obtain, if $s \notin \mathbb{Z}$,

$$\begin{aligned}\zeta(s) &= \sum_{n\leq m} n^{-s} + \frac{I(s)}{\Gamma(s)(\mathrm{e}^{2\pi i s} - 1)} \\ &= \sum_{n\leq m} n^{-s} + \frac{\mathrm{e}^{-\pi i s}\Gamma(1-s)}{2\pi i} \int_C \frac{w^{s-1}\mathrm{e}^{-mw}}{\mathrm{e}^w - 1} \mathrm{d}w, \qquad (4.14)\end{aligned}$$

if we use the identity

$$\Gamma(s)\Gamma(1-s) = \frac{\pi}{\sin(\pi s)}.$$

Now let $x \leq y$, so that

$$1 \ll x \leq (t/(2\pi))^{1/2}, \quad m = [x], \quad y = t/(2\pi x), \quad q = [y], \quad \eta = 2\pi y.$$

The contour $\mathcal{C}$ in (4.14) is replaced by the straight lines C_1, C_2, C_3, C_4 joining the points

$$\infty,\ c\eta + i\eta(1+c),\ -c\eta + i\eta(1-c),\ -c\eta - (2q+1)\pi i,\ \infty,$$

where $0 < c \le \frac{1}{2}$ is an absolute constant, and if y is an integer a small indentation is made above the pole $w = i\eta$ of the integrand in (4.14). The residue theorem gives

$$\zeta(s) = \sum_{n\le m} n^{-s} - 2\pi i \sum \mathrm{Res} + \frac{\mathrm{e}^{-\pi i s}\Gamma(1-s)}{2\pi i}\Big(\int_{C_1} + \int_{C_2} + \int_{C_3} + \int_{C_4}\Big),$$

where summation is over the residues of

$$F(w) := \frac{\mathrm{e}^{-\pi i s}\Gamma(1-s)}{2\pi i(\mathrm{e}^w - 1)}\mathrm{e}^{-mw}w^{s-1}$$

at the points $w = \pm 2n\pi i$ ($n = 1, 2, \ldots, q$). The residues at $\pm 2n\pi i$ contribute together, on using (1.7),

$$\begin{aligned}
&\frac{\mathrm{e}^{-\pi i s}\Gamma(1-s)}{2\pi i}\left\{(2ni\pi)^{s-1} + (-2ni\pi)^{s-1}\right\}\\
&= \frac{(2n\pi)^{s-1}\mathrm{e}^{-\pi i s}\Gamma(1-s)}{2\pi i}\left(\mathrm{e}^{\pi i(s-1)/2} + \mathrm{e}^{3\pi i(s-1)/2}\right)\\
&= \frac{(2n\pi)^{s-1}\Gamma(1-s)}{2\pi i}(-i\mathrm{e}^{-\pi i s/2} + i\mathrm{e}^{\pi i s/2})\\
&= -\frac{2^s(n\pi)^{s-1}\Gamma(1-s)}{2\pi i}\sin\frac{\pi s}{2} = -(2\pi i)^{-1}n^{s-1}\chi(s).
\end{aligned}$$

Therefore we have

$$\zeta(s) = \sum_{n\le m} n^{-s} + \chi(s)\sum_{n\le q} n^{s-1} + \frac{\mathrm{e}^{-\pi i s}\Gamma(1-s)}{2\pi i}\Big(\int_{C_1} + \int_{C_2} + \int_{C_3} + \int_{C_4}\Big), \tag{4.15}$$

and the proof reduces to showing that the last expression in (4.15) is

$$\ll x^{-\sigma} + t^{1/2-\sigma}y^{\sigma-1}.$$

Let $w = u + iv = \rho\mathrm{e}^{i\varphi}$, $0 < \varphi < 2\pi$, so that $|w^{s-1}| = \rho^{\sigma-1}\mathrm{e}^{-t\varphi}$. On C_4 we have $\varphi \ge \frac{5}{4}$, $\rho > A\eta$, $|\mathrm{e}^w - 1| > A$, where $A > 0$ denotes (possibly different) absolute constants. Hence

$$\int_{C_4} \ll \eta^{\sigma-1}\mathrm{e}^{-5\pi t/4}\int_{-c\eta}^{\infty}\mathrm{e}^{-mu}\,du \ll \exp(mc\eta - 5\pi t/4) \ll \exp(tc - 5\pi t/4).$$

On C_3 we have

$$\begin{aligned}\varphi &\geq \frac{\pi}{2} + \arctan\frac{c}{1-c} = \frac{\pi}{2} + \int_0^{c/(1-c)} \frac{\mathrm{d}x}{1+x^2}\\ &> \frac{\pi}{2} + A + \int_0^{c/(1-c)} \frac{\mathrm{d}x}{(1+x)^2} = \frac{\pi}{2} + A + c.\end{aligned}$$

Hence on C_3

$$w^{s-1}\mathrm{e}^{-mw} \ll \eta^{\sigma-1}\exp\Big\{-t(\pi/2 + A + c) + mc\eta\Big\} \ll \eta^{\sigma-1}\mathrm{e}^{-t(\pi/2+A)}.$$

Since $|\mathrm{e}^w - 1| > A$ we obtain

$$\int_{C_3} \ll \eta^{\sigma}\mathrm{e}^{-t(\pi/2+A)}.$$

On C_1 we have $|\mathrm{e}^w - 1| > A\mathrm{e}^u$. Therefore

$$\frac{w^{s-1}\mathrm{e}^{-mw}}{\mathrm{e}^w - 1} \ll \eta^{\sigma-1}\exp\left(-t\arctan\frac{\eta + c\eta}{u} - mu - u\right).$$

Since $m + 1 \geq x = t/\eta$ and

$$\frac{\mathrm{d}}{\mathrm{d}u}\left\{\arctan\frac{\eta+c\eta}{u} + \frac{u}{\eta}\right\} = -\frac{(1+c)u}{u^2 + (1+c)^2\eta^2} + \frac{1}{\eta} > 0,$$

we obtain

$$\begin{aligned}\arctan\frac{\eta+c\eta}{u} + \frac{u}{\eta} &\geq \arctan\frac{1+c}{c} + c\\ &= \frac{\pi}{2} + c - \arctan\frac{c}{1+c} = \frac{\pi}{2} + A,\end{aligned}$$

since for $0 < \theta < 1$

$$\arctan\theta < \int_0^{\theta} \frac{\mathrm{d}x}{(1-x)^2} = \frac{\theta}{1-\theta}.$$

Hence

$$\begin{aligned}\int_{C_1} &\ll \eta^{\sigma-1}\int_0^{\pi\eta} \mathrm{e}^{-(\pi/2+A)t}\,\mathrm{d}u + \eta^{\sigma-1}\int_{\pi\eta}^{\infty} \mathrm{e}^{-xu}\,\mathrm{d}u\\ &\ll \eta^{\sigma}\mathrm{e}^{-(\pi/2+A)t} + \eta^{\sigma-1}\mathrm{e}^{-\pi\eta x} \ll \eta^{\sigma}\mathrm{e}^{-(\pi/2+A)t}.\end{aligned}$$

Recalling that

$$\mathrm{e}^{-\pi i s}\Gamma(1-s) \ll t^{1/2-\sigma}\mathrm{e}^{\pi t/2} \tag{4.16}$$

we obtain

$$\mathrm{e}^{-\pi i s}\Gamma(1-s)\Big(\int_{C_1} + \int_{C_3} + \int_{C_4}\Big) \ll \mathrm{e}^{-At} \ll x^{-\sigma} + t^{1/2-\sigma}y^{\sigma-1},$$

and therefore it remains only to deal with $\int_{C_2}$. Here we have

$$w = i\eta + \lambda \mathrm{e}(1/8), \quad -\sqrt{2}c\eta \le \lambda \le \sqrt{2}c\eta.$$

Hence

$$\begin{aligned} w^{s-1} &= \exp\Big\{(s-1)\Big(\frac{\pi i}{2} + \log(\eta + \lambda \mathrm{e}^{-\pi i/4})\Big)\Big\} \\ &= \exp\left\{(s-1)\left(\frac{\pi i}{2} + \log\eta + \frac{\lambda}{\eta}\mathrm{e}^{-\pi i/4} - \frac{\lambda^2}{2\eta^2}\mathrm{e}^{-\pi i/2} + O(\lambda^3\eta^{-3})\right)\right\} \\ &\ll \eta^{\sigma-1}\exp\left\{\left(-\frac{\pi}{2} + \frac{\lambda}{\sqrt{2}\eta} - \frac{\lambda^2}{2\eta^2} + O(\lambda^3\eta^{-3})\right)t\right\}. \end{aligned}$$

Note that for $u \ge 0$ we have

$$\frac{\mathrm{e}^{-mw+xw}}{\mathrm{e}^w - 1} \ll \frac{\mathrm{e}^{(x-m-1)u}}{1 - \mathrm{e}^{-u}},$$

while for $u < 0$

$$\frac{\mathrm{e}^{-mw+xw}}{\mathrm{e}^w - 1} \ll \frac{\mathrm{e}^{(x-m)u}}{\mathrm{e}^u - 1},$$

which is bounded for $|u| > \pi/2$. Since

$$|\mathrm{e}^{-xw}| = \mathrm{e}^{-\lambda t/(\eta\sqrt{2})},$$

we see that the part of $\int_{C_2}$ with $|u| > \pi/2$ is

$$\begin{aligned} &\ll \eta^{\sigma-1}\mathrm{e}^{-\pi t/2}\int_{-c\eta\sqrt{2}}^{c\eta\sqrt{2}} \exp\left(-\frac{\lambda^2 t}{2\eta^2} + O(\lambda^3 t\eta^{-3})\right) \mathrm{d}\lambda \\ &\ll \eta^{\sigma-1}\mathrm{e}^{-\pi t/2}\int_{-\infty}^{\infty} \exp\left(-\frac{A\lambda^2 t}{\eta^2}\right) \mathrm{d}\lambda \\ &\ll \eta^{\sigma} t^{-1/2}\mathrm{e}^{-\pi t/2}. \end{aligned}$$

The arguments used above apply also if $|u| \le \pi/2$ and $|\mathrm{e}^w - 1| > A$. Otherwise suppose that the contour goes too near to the pole at $w = 2q\pi i$. Take it around the arc of the circle

$$w = 2q\pi i + \frac{\pi}{2}\mathrm{e}^{i\theta} \qquad (0 \le \theta \le 2\pi).$$

Therefore

$$\begin{aligned} \log(w^{s-1}\mathrm{e}^{-mw}) &= -\frac{m}{2}\pi\mathrm{e}^{i\theta} + (s-1)\left\{\frac{\pi i}{2} + \log\left(2q\pi + \frac{1}{2i}\pi\mathrm{e}^{i\theta}\right)\right\} \\ &= -\frac{m}{2}\pi\mathrm{e}^{i\theta} - \frac{\pi}{2}t + (s-1)\log(2q\pi) + \frac{t\mathrm{e}^{i\theta}}{4q} + O(1) \\ &= -\frac{\pi}{2}t + (s-1)\log(2q\pi) + O(1), \end{aligned}$$

since for $x \le y, q = [y]$ we have

$$m\pi - \frac{t}{2q} = \frac{2mq\pi - t}{2q}$$
$$= \frac{2\pi}{2q}\Big\{[x][y] - ([x] + O(1))([y] + O(1))\Big\} = O(1).$$

Hence

$$w^{s-1}\mathrm{e}^{-mw} \ll q^{\sigma-1}\mathrm{e}^{-\pi t/2},$$

and the contribution of this part is

$$\ll \eta^{\sigma-1}\mathrm{e}^{-\pi t/2}.$$

Using (4.16) again we finally have

$$\mathrm{e}^{-\pi is}\Gamma(1-s)\int_{C_2} \ll t^{1/2-\sigma}(\eta^\sigma t^{-1/2} + \eta^{\sigma-1}) \ll x^{-\sigma} + t^{-1/2}x^{1-\sigma},$$

which in view of (4.15) proves the theorem for $x \le y$.

To deal with the case $x \ge y$ change s into $1-s$ in (4.5). Then for $x \le y$

$$\zeta(1-s) = \sum_{n\le x} n^{s-1} + \chi(1-s)\sum_{n\le y} n^{-s} + O(x^{\sigma-1}) + O(t^{\sigma-1/2}y^{-\sigma}).$$

Hence by the functional equation for $\zeta(s)$ and (2.16)

$$\zeta(s) = \chi(s)\zeta(1-s) = \chi(s)\sum_{n\le x} n^{s-1} + \chi(s)\chi(1-s)\sum_{n\le y} n^{-s}$$
$$+ O(t^{1/2-\sigma}x^{\sigma-1}) + O(y^{-\sigma}).$$

Using $\chi(s)\chi(1-s) = 1$ we have

$$\zeta(s) = \sum_{n\le y} n^{-s} + \chi(s)\sum_{n\le x} n^{s-1} + O(t^{1/2-\sigma}x^{\sigma-1}) + O(y^{-\sigma}), \tag{4.17}$$

for $x \le y$. Interchanging x and y in (4.17) we obtain (4.5) with $x \ge y$.

4.3 The AFE for the powers of $\zeta(s)$

We pass now to the analog of Theorem 4.3 for $\zeta^k(s)$ when $k \ge 2$ is a fixed integer. When $k = 2$ we have Hardy and Littlewood's theorem 4.4, namely

$$\zeta^2(s) = \sum_{n\le x} d(n)n^{-s} + \chi^2(s)\sum_{n\le y} d(n)n^{s-1} + O(x^{1/2-\sigma}\log t) \tag{4.18}$$

when $x, y, t > C > 0, 4\pi^2 xy = t^2$. The result holds uniformly for $0 < \sigma < 1$ with the number of divisors function $d(n)$. We shall formulate and prove the general result (which does not have the error term as good as the one in (4.18)), and the interested reader may look for a proof of (4.18) in [Mot1], who relates the error term to the error term in the Dirichlet divisor problem, or [Iv1, chapter 4]. The general result, proved by the present author in [Iv1, chapter 4], is based on the method of Hardy and Littlewood [HaLi3], in the form given by R. Wiebelitz [Wie]. Best known estimates for power moments of $\zeta(s)$ lead to overall improvements of Wiebelitz's results, but as k grows the order of the error terms in the AFE becomes rather large, which is to be only expected. Thus for practical reasons the detailed analysis is carried out only for $3 \le k \le 12$. The method, where the functional equation for $\zeta(s)$ plays a crucial rôle, obviously generalizes to the functions of Selberg's class.

To formulate the results, first we introduce some notation. Let

$$X(s) := \chi^k(s), \quad \log T := -\frac{X'(\frac{1}{2}+it)}{X(\frac{1}{2}+it)} = -k\frac{\chi'(\frac{1}{2}+it)}{\chi(\frac{1}{2}+it)}, \tag{4.19}$$

and furthermore as in the proof of Theorem 4.3 we may assume, without loss of generality, that $t > 0$. From the functional equation (1.5) and the definition of $\chi(s)$ in (1.7), it follows that

$$\frac{\chi'(s)}{\chi(s)} = \log\pi - \frac{1}{2}\frac{\Gamma'(\frac{1}{2}(1-s))}{\Gamma(\frac{1}{2}(1-s))} - \frac{1}{2}\frac{\Gamma'(\frac{1}{2}s)}{\Gamma(\frac{1}{2}s)}.$$

Hence for $s = 1/2 + it$ the above equation shows that T, as defined by (4.19), is real and moreover using (2.16) we have

$$-\frac{\chi'(\frac{1}{2}+it)}{\chi(\frac{1}{2}+it)} = -\log 2\pi + \log t + O(t^{-2}).$$

Combining this relation with (4.19) we obtain

$$T = \left(\frac{t}{2\pi}\right)^k \left(1 + O\left(\frac{1}{t^2}\right)\right). \tag{4.20}$$

Further we suppose that $xy = (t/(2\pi))^k$, $0 < \sigma < 1$, and we define

$$R_x(s) = \frac{x^{1-s}}{1-s}\sum_{\nu=0}^{k-1}\sum_{\rho=\nu+1}^{k} a_{-\rho,k} r_{\rho,\nu}(1-s)^{1+\nu-\rho}\log^\nu x, \tag{4.21}$$

where $a_{j,k}$ is the coefficient of $(s-1)^j$ in the Laurent expansion of $\zeta^k(s)$ at $s=1$ and

$$r_{j,m} = \frac{1}{m!}\sum_{i=0}^{j-m-1}(-1)^i\binom{k+i-1}{i}\binom{k-1}{j-i-m-1}.$$

Therefore in the general case we have

$$R_x(s) + \chi^k(s)R_y(1-s) \ll x^{1/2-\sigma}t^{-1}(x+y)^{1/2}\log^{k-1}t, \tag{4.22}$$

while for $k=3$ we may write, for some absolute constant D,

$$R_x(s) = \frac{x^{1-s}}{1-s}\left(\tfrac{1}{2}\log^2 x + 3\gamma\log x + D\right) + O(x^{1-\sigma}t^{-2}\log t). \tag{4.23}$$

Finally let c be an upper bound for

$$\mu(1/2) = \limsup_{t\to\infty}\frac{\log|\zeta(\frac{1}{2}+it)|}{\log t}.$$

We shall now give a proof of the AFE for $\zeta^k(s)$ for $3 \le k \le 12$, although it has been already remarked that the method of proof may be used both when $k=2$ and when $k > 13$, but in the latter case the error terms tend to be large as k increases. The AFE that will be proved is contained in the following theorem.

Theorem 4.10 *With the notation introduced above we have*

$$\zeta^k(s) = \sum_{n\le x} d_k(n)n^{-s} + \chi^k(s)\sum_{n\le y} d_k(n)n^{s-1} - R_x(s)$$
$$-\chi^k(s)R_y(1-s) + \Delta_k(x,y), \tag{4.24}$$

where

$$xy = (t/(2\pi))^k, 0 < \sigma < 1; x, y, t > C > 0$$

and $\Delta_k(x,y) = \Delta_k(x,y;t)$ *may be considered as an error term which depends on* k, x, y *and* t. *We have uniformly in* σ

$$\Delta_3(x,y) \ll_\varepsilon x^{1/2-\sigma}t^{1/12+\varepsilon}, \quad \Delta_4(x,y) \ll_\varepsilon x^{1/2-\sigma}t^{1/6+\varepsilon}, \tag{4.25}$$

and

$$\Delta_k(x,y) \ll_\varepsilon t^\varepsilon\Big\{x^{1/2-\sigma}\min(x^{1/2},y^{1/2})t^{-2}$$
$$+(x+y)^{1/2}x^{1/2-\sigma}t^{-\beta} + x^{1/2-\sigma}t^{\frac{1}{3}+\frac{461}{3280}(k-4)}\Big\} \tag{4.26}$$

for $5 \le k \le 12$, *where* (see (4.30))

$$\beta = \frac{8}{3(k-4)} + \frac{8c+1}{2}, \quad c = \frac{32}{205}.$$

Remark 4.11 In case $k = 3$ or $k = 4$ we obtain the following expressions, on taking in both cases $x = y$, $\sigma = \frac{1}{2}$:

$$\begin{aligned} Z^3(t) &= 2 \sum_{n \le (t/(2\pi))^{3/2}} d_3(n) n^{-1/2} \cos\left(t \log\left(\frac{(\frac{t}{2\pi})^{3/2}}{n}\right) - \frac{3t}{2} - \frac{3\pi}{8}\right) \\ &\quad + O_\varepsilon(t^{1/12+\varepsilon}), \\ Z^4(t) &= 2 \sum_{n \le (t/(2\pi))^{2}} d_4(n) n^{-1/2} \cos\left(t \log\left(\frac{(\frac{t}{2\pi})^{2}}{n}\right) - 2t - \frac{\pi}{2}\right) \\ &\quad + O_\varepsilon(t^{1/6+\varepsilon}). \end{aligned} \tag{4.27}$$

In general, one expects to have

$$\begin{aligned} Z^k(t) &= 2 \sum_{n \le (t/(2\pi))^{k/2}} d_k(n) n^{-1/2} \cos\left(t \log\left\{\frac{(\frac{t}{2\pi})^{k/2}}{n}\right\} - \frac{kt}{2} - \frac{k\pi}{8}\right) \\ &\quad + \text{“error term”}, \end{aligned}$$

although it is hard even to speculate on the size of the “error term” in such a formula. Even for the error terms in (4.27) (cases $k = 3$ and $k = 4$), where we have fairly good upper bounds, it is unclear what should be lower bounds (i.e. omega results).

Remark 4.12 Theorem 4.10 is a new result. It improves on theorem 4.4 of [Iv1], due primarily to the use of better estimates for the mean values of $|\zeta(\frac{1}{2} + it)|$ which were not available at the time of writing of [Iv1].

Proof of Theorem 4.10 We begin the proof of (4.24) by remarking that for technical reasons the condition $xy = (t/(2\pi))^k$ is replaced by $xy = T$ (see (4.20)). The error term that is made in this process is seen to be

$$\ll_\varepsilon x^{1/2-\sigma} \min\left(x^{1/2}, y^{1/2}\right) t^{\varepsilon - 2},$$

which is negligible in (4.25) and present in (4.26). We shall begin the proof with the general (4.24), but at a suitable point we shall distinguish between the cases $k = 3$, $k = 4$ and $k > 4$. For notational convenience we define

$$\Phi_1(u) = \frac{\Phi(u)}{(u-s)^2} = \frac{T^{u-s} X(u) - X(s)}{(u-s)^2} \qquad \left(\tfrac{1}{2} \le \sigma \le \tfrac{3}{2}\right). \tag{4.28}$$

Thus for $\mathbb{R}\mathrm{e}\, u \le \frac{1}{2}$ and also for $\mathbb{R}\mathrm{e}\, u < \min(\sigma, 1)$ the function $\Phi_1(u)$ is seen to be regular. Moreover uniformly in s, for $\mathbb{R}\mathrm{e}\, u = \frac{1}{2}$, we have

$$\Phi(u) \ll t^{k/2 - k\sigma} \min\left(1, t^{-1}|s-u|^2\right). \tag{4.29}$$

To see this, observe that by Taylor's formula

$$T^{u-s}X(u) = X(s) + (u-s)(X'(s) + X(s)\log T) + \frac{1}{2!}(X''(s) + \cdots)(u-s)^2 + \cdots,$$

and since by (4.19) we have

$$X'(\tfrac{1}{2} + it) + X(\tfrac{1}{2} + it)\log T = 0,$$

it is seen that $\Phi_1(u)$ is regular for $\mathbb{R}\mathrm{e}\, u = \frac{1}{2}$, when the double zeros $u = s = \frac{1}{2} + it$ of the numerator and denominator cancel each other, and the other ranges for u are easy. This discussion incidentally shows why the definition of T in (4.19), which may have looked mysterious at first, is a natural one to make. To see that (4.29) holds observe that, for $\mathbb{R}\mathrm{e}\, u = \frac{1}{2}$, one has $|X(u)| = 1$, so that by (2.17)

$$|\Phi(u)| \le |T^{u-s}| + |X(s)| \ll t^{k(1/2-\sigma)}.$$

This proves the first bound in (4.29). For the second one note that if $|s-u|^2 \ll t$, then $\mathbb{R}\mathrm{e}\, u \asymp t$, and since we have

$$\frac{\mathrm{d}^2}{\mathrm{d}s^2}\big(\log \chi(s)\big) \ll \frac{1}{t} \qquad (s = \sigma + it, t > 1),$$

we may write

$$\Phi(u) = T^{u-s}X(u)\left(1 - \frac{X(s)}{X(u)}T^{s-u}\right)$$

and use Taylor's formula and (4.19) with $u = \frac{1}{2} + iv,\ v \in \mathbb{R}$. Therefore

$$\begin{aligned} 1 - \frac{X(s)}{X(u)}T^{s-u} &= 1 - \exp\Big\{-k(s-u)\frac{\chi'}{\chi}(\tfrac{1}{2} + it) + k\log\chi(\sigma + it) \\ &\quad - k\log\chi(\tfrac{1}{2} + iv)\Big\} \\ &= 1 - \exp\Big\{-k(s-u)\left(\frac{\chi'}{\chi}(\tfrac{1}{2} + it) - \frac{\chi'}{\chi}(\tfrac{1}{2} + iv)\right) \\ &\quad + O\left(\frac{|s-u|^2}{t}\right)\Big\} \\ &= 1 - \exp\Big\{k\left(\sigma + it - iv + \tfrac{1}{2}\right)^2 O(t^{-1})\Big\} \ll \frac{|s-u|^2}{t}. \end{aligned}$$

This gives the second bound in (4.29).

In the course of the proof we shall need the power moment estimate

$$\int_{T-G}^{T+G} |\zeta(\tfrac{1}{2}+it)|^k \,\mathrm{d}t \ll_\varepsilon GT^{(k-4)c+\varepsilon} \qquad (T^{2/3} \le G \ll T, k \ge 4). \tag{4.30}$$

where $\zeta(\frac{1}{2}+it) \ll t^{c+\varepsilon}$, so we may take $c = 32/205 = 0.156\,09\ldots$, which is the best known value due to M. N. Huxley [Hux2] (see also [HuIv]). We shall also need

$$\int_0^T |\zeta(\tfrac{1}{2}+it)|^k \,\mathrm{d}t \ll_\varepsilon T^{1+(k-4)/8+\varepsilon} \qquad (4 \le k \le 12). \tag{4.31}$$

The estimate (4.30) follows trivially from (4.55). The estimate (4.31) is contained in [Iv1, theorem 8.3].

The first step in the proof is to use the inversion formula (A.11) of [Iv1] to obtain

$$\begin{aligned} I &:= \frac{1}{2\pi i}\int_{2-i\infty}^{2+i\infty} \zeta^k(s+w)x^w w^{-k}\,\mathrm{d}w \\ &= \sum_{n=1}^{\infty} d_k(n)n^{-s}\left\{\frac{1}{2\pi i}\int_{2-i\infty}^{2+i\infty}\left(\frac{x}{n}\right)^w w^{-k}\,\mathrm{d}w\right\} \\ &= \frac{1}{(k-1)!}\sum_{n\le x} d_k(n)n^{-s}\log^{k-1}(x/n) := S_x, \end{aligned}$$

say. The basic idea, which originated with Hardy and Littlewood, is to use a differencing argument to recover $\sum_{n\le x} d_k(n)n^{-s}$ from the same sum weighted by $\log^{k-1}(x/n)$. To achieve this, first we move the line of integration in I to $\mathbb{R}\mathrm{e}\, w = -\gamma$, where

$$0 < \gamma < 3/4, \quad \sigma - 1 < \gamma, \quad \gamma \ne \sigma, \quad \gamma \ne \sigma - 1/2.$$

In doing this we pass over the poles $w = 0$ and $w = 1 - s$ of the integrand, with the respective residues

$$F_x := \sum_{m=0}^{k-1} \frac{(\zeta^k(s))^{(m)}(s)}{m!(k-1-m)!}(\log x)^{k-1-m}$$

and

$$Q_x := \frac{x^{1-s}}{(k-1)!(1-s)^k}\sum_{m=0}^{k-1}\frac{(-1)^m(k+m-1)!}{m!(1-s)^m}\sum_{r=m+1}^{k} a_{-r,k}\frac{\log^{r-m-1} x}{(r-m-1)!}.$$

Hence by the residue theorem we obtain

$$J_0 := \frac{1}{2\pi i}\int_{-\gamma-i\infty}^{-\gamma+i\infty} \zeta^k(s+w)x^w w^{-k}\,\mathrm{d}w = I - F_x - Q_x = S_x - F_x - Q_x. \tag{4.32}$$

In the integral in (4.32) set $z = s + w$, replace x by T/y, use the functional equation for $\zeta^k(s)$ and (4.19) in the form

$$T^{u-s}X(u) = X(s) + \Phi(u; s, T)$$

to obtain

$$\begin{aligned} J_0 &= X(s)\frac{1}{2\pi i}\int_{\sigma-\gamma-i\infty}^{\sigma-\gamma+i\infty} \zeta^k(1-z)y^{s-z}(z-s)^{-k}\,\mathrm{d}z \\ &\quad + \frac{1}{2\pi i}\int_{\sigma-\gamma-i\infty}^{\sigma-\gamma+i\infty} \zeta^k(1-z)\Phi(z; s, T)y^{s-z}(z-s)^{-k}\,\mathrm{d}z \\ &= X(s)J_1 + J_2, \end{aligned}$$

say. This is a crucial point which explains the definition of the function Φ. We use again (A.12) of [Iv1] to deduce that, for $\sigma < \gamma$,

$$J_1 = \frac{(-1)^k}{(k-1)!}\sum_{n\le y} d_k(n)n^{s-1}\log^{k-1}(x/n) := S_y, \tag{4.33}$$

similarly to the notation used in evaluating I. For $\sigma > \gamma$ we must take into account the pole $z = 0$, where the integrand has a residue

$$Q_y = \frac{(-1)^k y^s}{s^k(k-1)!}\sum_{m=0}^{k-1}\frac{(-1)^m(k+m-1)!}{m!s^m}\sum_{r=m+1}^{k} a_{-r,k}\frac{\log^{r-m-1} x}{(r-m-1)!},$$

so that altogether

$$J_1 = S_y - \varepsilon_\gamma Q_y,$$

where

$$\varepsilon_\gamma = \begin{cases} 0 & \text{if } \sigma < \gamma, \\ 1 & \text{if } \sigma > \gamma. \end{cases}$$

The line of integration in J_2 is moved to $\mathrm{Re}\, z = 1/4$, and for $\sigma < \gamma$ the pole $z = 0$ of the integrand is passed. In calculating the residue note that

$$X(0) = X'(0) = \cdots = X^{(k-1)}(0) = 0,$$

since in $X(u)$ and its first $k-1$ derivatives the factor $\sin(\frac{1}{2}\pi u)$ comes in. This leads to

$$\begin{aligned} J_2 &= \frac{1}{2\pi i}\int_{1/4-i\infty}^{1/4+i\infty} \zeta^k(1-z)\Phi(z)(z-s)^{-k}y^{s-z}\,\mathrm{d}z - (1-\varepsilon_\gamma)X(s)Q_y \\ &= J_y - (1-\varepsilon_\gamma)X(s)Q_y. \end{aligned} \tag{4.34}$$

Inserting the expressions (4.32) and (4.34) in (4.30) we obtain

$$F_x - S_x + Q_x = -X(s)J_1 - J_2 = -X(s)(S_y - Q_y) - J_y. \tag{4.35}$$

At this stage of the proof we shall use a differencing argument. The underlying idea is that (4.35) remains true if x and y are replaced by $x\mathrm{e}^{\nu h}$ and $y\mathrm{e}^{-\nu h}$, respectively, where $0 < h \le 1$ and ν is an integer satisfying $0 \le \nu \le k-1$, which will be suitably chosen later. Thus we see that the condition $x\mathrm{e}^{\nu h}\cdot y\mathrm{e}^{-\nu h} = (t/2\pi)^k$ is preserved. Now we shall sum (4.35) with weight $(-1)^\nu\binom{k-1}{\nu}$ for $0 \le \nu \le k-1$ to recover the AFE by means of the elementary identity

$$\sum_{\nu=0}^{m}(-1)^\nu\binom{m}{\nu}\nu^p = \begin{cases} m! & \text{if } p = m \\ 0 & \text{if } p < m, \end{cases} \tag{4.36}$$

whenever $p \in \mathbb{N}$. One obtains (4.36), for example, starting from

$$(1+x)^m = \binom{m}{0} + \binom{m}{1}x + \binom{m}{2}x^2 + \cdots + \binom{m}{m}x^m$$

and differentiating, taking eventually $x = -1$. In the first step we have

$$m(1+x)^{m-1} = \binom{m}{1} + 2\binom{m}{2}x + \cdots + m\binom{m}{m}x^{m-1},$$

and then the proof is finished if $p = 1$ by taking $x = -1$. Otherwise the above equation is multiplied by x and differentiated again and the process is repeated sufficiently many times. Finally, for $p = m$, we obtain

$$\sum_{\nu=0}^{m}(-1)^\nu\binom{m}{\nu}\nu^p = m!,$$

since we arrive at an expression whose left-hand side is $m!$ plus a polynomial in $x+1$, and taking $x = -1$ we obtain (4.36).

We also need the estimate

$$\mathrm{e}^z = \sum_{n=0}^{M}\frac{z^n}{n!} + O(|z|^{M+1}) \qquad (M \ge 1,\ a \le \operatorname{Re} z \le b), \tag{4.37}$$

where a and b are fixed. To distinguish better the sums which will arise in this process we introduce left indices to obtain from (4.35)

$$\sum_{\nu=0}^{k-1}(-1)^{\nu}\binom{k-1}{\nu}\Big({}_{\nu}F_x - {}_{\nu}S_x + {}_{\nu}Q_x + X(s)({}_{\nu}S_y - {}_{\nu}Q_y) + {}_{\nu}J_y\Big) = 0,$$

or abbreviating

$$\bar{F}_x - \bar{S}_x + \bar{Q}_x + X(s)\bar{S}_y - X(s)\bar{Q}_y + \bar{J}_y = 0. \tag{4.38}$$

Each term in (4.38) will be evaluated or estimated separately. We have

$$\bar{F}_x = \sum_{m=0}^{k-1}\frac{(\zeta^k(s))^{(m)}}{m!(k-1-m)!}A_m(x)$$

with

$$\begin{aligned} A_m(x) &:= \sum_{\nu=0}^{k-1}(-1)^{\nu}\binom{k-1}{\nu}(\log x + \nu h)^{k-1-m} \\ &= \sum_{r=0}^{k-1-m}\binom{k-1-m}{r}h^r \log^{k-1-m-r} x \sum_{\nu=0}^{k-1}(-1)^{\nu}\binom{k-1}{\nu}\nu^r \\ &= (k-1)!h^{k-1} \end{aligned}$$

for $m = 0$, and otherwise $A_m(x) = 0$, where we used (4.36). Therefore

$$\bar{F}_x = h^{k-1}\zeta^k(s),$$

and this is exactly what is needed for the approximate functional equation that will follows on dividing (4.35) by h^{k-1} with a suitably chosen h. Consider next

$$\begin{aligned} \bar{S}_x &= \frac{1}{(k-1)!}\sum_{n\le x} d_k(n)n^{-s}\sum_{\nu=0}^{k-1}\binom{k-1}{\nu}(-1)^{\nu}\Big(\nu h + \log(x/n)\Big)^{k-1} \\ &\quad + \frac{1}{(k-1)!}\sum_{\nu=0}^{k-1}\binom{k-1}{\nu}(-1)^{\nu}\sum_{x<n\le xe^{\nu h}} d_k(n)n^{-s}\Big(\nu h + \log(x/n)\Big)^{k-1} \\ &= \sum\nolimits_1 + \sum\nolimits_2, \end{aligned} \tag{4.39}$$

say. Analogously to the evaluation of $\bar{F}_x$ it follows that

$$\sum\nolimits_1 = h^{k-1}\sum_{n\le x} d_k(n)n^{-s}. \tag{4.40}$$

We estimate $\sum_2$ trivially, on using the elementary bound $d_k(n) \ll_{k,n} n^{\varepsilon}$, to obtain

$$\Big|\sum\nolimits_2\Big| \le \frac{1}{(k-1)!}\sum_{\nu=0}^{k-1}\binom{k-1}{\nu}(2\nu h)^{k-1}x^{-\sigma}\sum_{x<n\le x\mathrm{e}^{(k-1)h}} d_k(n)$$
$$\ll_{\varepsilon} h^{k-1}x^{-\sigma}t^{\varepsilon}\Big(1+x(\mathrm{e}^{(k-1)h}-1)\Big) \ll_{\varepsilon} t^{\varepsilon}(h^{k-1}x^{-\sigma}+h^k x^{1-\sigma}). \tag{4.41}$$

In a similar way it follows that

$$-X(s)\bar{S}_y = h^{k-1}X(s)\sum_{n\le y} d_k(n)n^{s-1} + O_{\varepsilon}\Big(t^{\varepsilon}x^{1/2-\sigma}(h^{k-1}y^{-1/2}+h^k y^{1/2})\Big). \tag{4.42}$$

Next we have

$$\bar{Q}_x = \frac{x^{1-s}}{(k-1)!(1-s)^k}\sum_{\mu=0}^{k-1}\frac{(-1)^{\mu}(k+\mu-1)!}{\mu!(1-s)^{\mu}}\sum_{\rho=\mu+1}^{k}\frac{a_{-\rho,k}}{(\rho-\mu-1)!}B_{\mu\rho},$$

where

$$B_{\mu\rho} = \sum_{\nu=0}^{k-1}(-1)^{\nu}\binom{k-1}{\nu}\mathrm{e}^{\nu h(1-s)}(\nu h+\log x)^{\rho-\mu-1}.$$

Using (4.36) and (4.37) with $M = k-1$ it follows that

$$B_{\mu\rho} = (k-1)!h^{k-1}\sum_{n+m=k-1,0<m<\rho-\mu-1}\frac{(1-s)^n}{n!}\binom{\rho-\mu-1}{m}\log^{\rho-\mu-1-m} x$$
$$+ O(h^k t^k \log^{\rho-\mu-1} x).$$

If we set $\nu = \rho-\mu-1-m$, change the order of summation and collect the constants, we obtain

$$\bar{Q}_x = \frac{x^{1-s}}{1-s}h^{k-1}\sum_{\nu=0}^{k-1}\sum_{\rho=\nu+1}^{k} a_{-\rho,k}r_{\rho,\nu}(1-s)^{1+\nu-\rho}\log^{\nu} x + O_{\varepsilon}(h^k x^{1+\varepsilon-\sigma})$$
$$= h^{k-1}R_x(s) + O_{\varepsilon}(h^k x^{1+\varepsilon-\sigma}). \tag{4.43}$$

The same argument applies to $\bar{Q}_y$ and yields

$$\bar{Q}_y = -h^{k-1}R_y(1-s) + O_{\varepsilon}(h^k x^{1/2+\varepsilon-\sigma}y^{1/2}). \tag{4.44}$$

Therefore we are left with the evaluation of

$$\bar{J}_y = \frac{1}{2\pi i}\int_{1/4-i\infty}^{1/4+i\infty} \zeta^k(1-z)\Phi(z)y^{z-s}(z-s)^{-k} \\ \times \left(\sum_{m=0}^{k-1}(-1)^m\binom{k-1}{m}\mathrm{e}^{-hm(s-z)}\right)\mathrm{d}z.$$

Observing that $\Phi(z)$ has a double zero at $z = s$ and that

$$\sum_{m=0}^{k-1}(-1)^m\binom{k-1}{m}\mathrm{e}^{-hm(s-z)} = (1-\mathrm{e}^{-hs+hz})^{k-1}$$

has a zero of order $k-1$ at $z-s$, we can move the line of integration in $\bar{J}_y$ to $\mathbb{R}\mathrm{e}\, z = 1/2$ to obtain with $w = u + iv\ (u, v \in \mathbb{R})$

$$\begin{aligned} \bar{J}_y &= \frac{1}{2\pi i}\int_{1/2-i\infty}^{1/2+i\infty} \zeta^k(1-w)\Phi(w)y^{s-w}(w-s)^{-k} \\ &\quad\times\left(\sum_{m=0}^{k-1}(-1)^m\binom{k-1}{m}\mathrm{e}^{-hm(s-w)}\right)\mathrm{d}w \\ &= \frac{1}{2\pi i}\left(\int_{|v-t|\le G}\cdots\,\mathrm{d}w + \int_{|v-t|>G,|v|\le 2t^\beta}\cdots\,\mathrm{d}w + \int_{|v|\ge 2t^\beta}\cdots\,\mathrm{d}w\right) \\ &= j_1 + j_2 + j_3, \end{aligned} \tag{4.45}$$

say. Here $\beta\,(\ge 1)$ is the number appearing in (4.26), and $t^\varepsilon < G \le t^{2/3}$ is a parameter that will be suitably chosen. We distinguish now between the cases $3 \le k \le 4$ and $k > 4$, and treat first the latter case. For j_1 we use (4.29) in the form

$$\Phi(w) \ll t^{k/2-k\sigma-1}|s-w|^2,$$

and majorize the sum over m in (4.45) by $O(h^{k-1}|s-w|^{k-1})$, which follows when we combine (4.36) and (4.37). Therefore we have, since $xy \asymp t^k$,

$$\begin{aligned} j_1 &\ll h^{k-1}y^{\sigma-1/2}t^{k(1/2-\sigma)-1}G\int_{t-G}^{t+G}|\zeta(\tfrac{1}{2}+iv)|^k\,\mathrm{d}v \\ &\ll h^{k-1}x^{1/2-\sigma}Gt^{-1}\int_{t-t^{2/3}}^{t+t^{2/3}}|\zeta(\tfrac{1}{2}+iv)|^k\,\mathrm{d}v \\ &\ll_\varepsilon h^{k-1}x^{1/2-\sigma}Gt^{(k-4)c+\varepsilon-1/3}, \end{aligned} \tag{4.46}$$

where (4.30) was used. To estimate j_2 we use $\Phi(w) \ll t^{k/2-k\sigma}$ and the same majorization for the sum over m as above to obtain

$$j_2 \ll h^{k-1}x^{1/2-\sigma}\int_{|v-t|>G,|v|\le 2t^\beta} |\zeta(\tfrac{1}{2}+iv)|^k|v-t|^{-1}\,\mathrm{d}v. \tag{4.47}$$

The integral in (4.47) is split into subintegrals $j_{21}, j_{22}, j_{23}, j_{24}, j_{25}$ over the intervals

$$[-2t^\beta,\,-2t],\ [-2t,\,t/2],\ [t/2,\,t-G],\ [t+G,\,2t],\ [2t,\,2t^\beta],$$

respectively, since $\beta \ge 1$. Using (4.31) it follows at once that

$$j_{22} \ll_\varepsilon t^{(k-4)/8+\varepsilon},$$

and the remaining integrals $j_{2\ell}$ ($\ell \ne 2$) are integrated by parts and then estimated. For example, for j_{23} we have with

$$H(v) := -\int_v^{t-G} |\zeta(\tfrac{1}{2}+ix)|^k\,\mathrm{d}x$$

and (4.31) that

$$\begin{aligned} j_{23} &= H(v)(t-v)^{-1}\Big|_{t/2}^{t-G} + \int_{t/2}^{t-G} H(v)(t-v)^{-2}\,\mathrm{d}v \\ &\ll_\varepsilon t^{(k-4)/8+\varepsilon} + G^{-1}t^{1+(k-4)/8+\varepsilon} \ll_\varepsilon G^{-1}t^{1+(k-4)/8+\varepsilon} \end{aligned} \tag{4.48}$$

since $G \le t^{2/3}$, and the same bound similarly holds for j_{24}. Again using (4.31) we have

$$j_{21} + j_{25} \ll_\varepsilon t^{\beta((k-4)/8+\varepsilon)},$$

and so

$$\begin{aligned} j_1 + j_2 \ll_\varepsilon t^\varepsilon(&h^{k-1}x^{1/2-\sigma}Gt^{(k-4)c-1/3} + h^{k-1}x^{1/2-\sigma}G^{-1}t^{1+(k-4)/8} \\ &+ h^{k-1}x^{1/2-\sigma}t^{\beta(k-4)/8}). \end{aligned} \tag{4.49}$$

We choose now G in such a way that the first two terms on the right-hand side of (4.49) are equal. Thus with $c = 32/205$ we let

$$G = t^{2/3+(k-4)(1/8-c)/2}.$$

This choice of G obviously satisfies the condition $t^\varepsilon < G \le t^{2/3}$, since $k > 4$ and the known values for c satisfy $c > 1/8$. We obtain then

$$j_1 + j_2 \ll_\varepsilon t^\varepsilon h^{k-1}x^{1/2-\sigma}\Big(t^{\frac{1}{3}+\frac{1}{16}(k-4)(8c+1)} + t^{\frac{1}{8}\beta(k-4)}\Big).$$

If we choose

$$\beta = \frac{8}{3(k-4)} + \frac{8c+1}{2} \tag{4.50}$$

then $\beta > 1$, since $c > 1/8$. Hence above estimate reduces to

$$j_1 + j_2 \ll_\varepsilon t^\varepsilon h^{k-1} x^{1/2-\sigma} t^{\frac{1}{8}\beta(k-4)} = h^{k-1} x^{1/2-\sigma} t^{\frac{1}{3}+\frac{461}{3280}(k-4)+\varepsilon}, \tag{4.51}$$

with $461/3280 = 0.140\,5487\ldots$. Integration by parts and (4.31) give

$$j_3 \ll x^{1/2-\sigma} \int_{2t^\beta}^\infty |\zeta(\tfrac{1}{2}+iv)|^k v^{-k}\,dv \ll_\varepsilon x^{1/2-\sigma} t^{\beta(4-7k+\varepsilon)/8},$$

where we used $\Phi(w) \ll t^{k/2-k\sigma}$ and $\sum_{m=0}^{k-1} \ll 1$ for the sum appearing in (4.45), since

$$\left|e^{-hm(s-w)}\right| = e^{-hm(\sigma-1/2)} \ll 1.$$

Finally, combining all the estimates (4.38)-(4.51) we obtain

$$h^{k-1}\Big(\zeta^k(s) - \sum_{n\le x} d_k(n)n^{-s} - \chi^k(s)\sum_{n\le y} d_k(n)n^{s-1} + R_x(s) + \chi^k(s)R_y(1-s)\Big)$$
$$\ll_\varepsilon t^\varepsilon x^{1/2-\sigma}\left(h^k(x+y)^{1/2} + h^{k-1}t^{\frac{1}{8}\beta(k-4)} + t^{\beta(4-7k)/8}\right). \tag{4.52}$$

Choosing $h = t^{-\beta}$, where β is defined by (4.50), it is seen that the last two terms in (4.52) are equal, and the desired AFE follows from (4.52) on dividing by h^{k-1}, if we note that

$$\frac{\beta(k-4)}{8} = \frac{1}{3} + \frac{461}{3280}(k-4),$$

and recall that the error term made by replacing the condition $xy = (t/(2\pi))^k$ by $xy = T$ is

$$\ll_\varepsilon x^{1/2-\sigma} t^{\varepsilon-2} \min(x^{1/2},\ y^{1/2}).$$

This settles the case $k > 4$, and we still have to consider the cases $k = 3$ and $k = 4$. The only changes in the proof will be in the integrals j_1, j_2, j_3 appearing in (4.45), where sharper estimates than those used for the general case $k > 4$ are available.

For $k = 3$ we choose G in (4.45) to satisfy

$$t^{131/416+\varepsilon} \le G \le t^{2/3}. \tag{4.53}$$

We shall use the strongest known mean value result

$$\int_0^T |\zeta(\tfrac{1}{2}+it)|^2\,\mathrm{d}t \;=\; T\log\Big(\frac{T}{2\pi}\Big) + (2\gamma-1)T + E(T),$$
$$E(T) \ll_\varepsilon T^{131/416+\varepsilon}, \quad \frac{131}{416} = 0.314\,903\ldots \tag{4.54}$$

and likewise

$$\int_0^T |\zeta(\tfrac{1}{2}+it)|^4\,\mathrm{d}t = TP_4(\log T) + E_2(T), \qquad E_2(T) \ll T^{2/3}\log^C T, \tag{4.55}$$

where γ is Euler's constant, $P_4(x)$ is a explicit polynomial in x with leading coefficient $1/(2\pi^2)$ and one can take, e.g., $C = 8$.

Then by (4.54), (4.55) and the Cauchy-Schwarz inequality for integrals we obtain, instead of (4.46),

$$\begin{aligned} j_1 &\ll h^2x^{1/2-\sigma}Gt^{-1}\int_{t-G}^{t+G}|\zeta(\tfrac{1}{2}+iv)|^3\,\mathrm{d}v \\ &\ll h^2x^{1/2-\sigma}Gt^{-1}\left(\int_{t-G}^{t+G}|\zeta(\tfrac{1}{2}+iv)|^2\,\mathrm{d}v\int_{t-2t^{2/3}}^{t+2t^{2/3}}|\zeta(\tfrac{1}{2}+iv)|^4\,\mathrm{d}v\right)^{1/2} \\ &\ll_\varepsilon h^2x^{1/2-\sigma}t^{-2/3+\varepsilon}G^{3/2}. \end{aligned}$$

For j_{23} we obtain similarly

$$\begin{aligned} j_{23} &\ll h^2x^{1/2-\sigma}\int_{t/2}^{t-G}|\zeta(\tfrac{1}{2}+iv)|^3\frac{\mathrm{d}v}{t-v} \\ &\ll h^2x^{1/2-\sigma}\sum_{n=1}^{O(t/G)}\int_{t-(n+1)G}^{t-nG}|\zeta(\tfrac{1}{2}+iv)|^3\frac{\mathrm{d}v}{t-v} \\ &\ll_\varepsilon h^2x^{1/2-\sigma}\sum_{n=1}^{O(t/G)}\frac{1}{nG}G^{1/2}t^{1/3+\varepsilon} \\ &\ll_\varepsilon h^2x^{1/2-\sigma}G^{-1/2}t^{1/3+\varepsilon}. \end{aligned}$$

We choose

$$G \;=\; t^{1/2}$$

to make the estimates for j_1 and j_{23} equal, namely $\ll_\varepsilon h^2x^{1/2-\sigma}t^{1/12+\varepsilon}$. This choice for G obviously satisfies the condition (4.53). At last using

$$\int_0^T |\zeta(\tfrac{1}{2}+it)|^3\,\mathrm{d}t \;\ll_\varepsilon\; T^{1+\varepsilon}$$

we obtain

$$j_1 + j_2 + j_3 \ll_\varepsilon t^\varepsilon x^{1/2-\sigma}(h^2 t^{1/12+\varepsilon} + t^{-2\beta}).$$

Thus for $k = 3$ we obtain (4.52), only the right-hand side will now be

$$\ll_\varepsilon t^\varepsilon x^{1/2-\sigma}\left(h^3(x+y)^{1/2} + h^2 t^{1/12} + t^{-2\beta}\right).$$

Now we set $\beta = 3/2$, $h = t^{-\beta} = t^{-3/2}$. Then from $x + y \ll t^3$ we infer that

$$h^3(x+y)^{1/2} \ll h^2,$$

and dividing (4.52) (with $k = 3$) by h^2 we obtain the desired AFE (4.24) for $\zeta^3(s)$ with the bound for the error term given by (4.25), namely

$$\begin{aligned}\zeta^3(s) = \sum_{n\le x} d_3(n)n^{-s} + \chi^3(s)\sum_{n\le y} d_3(n)n^{s-1} - R_x(s)\\ -\chi^3(s)R_y(1-s) + O_\varepsilon(x^{1/2-\sigma}t^{1/12+\varepsilon}).\end{aligned}\tag{4.56}$$

It remains to consider yet the case $k = 4$, where the estimation is identical with the general case up to equation (4.47), only now for $H(v)$ we shall use (4.55). Therefore for j_{23} in (4.47) with $k = 4$ we have

$$j_{23} \ll_\varepsilon t^\varepsilon\left(1 + \int_{t/2}^{t-G}(t - v - G + t^{2/3})\frac{\mathrm{d}v}{(t-v)^2}\right) \ll_\varepsilon G^{-1}t^{2/3+\varepsilon}.$$

We then have

$$j_1 + j_2 \ll_\varepsilon t^\varepsilon(h^3x^{1/2-\sigma}Gt^{-1/3} + h^3x^{1/2-\sigma}G^{-1}t^{2/3}) \ll_\varepsilon h^3x^{1/2-\sigma}t^{1/6+\varepsilon}$$

for $G = t^{1/2}$. Since $j_3 \ll_\varepsilon x^{1/2-\sigma}t^{\beta(\varepsilon-3)}$ we obtain (4.52), where the right-hand side will be

$$\ll_\varepsilon t^\varepsilon x^{1/2-\sigma}\left(h^4(x+y)^{1/2} + h^3t^{1/6+\varepsilon} + t^{-3\beta}\right).$$

The result given by (4.24) and (4.25) follows for $\beta = 2$, $h = t^{-\beta}$ on dividing (4.52) (with $k = 4$) by h^3, since

$$h(x+y)^{1/2}x^{1/2-\sigma} \ll t^{-2}(t^4)^{1/2}x^{1/2-\sigma} = x^{1/2-\sigma}.$$

From this discussion and (4.56) the formulas in (4.27) follows easily. This ends the proof of Theorem 4.10.

4.4 The reflection principle

The AFEs in Theorem 4.3, Theorem 4.4 and Theorem 4.10 all have a symmetric property if $x = y$, especially when $\sigma = 1/2$, which is very useful in applications. However, when one seeks estimates for averages of powers of moduli of $\zeta(s)$ in the critical strip, it turns out that the AFEs of the types just considered have two shortcomings. First, the lengths of the sums depend on t, and second, the error terms for $k \ge 3$ and $\sigma = 1/2$ already are not small (i.e. they are not $\ll_\varepsilon t^\varepsilon$). We shall proceed now to derive another type of AFE, which though lacking symmetry, is in many problems concerning averages of $\zeta(s)$ quite adequate. The idea, which has permeated the whole theory since the pioneering days of B. Riemann, is to use the functional equation in the form

$$\zeta^k(w) = \chi^k(w)\zeta^k(1-w) \tag{4.57}$$

for some w with $\mathbb{R}\mathrm{e}\, w < 0$, to split the absolutely convergent series

$$\zeta^k(1-w) \;=\; \sum_{n=1}^{\infty} d_k(n) n^{w-1}$$

at some suitable M and to estimate the terms satisfying $n > M$ trivially. This approach is very flexible, and the error terms that will arise will be small. The starting point is the classical Mellin inversion integral for the gamma-function, namely

$$\mathrm{e}^{-w} = \frac{1}{2\pi i}\int_{c-i\infty}^{c+i\infty} \Gamma(s) w^{-s}\,\mathrm{d}s \qquad (c > 0,\ \mathbb{R}\mathrm{e}\, w > 0). \tag{4.58}$$

In (4.58) we set $w = Y^h$, $s = w/h$ and suppose $Y, h > 0$. In view of the functional equation $\Gamma(z+1) = z\Gamma(z)$ we obtain, by moving the line of integration,

$$\mathrm{e}^{-Y^h} = \frac{1}{2\pi i}\int_{2-i\infty}^{2+i\infty} Y^{-w}\Gamma(1+w/h) w^{-1}\,\mathrm{d}w. \tag{4.59}$$

Now replacing Y by n/Y and using

$$\sum_{n=1}^{\infty} d_k(n) n^{-z} = \zeta^k(z) \qquad (\mathbb{R}\mathrm{e}\, z > 1,\ k \in \mathbb{N})$$

it follows from (4.59) when $\sigma \ge 0$ that

$$\sum_{n=1}^{\infty} d_k(n)\mathrm{e}^{-(n/Y)^h} n^{-s} = \frac{1}{2\pi i}\int_{2-i\infty}^{2+i\infty} \zeta^k(w+s) Y^w \Gamma(1+w/h) w^{-1}\,\mathrm{d}w. \tag{4.60}$$

Now we suppose that

$$s = \sigma + it,\ 0 \le \sigma \le 1,\ h^2 \le t \le T,\ h = \log^2 T,\ 1 \ll Y \ll T^c$$

for some fixed $c > 0$. We move the line of integration in (4.60) to $\mathbb{R}\mathrm{e}\,(s + w) = -1/2$. Using Stirling's formula in the form

$$|\Gamma(s)| = \sqrt{2\pi}\,|t|^{\sigma-1/2}\mathrm{e}^{-\pi|t|/2}\Big(1 + O(|t|^{-1})\Big) \qquad (|t| \ge t_0 > 0) \tag{4.61}$$

it is seen that the residue at the pole $w = 1 - s$ is $o(1)$, while the residue at the pole $w = 0$ is $\zeta^k(s)$. Using the functional equation (4.57) we have

$$\sum_{n=1}^{\infty} d_k(n)\mathrm{e}^{-(n/Y)^h} n^{-s} = \zeta^k(s) + o(1) + I_1 + I_2, \tag{4.62}$$

say, where for some M satisfying $1 \ll M \ll T^c$ we have

$$I_1 = \frac{1}{2\pi i}\int_{\mathbb{R}\mathrm{e}\,(s+w)=-1/2} \chi^k(s + w)\sum_{n\le M} d_k(n)n^{w+s-1}Y^w\Gamma(1 + w/h)\frac{\mathrm{d}w}{w}, \tag{4.63}$$

and

$$I_2 = \frac{1}{2\pi i}\int_{\mathbb{R}\mathrm{e}\,(s+w)=-1/2} \chi^k(s + w)\sum_{n> M} d_k(n)n^{w+s-1}Y^w\Gamma(1 + w/h)\frac{\mathrm{d}w}{w}. \tag{4.64}$$

In I_2 we move the line of integration to $\mathbb{R}\mathrm{e}\,(s + w) = -h/2$, noting that the integrand is regular for $-h/2 \le \mathbb{R}\mathrm{e}\,(s + w) \le -1/2$, and aiming to choose M in such a way that $I_2 = o(1)$ as $T \to \infty$. With $w = u + iv$ $(u, v \in \mathbb{R})$ we obtain, on using (2.17) and Stirling's formula (4.61),

$$\begin{aligned} I_2 &\ll \int_{-\infty}^{\infty} |\chi(-\tfrac{1}{2}h + iv + it)|^k \\ &\quad\times \sum_{n>M} d_k(n)n^{-1-h/2}Y^{-h/2}|\Gamma(\tfrac{1}{2} - \sigma/h + iv/h)|\,\mathrm{d}v \\ &\ll (MY)^{-h/2}\log^k T\int_0^T (t + v)^{k(1+h)/2}\,\mathrm{d}v + \int_T^{\infty} \mathrm{e}^{-v/h}\,\mathrm{d}v \\ &\ll (MY)^{-h/2}\log^k T(2T)^{k(1+h)/2} + o(1) = o(1) \end{aligned}$$

as $T \to \infty$ if

$$M \ \ge\ (3T)^k Y^{-1}. \tag{4.65}$$

The flexibility of this method is best seen in various possibilities for the estimation of I_1 in (4.63). The line of integration in I_1 may be moved to $\mathbb{R}\mathrm{e}\,(s + w) = \alpha$, $0 < \alpha < 1$ for a fixed α such that $\alpha \ne \sigma$, so that the sum appearing in (4.63) will be "reflected," hence the name the *reflection principle*.

Letting

$$\delta_\alpha = \begin{cases} 1 & \text{if } \alpha > \sigma, \\ 0 & \text{if } \alpha < \sigma, \end{cases} \tag{4.66}$$

we obtain by the residue theorem

$$I_1 = -\delta_\alpha \chi^k(s) \sum_{n \le M} d_k(n) n^{s-1} + \frac{1}{2\pi i} \int_{\mathbb{R}\mathrm{e}\,(s+w)=-1/2} \chi^k(s+w) \sum_{n \le M} d_k(n) n^{w+s-1} Y^w \Gamma(1 + w/h) \frac{\mathrm{d}w}{w}. \tag{4.67}$$

The terms with $n > 2Y$ in (4.62) are trivially $o(1)$, and the part of the integral in (4.67) with $|v| = |\mathbb{I}\mathrm{m}\, w| \ge h^2$ is also $o(1)$ by Stirling's formula. Therefore combining (4.62)-(4.67) we obtain the next theorem.

Theorem 4.13 *For* $s = \sigma + it, 0 \le \sigma \le 1, k \ge 1$ *a fixed integer,* $h = \log^2 T, h^2 \le t \le T,\ 1 \ll Y \ll T^c, 0 < \alpha < 1\,(\alpha \ne \sigma),\ \delta_\alpha$ *given by* (4.66), $M \ge (3T)^k Y^{-1}$, *we have uniformly in* σ *and* t, *as* $T \to \infty$,

$$\zeta^k(s) = \sum_{n \le 2Y} d_k(n) n^{-s} \mathrm{e}^{-(n/Y)^h} + \delta_\alpha \chi^k(s) \sum_{n \le M} d_k(n) n^{s-1} + o(1) - \frac{1}{2\pi i} \int_{\mathbb{R}\mathrm{e}\,(s+w)=-1/2, |\mathbb{I}\mathrm{m}\, w| \le h^2} \chi^k(s+w) \times \sum_{n \le M} d_k(n) n^{w+s-1} Y^w \Gamma(1 + w/h) \frac{\mathrm{d}w}{w}. \tag{4.68}$$

Remark 4.14 This result is theorem 4.10 of [Iv1]. This is the desired (unsymmetrical) type of AFE, where the lengths of the sums do not depend on t, but on Y, and where the error term is $o(1)$ as $T \to \infty$. The method of proof generalizes easily to elements of $\mathcal{S}$.

4.5 The AFEs with smooth weights

The last AFE that will be considered in this chapter will be symmetric, but it will contain smooth functions whose presence will enable us to obtain sharp error terms (see [Iv4, chapter 4]). Before we proceed to the formulation of the result we need a technical result which yields an explicit construction of the desired smooth functions. This is

Lemma 4.15 *Let $b > 1$ be a fixed constant. There exists a real-valued function $\rho(x)$ such that*

(i) $\rho(x) \in C^\infty(0, \infty)$,
(ii) $\rho(x) + \rho(1/x) = 1$ *for* $x > 0$,
(iii) $\rho(x) = 0$ *for* $x \geq b$.

Proof of Lemma 4.15 Let us define, for $\alpha > \beta > 0$,

$$\varphi(t) = \exp\left(t^2 - \beta^2)^{-1}\right) \left\{ \int_{-\beta}^{\beta} \exp\left((u^2 - \beta^2)^{-1}\right) \mathrm{d}u \right\}^{-1}$$

if $|t| < \beta$, and put $\varphi(t) = 0$ if $|t| \geq \beta$, and let

$$f(x) := \int_{x-\alpha}^{x+\alpha} \varphi(t)\,\mathrm{d}t = \int_{-\infty}^{x} \Big(\varphi(t+\alpha) - \varphi(t-\alpha)\Big)\,\mathrm{d}t.$$

Then $\varphi(t) \in C(-\infty, \infty)$, $\varphi(t) \geq 0$ $(\forall t)$, and from the definition of φ and f it follows that $f(x) \in C^\infty(-\infty, \infty)$, $f(x) \geq 0$ $(\forall x)$ and if $\varphi(t)$ is even, then $f(x)$ is also even. If $|x| < \alpha - \beta$, then

$$\int_{x-\alpha}^{x+\alpha} \varphi(t)\,\mathrm{d}t = \int_{-\beta}^{\beta} \varphi(t)\,\mathrm{d}t = 1.$$

Moreover, if $|x| \geq \alpha + \beta$, then $f(x) = 0$. Hence

$$f(x) = \begin{cases} 0 & \text{if } |x| \geq \alpha + \beta, \\ 1 & \text{if } |x| < \alpha - \beta. \end{cases}$$

Now choose $\alpha = \frac{1}{2}(b+1)$, $\beta = \frac{1}{2}(b-1)$. Then

$$f(x) = \begin{cases} 0 & \text{if } x \geq b, \\ 1 & \text{if } 0 \leq x \leq 1. \end{cases}$$

Finally we set

$$\rho(x) := \frac{1}{2}\left(1 + f(x) - f\left(\frac{1}{x}\right)\right). \tag{4.69}$$

Property (i) of Lemma 4.15 is obvious. Next by using (4.69) we have

$$\rho(x) + \rho\left(\frac{1}{x}\right) = \frac{1}{2}\left\{1 + f(x) - f\left(\frac{1}{x}\right)\right\} + \frac{1}{2}\left\{1 + f\left(\frac{1}{x}\right) - f(x)\right\} = 1,$$

which establishes (ii) of Lemma 4.15. Lastly, if $x \geq b$,

$$\rho(x) = \frac{1}{2}\left(1 - f\left(\frac{1}{x}\right)\right) = 0$$

since $1/x \leq 1/b < 1$ and $f(x) = 1$ for $0 \leq x \leq 1$.

The AFE with a smooth weight ρ is given by the following theorem.

Theorem 4.16 *For* $\frac{1}{2} \le \sigma < 1$ *fixed,* $1 \ll x, y \ll t^k$, $s = \sigma + it$, $xy = (t/(2\pi))^k$, $t \ge t_0 > 0$ *and* $k \ge 1$ *a fixed integer we have*

$$\zeta^k(s) = \sum_{n=1}^{\infty} \rho\Big(\frac{n}{x}\Big) d_k(n) n^{-s} + \chi^k(s) \sum_{n=1}^{\infty} \rho\Big(\frac{n}{y}\Big) d_k(n) n^{s-1}$$
$$+ O(t^{\frac{1}{3}k(1-\sigma)-1}) + O(t^{k(1/2-\sigma)-2} y^{\sigma} \log^{k-1} t), \tag{4.70}$$

where $\rho(x)$ *is a smooth function satisfying the conditions of Lemma* 4.15.

Proof of Theorem 4.16 Let

$$R(s) := \int_0^\infty \rho(x) x^{s-1}\,\mathrm{d}x \tag{4.71}$$

be the Mellin transform of $\rho(x)$. This is a regular function of s for $\mathbb{R}\mathrm{e}\, s > 0$, but it possesses analytic continuation to $\mathbb{C}$. Its only singularity is a simple pole at $s = 0$ with residue 1. For $\mathbb{R}\mathrm{e}\, s > 0$

$$R(s) = \int_0^b \rho(x) x^{s-1}\,\mathrm{d}x = \frac{x^s}{s}\rho(x)\Big|_0^b - \int_0^b \frac{x^s}{s}\rho'(x)\,\mathrm{d}x = \frac{1}{s}\int_0^\infty \rho'(x) x^s\,\mathrm{d}x,$$

and the last integral is an analytic function for $\mathbb{R}\mathrm{e}\, s > -1$. Since

$$-\int_0^b \rho'(x)\,\mathrm{d}x = 1,$$

the residue of $R(s)$ at $s = 0$ equals unity. In general, repeated integration by parts gives, for $N \ge 0$ an integer,

$$R(s) = \frac{(-1)^{N+1}}{s(s+1)\cdots(s+N-1)(s+N)} \int_0^b \rho^{(N+1)}(x) x^{s+N}\,\mathrm{d}x. \tag{4.72}$$

Taking N sufficiently large it is seen that (4.72) provides analytic continuation of $R(s)$ to $\mathbb{C}$. Moreover for any N and σ in a fixed strip ($s = \sigma + it$) we have

$$R(s) \ll_N |t|^{-N} \qquad (|t| \to \infty). \tag{4.73}$$

In fact, (4.72) shows that $R(s)$ is regular at all points except $s = 0$, since for $N \ge 1$

$$\int_0^b \rho^{(N+1)}(x)\,\mathrm{d}x = \rho^{(N)}(b) - \rho^{(N)}(0) = 0.$$

For $-1 < \mathbb{R}\mathrm{e}\, s < 1$ and $s \ne 0$ we have

$$R(s) = -\frac{1}{s}\int_0^b \rho'(x) x^s\,\mathrm{d}x = -\frac{1}{s}\int_0^\infty \rho'(x) x^s\,\mathrm{d}x = -\frac{1}{s}\int_0^\infty \rho'\Big(\frac{1}{t}\Big) t^{-s-2}\,\mathrm{d}t$$

after change of variable $x = 1/t$. But

$$\rho(t) + \rho\Big(\frac{1}{t}\Big) = 1, \quad \rho'(t) - t^{-2}\rho'\Big(\frac{1}{t}\Big) = 0, \quad \rho'\Big(\frac{1}{t}\Big) = t^2\rho'(t).$$

Hence

$$R(s) = -\frac{1}{s}\int_0^\infty \rho'(t)t^{-s}\,\mathrm{d}t = -R(-s)$$

for $-1 < \mathbb{R}\mathrm{e}\, s < 1$. Then, by analytic continuation, for all s

$$R(-s) \;=\; -R(s), \tag{4.74}$$

so that $R(s)$ is an odd function. Conversely, from (4.74) by the inverse Mellin transform formula we can deduce the functional equation $\rho(x) + \rho(1/x) = 1$. The fact that $R(s)$ is odd plays an important rôle in the proof of Theorem 4.16.

Having established the necessary analytic properties of $R(s)$ we pass to the proof of (4.70), supposing that

$$s = \sigma + it, \quad \tfrac{1}{2} \le \sigma < 1, \quad t \ge t_0, \quad d = \mathbb{R}\mathrm{e}\, z > 1 - \sigma.$$

Then for any $n (\in \mathbb{N})$ we have from (4.71), by the inversion formula (see, e.g., (A.3) of [Iv1]) for Mellin transforms

$$\rho\Big(\frac{n}{x}\Big) = \frac{1}{2\pi i}\int_{d-i\infty}^{d+i\infty} R(z)\Big(\frac{x}{n}\Big)^z\,\mathrm{d}z.$$

By the absolute convergence of the series for $\zeta^k(z)$ this gives

$$\sum_{n=1}^\infty \rho\Big(\frac{n}{x}\Big)d_k(n)n^{-s} = \frac{1}{2\pi i}\int_{d-i\infty}^{d+i\infty} R(z)x^z\zeta^k(s+z)\,\mathrm{d}z.$$

We shift the line of integration to $\mathbb{R}\mathrm{e}\, z = -d$, passing the poles of the integrand at $z = 1-s$ and $z = 0$. Using (4.73) it is seen that the first residue is $\ll t^{-A}$ for any fixed $A > 0$. Hence by the residue theorem, (4.74) and the functional equation $\zeta(w) = \chi(w)\zeta(1-w)$ we have

$$\begin{aligned}
\sum_{n=1}^\infty \rho\Big(\frac{n}{x}\Big)d_k(n)n^{-s} &= \zeta^k(s) + O(t^{-A}) + \frac{1}{2\pi i}\int_{-d-i\infty}^{-d+i\infty} R(z)x^z\zeta^k(s+z)\,\mathrm{d}z \\
&= \zeta^k(s) + O(t^{-A}) + \frac{1}{2\pi i}\int_{-d-i\infty}^{-d+i\infty} R(z)x^z\chi^k(s+z)\zeta^k(1-s-z)\,\mathrm{d}z \\
&= \zeta^k(s) + O(t^{-A}) - \frac{1}{2\pi i}\int_{d+i\infty}^{d-i\infty} R(-w)x^{-w}\chi^k(s-w)\zeta^k(1-s+w)\,\mathrm{d}w \\
&= \zeta^k(s) + O(t^{-A}) - \frac{1}{2\pi i}\int_{d-i\infty}^{d+i\infty} R(w)x^{-w}\chi^k(s-w)\zeta^k(1-s+w)\,\mathrm{d}w.
\end{aligned}$$

Here we supposed that $1 \ll x, y \ll t^k$ and (similarly as in the proof of Theorem 4.10) that $xy = T$, where T is given by (4.19), hence by (4.20)

$$T = \left(\frac{t}{2\pi}\right)^k \left(1 + O\left(\frac{1}{t^2}\right)\right).$$

The idea is to derive first the AFE with x, y satisfying $xy = T$, and then to replace this by the condition $xy = (t/(2\pi))^k$, estimating the ensuing error term trivially. So far we have shown that

$$\zeta^k(s) = \sum_{n=1}^{\infty} \rho\Big(\frac{n}{x}\Big) d_k(n) n^{-s} + O(t^{-A}) + \frac{1}{2\pi i}\int_{d-i\infty}^{d+i\infty} R(w)\Big(\frac{y}{T}\Big)^w \chi^k(s-w)\zeta^k(1-s+w)\,\mathrm{d}w. \tag{4.75}$$

We choose d to satisfy $d > \sigma$, so that the series for $\zeta^k(1-s+w)$ in (4.75) is absolutely convergent and thus may be integrated termwise. Since $R(w) \ll |v|^{-A}$ for any fixed $A > 0$ if $v = \mathbb{I}\mathrm{m}\, w$, $|v| \ge v_0$, it is seen that the portion of the integral in (4.75) for which $|v| \ge t^\varepsilon$ makes a negligible contribution. Suppose now that N is a (large) fixed integer. For $|v| \le t^\varepsilon$

$$T^{-w}\chi^k(s-w) = \exp\Big\{kw\frac{\chi'(\frac12+it)}{\chi(\frac12+it)} + k\log\chi(s) - kw\frac{\chi'(s)}{\chi(s)} + k\sum_{j=2}^{N}\frac{(-w)^j}{j!}\cdot\frac{\mathrm{d}^j}{\mathrm{d}s^j}\log\chi(s))\Big\}\cdot(1 + O(t^{\varepsilon-N})) \tag{4.76}$$

because

$$\frac{\mathrm{d}^j}{\mathrm{d}s^j}(\log\chi(s)) \ll_j t^{-j+1} \qquad (j \ge 2). \tag{4.77}$$

Here we used (see (4.19))

$$\frac{\chi'(\frac12+it)}{\chi(\frac12+it)} = -\log t + \log(2\pi) + O(t^{-2}),$$
$$\frac{\chi'(\sigma+it)}{\chi(\sigma+it)} = -\log t + \log(2\pi) + O(t^{-1}),$$

which follows by logarithmic differentiation of the functional equation in the form $\zeta(s) = \chi(s)\zeta(1-s)$ and Stirling's formula. Therefore we have, for the expression in (4.76),

$$\exp(kw\cdots) = \chi^k(s)\Big(1 + G(w,s)\Big),$$

say, with $G(w, s) \ll_\varepsilon t^{\varepsilon-1}$ for $|v| \le t^\varepsilon$. For $\delta > 0$ and N sufficiently large we obtain

$$\begin{aligned}
&\frac{1}{2\pi i}\int_{d-i\infty}^{d+i\infty} R(w)y^w T^{-w}\chi^k(s-w)\zeta^k(1-s+w)\,\mathrm{d}w \\
&= O(t^{-A}) + \frac{1}{2\pi i}\int_{d-i\infty}^{d+i\infty} R(w)y^w\chi^k(s)\zeta^k(1-s+w)\,\mathrm{d}w \\
&\quad + \frac{1}{2\pi i}\int_{\delta-i\infty}^{\delta+i\infty} R(w)y^w\chi^k(s)\zeta^k(1-s+w)G(w,s)\,\mathrm{d}w.
\end{aligned}$$

Further we have, since $d > \sigma$,

$$\begin{aligned}
&\frac{1}{2\pi i}\int_{d-i\infty}^{d+i\infty} R(w)y^w\chi^k(s)\zeta^k(1-s+w)\,\mathrm{d}w \\
&= \chi^k(s)\sum_{n=1}^{\infty} d_k(n)n^{-s}\left\{\frac{1}{2\pi i}\int_{d-i\infty}^{d+i\infty} R(w)\Big(\frac{n}{y}\Big)^w\,\mathrm{d}w\right\} \\
&= \chi^k(s)\sum_{n=1}^{\infty} d_k(n)\rho\Big(\frac{n}{y}\Big)n^{-s}.
\end{aligned}$$

For $\delta > 0$ sufficiently small

$$\begin{aligned}
&\frac{1}{2\pi i}\int_{\delta-i\infty}^{\delta+i\infty} R(w)y^w\chi^k(s)\zeta^k(1-s+w)G(w,s)\,\mathrm{d}w \\
&\ll \left|\int_{|\mathbb{I}\mathrm{m}\, w|\le t^\varepsilon, \mathbb{R}\mathrm{e}\, w=\delta} R(w)y^w\zeta^k(s-w)\left(\chi^k(s)\chi^k(1-s+w)\right)\right) \\
&\quad \times G(w,s)\,\mathrm{d}w\Big| + t^{-A} \\
&\ll_\varepsilon t^{2\varepsilon-1}\int_{-t^\varepsilon}^{t^\varepsilon} |\zeta(\sigma+it-\delta+iv)|^k\,\mathrm{d}v + t^{-A}\,\mathrm{d}v \\
&\ll_\varepsilon t^{k\mu(\sigma)+\varepsilon_1-1} + t^{-A},
\end{aligned} \tag{4.78}$$

where $\lim_{\varepsilon\to 0}\varepsilon_1 = 0$ and

$$\mu(\sigma) = \limsup_{t\to\infty}\frac{\log|\zeta(\sigma+it)|}{\log t}$$

is the so-called *Lindelöf function* (see (3.8)-(3.9) and the work of M. N. Huxley [Hux2] for the sharpest known bound of $\mu(1/2)$, namely 32/205). For our purposes the standard bound

$$\mu(\sigma) < \frac{1-\sigma}{3} \qquad (\tfrac{1}{2} \le \sigma < 1) \tag{4.79}$$

will suffice. The bound (4.79) comes from $\mu(\frac{1}{2}) < \frac{1}{6}$, $\mu(1) = 0$ and convexity, while the use of Huxley's bound $\mu(\frac{1}{2}) \le 32/205$ would lead to small improvements. Thus, if $\varepsilon = \varepsilon(k)$ is sufficiently small, (4.78) and (4.79) yield

$$\zeta^k(s) = \sum_{n=1}^{\infty} \rho\Big(\frac{n}{x}\Big) d_k(n) n^{-s} + \chi^k(s) \sum_{n=1}^{\infty} \rho\Big(\frac{n}{y}\Big) d_k(n) n^{s-1} + O\big(t^{\frac{k(1-\sigma)}{3}-1}\big). \tag{4.80}$$

It is the use of (4.79) that required the range $\frac{1}{2} \le \sigma < 1$. We could also have considered the range $0 < \sigma \le \frac{1}{2}$. The analysis is, of course, quite similar but instead of (4.79) we would use

$$\mu(\sigma) \le \frac{1}{2} - \frac{2\sigma}{3} \qquad (0 \le \sigma \le 1/2),$$

which follows from (4.79), the functional equation and (2.17). This means that we would have obtained (4.80) with $O\big(t^{\frac{k(1-\sigma)}{3}-1}\big)$ replaced by $O\big(t^{k(\frac{1}{2}-\frac{2\sigma}{3})+\varepsilon-1}\big)$. As the final step of the proof of (4.70) we replace $y = Tx^{-1}$ by y, where we set $Y = x^{-1}(t/(2\pi))^k$. Then, for $n \ll y$, we have from (4.20)

$$\begin{aligned} Y - y &= O(t^{k-2}x^{-1}),\\ \rho\Big(\frac{n}{y}\Big) - \rho\Big(\frac{n}{Y}\Big) &\ll \frac{|Y-y|n}{y^2} \ll t^{k-2}x^{-1}y^{-1} \ll t^{-2}. \end{aligned}$$

Since $\rho(n/y) = 0$ for $n \ge by$, this means that if in (4.80) we replace y by Y the total error is

$$\ll t^{k(1/2-\sigma)} \sum_{n \le by} t^{-2} d_k(n) n^{\sigma-1} \ll t^{k(1/2-\sigma)-2} Y^{\sigma} \log^{k-1} t.$$

Then writing y for Y we obtain (4.70).

Remark 4.17 Theorem 4.16 can be generalized to a large class of L-functions. See, for example, the present author's work [Iv6] and G. Harcos [Har].

Remark 4.18 Note that, for ε and A fixed, the integral in (4.78) is bounded by

$$G(t) := t^{2\varepsilon-1} \int_{-t^{\varepsilon}}^{t^{\varepsilon}} |\zeta(\sigma + it - \delta + iv)|^k \, dv + t^{-A}.$$

In the above proof we have estimated $G(t)$ pointwise by the use of bounds for $\mu(\sigma)$, since in the above integral the interval of integration over t is short. However, if we seek an average of $G(t)$, we can get much better estimates by using bounds power moments of $|\zeta(\frac{1}{2}+it)$, since

$$\int_T^{2T} t^{2\varepsilon-1}\int_{-t^\varepsilon}^{t^\varepsilon} |\zeta(\sigma+it-\delta+iv)|^k \,dv\,dt$$
$$\ll_\varepsilon T^{2\varepsilon-1}\int_{-(2T)^\varepsilon}^{(2T)^\varepsilon}\left\{\int_T^{2T}|\zeta(\sigma+it-\delta+iv)|^k\,dt\right\}dv,$$

and for the integral in curly braces one can use the estimates of [Iv1, chapter 8]). This leads to a much better estimate for $\int_T^{2T} G(t)\,dt$ than by using the pointwise estimate for $G(t)$.

Remark 4.19 Taking $x = y = (t/(2\pi))^{k/2}, \sigma = 1/2$ in Theorem 4.16 we obtain

$$Z^k(t) = 2\sum_{n=1}^{\infty} d_k(n)\rho\left(\frac{n}{(\frac{t}{2\pi})^{k/2}}\right)n^{-1/2}\cos\left(t\log\left\{\frac{(\frac{t}{2\pi})^{k/2}}{n}\right\}-\frac{kt}{2}-\frac{k\pi}{8}\right)$$
$$+\,O\left(t^{\frac{1}{4}k-1}\log^{k-1}t\right). \tag{4.81}$$

Here we used $Z^k(t) = \zeta^k(\frac{1}{2}+it)(\chi(\frac{1}{2}+it))^{-k/2}$, and the asymptotic formula (2.17) to simplify $(\chi(\frac{1}{2}+it))^{-k/2}$, so that the ensuing error will be

$$\ll t^{-1}\sum_{n\ll t^{k/2}} d_k(n)n^{-1/2} \ll t^{k/4-1}\log^{k-1}t.$$

We also used implicitly the fact that the sum over n vanishes for $(\frac{t}{2\pi})^{k/2}/n > b$, that is for

$$n > \frac{1}{b}\left(\frac{t}{2\pi}\right)^{k/2} \qquad (b > 1). \tag{4.82}$$

The error term in (4.81) is better than the one that follow from (4.25) and (4.26) if we evaluate $Z^k(t)$ from these formulas. However, in applications sometimes it is not easy to get rid of the ρ-function, so that Theorem 4.10 certainly has its advantages.

Notes

We do not know exactly what motivated Riemann to conjecture the RH. Some mathematicians, like Felix Klein (Christian Felix Klein, April 25, 1849-June 22, 1925, German mathematician, known for his work in group theory, function theory, non-Euclidean geometry, and on the connections between geometry and group theory) thought that he was inspired by a sense of general beauty and symmetry in Mathematics. Although doubtlessly the truth of the RH would provide such harmonious symmetry, we also know now that Riemann undertook rather extensive numerical calculations concerning $\zeta(s)$ and its zeros. C. L. Siegel [Sie] studied Riemann's unpublished notes, kept in the University library at Göttingen. It turned out that Riemann had computed several zeros of the zeta-function and had a deep understanding of its analytic behavior. It is very fortunate that Siegel in 1932 (*op. cit.*) provided rigorous proof of the formula that had its genesis in Riemann's work. It came to be known later as *the Riemann-Siegel formula* and plays a fundamental rôle in zeta-function theory. One can quote [Edw], p. 136, who says: "Anyone who has read Siegel's paper is unlikely to assert, as Hardy did in 1915, that Riemann 'could not prove' the statements he made about the zeta-function, or to call them, as Landau did in 1908, 'conjectures'."

R. R. Hall in [Hal6] obtains the approximate functional equation

$$\zeta(s)\zeta(s+i\theta) = \sum_{n\le t/(2\pi)} \tau(n,-\theta)n^{-s} + \chi(s)\chi(s+\theta)\sum_{n\le t/(2\pi)} \tau(n,\theta)n^{s-1} + O(t^{1/2-\sigma}\log t).$$

Here $s = \sigma + it, 0 < \sigma < 1, 0 \le \theta \ll \sqrt{t}, t \ge t_0$, and $\tau(n,\theta) := \sum_{d|n} d^{i\theta}$ is the divisor function that generalizes $d(n)$ $(= \tau(n,0))$. A more general result of this shape was obtained by J. R. Wilton [Wil], but his error term is weaker by a log-factor. He follows the elegant proof of the AFE for $\zeta^2(s)$ by Y. Motohashi [Mot1]. Hall also proves (*op. cit.*) that

$$\int_0^T |Z(t)Z(t+\alpha)|^2\,\mathrm{d}t = 2T\sum_{n\le T/(2\pi)} |\tau(n,\alpha)|^2 n^{-1} + O(T\log^3 T)$$

uniformly for $0 < \alpha \le 1$ and $T \ge T_0$. In [Hal1] Hall proved that

$$\int_0^T |Z(t)Z'(t)|^2\,\mathrm{d}t = \frac{1}{120\pi^2}T\log^4 T + O(T\log^3 T),$$

$$\int_0^T \Big(Z'(t)\Big)^4\,\mathrm{d}t = \frac{1}{1120\pi^2}T\log^8 T + O(T\log^7 T).$$

This follows by using the method of A. E. Ingham [Ing] (Albert Edward Ingham, April 3, 1900-September 6, 1967, English mathematician), who proved in 1926 the weak asymptotic formula for the fourth moment of $Z(t)$ (see [Iv1] for a proof), namely

$$\int_0^T Z^4(t)\,\mathrm{d}t = \frac{1}{2\pi^2}T\log^4 T + O(T\log^3 T).$$

Additional mean value results for $Z(t)$ and its derivatives are obtained by R. R. Hall [Hal2]; for example, he proved that

$$\int_0^T Z^2(t)\Big(Z''(t)\Big)^2\,\mathrm{d}t = \frac{1}{672\pi^2}T\log^8 T + O(T\log^7 T).$$

To show that the error term in (4.6) of Theorem 4.4 is the best possible (the argument is due to M. Jutila) one may proceed as follows. Take $\sigma = \frac{1}{2}, t = 2\pi N, x = N - \delta t^{1/2}$, where $N \in \mathbb{N}$ and $\delta > 0$ is sufficiently small. We shall suppose that the error term in (4.6) is $o(\log t)$ as $t \to \infty$ and derive a contradiction. Thus for $s = \frac{1}{2} + it, t \to \infty$ we have

$$\zeta^2(s) = \sum_{n\le x} d(n)n^{-s} + \chi^2(s)\sum_{n\le y} d(n)n^{s-1} + o(\log t),$$

$$\zeta^2(s) = \sum_{n\le y} d(n)n^{-s} + \chi^2(s)\sum_{n\le x} d(n)n^{s-1} + o(\log t).$$

Multiplying by N^{it} and subtracting we obtain

$$\chi^2(\tfrac{1}{2}+it)\sum_{x<n\le y} d(n)n^{-1/2}(Nn)^{it} - \sum_{x<n\le y} d(n)n^{-1/2}(N/n)^{it} = o(\log t). \tag{4.83}$$

Our choice for t gives $e^{2it} = 1$, hence using (4.2) formula (4.83) becomes

$$\sum_{x<n\le y} d(n)n^{-1/2}\left\{-(n/N)^{-it} + i(n/N)^{it}\right\} = o(\log t). \tag{4.84}$$

Now for $x \le n \le y$ we have

$$\begin{aligned}(n/N)^{\pm it} &= \exp\left\{\pm 2\pi i N \log(1 + (n-N)N^{-1})\right\}\\ &= 1 + O\left((n-N)^2 N^{-1}\right) = 1 + O(\delta^2).\end{aligned}$$

Therefore a suitable choice of δ gives

$$\left|\sum_{x<n\le y} d(n)n^{-1/2}\left\{-(n/N)^{-it} + i(n/N)^{it}\right\}\right| \gg \log t, \tag{4.85}$$

since $y = N + \delta t^{1/2} + O(1)$, when we use the bound (see [Iv1])

$$\Delta(x) := \sum_{n\le x} d(n) - x(\log x + 2\gamma - 1) = O(x^{1/3}) \tag{4.86}$$

for the error term in *the Dirichlet divisor problem.* Coupled with the trivial estimation for the sum in (4.85), we obtain a contradiction with (4.84). This shows that, in general, the error term in (4.6) cannot be $o(\log t)$ when $\sigma = \frac{1}{2}$.

For a proof of a sharper version of Theorem 4.4, see the works of Y. Motohashi [Mot1], [Mot2]. Motohashi expresses the error term in the AFE for $\zeta^2(s)$ in terms of $\Delta(x)$. Namely he writes the AFE for $\zeta^2(s)$ (see (4.6)) as

$$\zeta^2(s) = \sum_{n\le X} d(n)n^{-s} + \chi^2(s)\sum_{n\le Y} d(n)n^{s-1} + D(s, X),$$

where $0 \le \sigma \le 1,\ t \ge 2,\ XY = t^2/(4\pi^2)$. Then in the first part of [Mot1] he proves that

$$D\left(s, \frac{t}{2\pi}\right) = -2\left(\frac{\pi}{t}\right)^{1/2}\Delta\left(\frac{t}{2\pi}\right)\chi(s) + O(t^{1/4-\sigma}).$$

In particular, from his formula and (4.86) one obtains

$$\zeta^2(s) = \sum_{n\le t/(2\pi)} d(n)n^{-s} + \chi^2(s)\sum_{n\le t/(2\pi)} d(n)n^{s-1} + O(t^{1/3-\sigma}),$$

which cannot be deduced from Theorem 4.4. In [Mot1, part II], Motohashi provides an asymptotic expansion of $D\left(s, \frac{t}{2\pi}\right)$. In the most important case $X = Y = t/(2\pi)$ he shows that the bound (see (4.6))

$$D\left(s, \frac{t}{2\pi}\right) \ll t^{1/2-\sigma}\log t \tag{4.87}$$

can be obtained rather simply from (4.5), where the error term is denoted by $E(s, x)$. Namely, by the classical *Dirichlet hyperbola method* one has

$$\sum_{n\le N} d(n)a_n = 2\sum_{n\le\sqrt{N}}\sum_{m\le N/m} a_{nm} - \sum_{n\le\sqrt{N}}\sum_{m\le\sqrt{N}} a_{nm}.$$

This gives, with $U = t/(2\pi)$, $u = \sqrt{U}$,

$$\begin{aligned}\sum_{n\le U} d(n)n^{-s} &+ \chi^2(s)\sum_{n\le U} d(n)n^{s-1}\\ &= 2\sum_{m\le u} m^{-s}\sum_{n\le U/m} n^{-s} + 2\chi^2(s)\sum_{m\le u} m^{s-1}\sum_{n\le U/m} n^{s-1}\\ &\quad -\Big(\sum_{m\le u} m^{-s}\Big)^2 - \chi^2(s)\Big(\sum_{m\le u} m^{s-1}\Big)^2.\end{aligned}$$

By the AFE for $\zeta(s)$ (see (4.5)) this is equal to

$$\begin{aligned}&2\sum_{m\le u} m^{-s}\Big\{\zeta(s) - \chi(s)\sum_{n\le m} n^{s-1} - E(s, U/m)\Big\}\\ &+2\chi^2(s)\sum_{m\le u} m^{s-1}\Big\{\zeta(1-s) - \chi(1-s)\sum_{n\le m} n^{-s} - E(1-s, U/m)\Big\}\\ &+2\chi(s)\sum_{m\le u} m^{-s}\sum_{n\le u} n^{s-1} - (\zeta(s) - E(s,u))^2.\end{aligned}$$

Using $\chi(s)\chi(1-s) = 1$ and rearranging, we obtain

$$\begin{aligned}\zeta^2(s) = &\sum_{n\le U} d(n)n^{-s} + \chi^2(s)\sum_{n\le U} d(n)n^{s-1} + 2\chi(s)\sum_{n\le u} 1/n\\ &+2\sum_{m\le u} m^{-s}E(s, U/m) + 2\chi^2(s)\sum_{m\le u} m^{s-1}E(1-s, U/m) + E^2(s,u).\end{aligned}$$

Finally by using the bounds

$$\chi(s) \ll t^{1/2-\sigma}, \quad E(s, U/m) \ll (m/t)^\sigma + t^{1/2-\sigma}m^{\sigma-1}, \quad E(s,u) \ll t^{-\sigma/2}$$

it is seen that (4.87) follows.

The bound for $E(T)$ in (4.54) is due to N. Watt [Watt] (see also the survey [HuIv]), while for the independent evaluation of the coefficients of $P_4(x)$ see J. B. Conrey [Con3] and the present author's paper [Iv5]. If the main term for the fourth moment in (4.55) is written as $Tp_4(L)$, $L = \log(T/(2\pi))$, then Conrey has shown that $p_4(x) = g_0(x) + g_1(x)$, where (γ is Euler's constant)

$$g_0(x) = \operatorname*{Res}_{s=0} \frac{2x^s\zeta^4(s+1)}{s(s+1)\zeta(2s+2)}$$

and

$$g_1(x) = \left(\frac{\mathrm{d}}{\mathrm{d}s}\right)^2 \frac{(x\mathrm{e}^{2\gamma})^s\left\{\frac12\zeta^2(s+1) - s^{-1}\zeta(2s+1) - \zeta(2s+2)\right\}}{(s+1)\zeta(s+2)}\Bigg|_{s=0}.$$

His analysis is based on the work of D. R. Heath-Brown [Hea3], where he obtained in closed form the first two leading coefficients of $p_4(x)$ and proved that $E_2(T) \ll_\varepsilon T^{7/8+\varepsilon}$. The present author [Iv5] also follows the argument of [Hea3], and obtains the coefficients of $p_4(x)$ also in closed form.

Numerical calculation shows that

$$Tp_4(L) = T\Big(0.050\,660L^4 + 0.496\,227L^3 + 0.937\,279L^2 + 1.353\,34L - 0.040\,924\Big),$$

where the coefficients are accurate within six decimal places. The forthcoming papers of G. A. Hiary and A. M. Odlyzko [HiOd1], and M. O. Rubinstein and S. Yamagishi [RuYa] contain many numerical results concerning various moments of $|\zeta(\frac12 + it)|$.

The bound in (4.55) for $E_2(T)$ is due to Y. Motohashi and the author [IvMo3] (the value $C = 8$ is worked out by Y. Motohashi in his monograph [Mot3]).

The Lindelöf hypothesis (LH) (see [Tit3]) is equivalent to the statement that $\mu(\sigma) = 1/2 - \sigma$ for $\sigma < 1/2$, and $\mu(\sigma) = 0$ for $\sigma \geq 1/2$. The behavior of $\mu(\sigma)$ in the critical strip is an important, unsolved problem. If the LH fails, it is perhaps true that $\mu(1/2) = 1/8$; in any case $\mu(1/2) \leq 32/205 = 0.156\,09\ldots$ is at present the best unconditional result of M. N. Huxley [Hux2] for the order of $\zeta(\frac{1}{2} + it)$. Note that $\mu(\sigma)$ is (unconditionally) a non-increasing, convex function of σ,

$$\mu(\sigma) = \frac{1}{2} - \sigma \quad (\sigma \leq 0), \qquad \mu(\sigma) = 0 \quad (\sigma \geq 1),$$

and by the functional equation one has

$$\mu(\sigma) = \frac{1}{2} - \sigma + \mu(1 - \sigma) \qquad (0 < \sigma < 1).$$

Theorem 4.13 is to be found in [Iv1], and Theorem 4.16 in [Iv4].

5

The derivatives of $Z(t)$

5.1 The θ and Γ functions

In applications it is sometimes useful not to have only an expression like (2.3), but to have similar expressions for the derivatives $Z^{(k)}(t)$ as well. Results of this type have been obtained by A. A. Karatsuba and S. M. Voronin [KaVo] and A. A. Lavrik [Lav1]. We shall follow the former work and derive an AFE for $Z^{(k)}(t)$ that is valid for all non-negative integers k. Before we formulate the result, we need the following lemma, which incidentally may serve as the basis for the proof of Stirling's formula for $\Gamma(s)$.

Lemma 5.1 *Let $\theta(t)$ be defined by* (1.19). *Then for $t \geq 2$ we have*

$$\theta(t) = \frac{t}{2}\log\frac{t}{2\pi} - \frac{t}{2} - \frac{\pi}{8} + \Delta(t) \tag{5.1}$$

with

$$\Delta(t) := \frac{t}{4}\log\left(1+\frac{1}{4t^2}\right) + \frac{1}{4}\arctan\frac{1}{2t} + \frac{t}{2}\int_0^\infty \frac{\psi(u)}{(u+\frac{1}{4})^2+(\frac{t}{2})^2}\,du, \tag{5.2}$$

where $\psi(x) = x - [x] - \frac{1}{2}$.

Proof of Lemma 5.1 From the product definition (1.23) of $\Gamma(s)$ we have

$$\log\Gamma(s) = -\gamma s - \log s + \sum_{n=1}^{\infty}\left(\frac{s}{n} - \log\left(1+\frac{s}{n}\right)\right). \tag{5.3}$$

For $N \in \mathbb{N}$ we set

$$\sum\nolimits_1 := \sum_{1/2<n<N+1/2} \frac{1}{n},$$
$$\sum\nolimits_2 := \sum_{0<n<N+1/2} \log(n+s),$$
$$\sum\nolimits_3 := \sum_{1/2<n<N+1/2} \log n.$$

From the familiar *Euler-Maclaurin summation formula* (see, e.g., (A.23) of [Iv1]) we have

$$\sum\nolimits_1 = \log N + \gamma + O\left(\frac{1}{N}\right),$$
$$\sum\nolimits_2 = \int_0^{N+1/2} \log(x+s)\,\mathrm{d}x + \int_0^{N+1/2} \frac{\psi(x)}{x+s}\,\mathrm{d}x - \tfrac{1}{2}\log s,$$
$$\sum\nolimits_3 = \int_{1/2}^{N+1/2} \log x\,\mathrm{d}x + \int_{1/2}^{N+1/2} \frac{\psi(x)}{x}\,\mathrm{d}x.$$

Inserting these expression in (5.3) and letting $N \to \infty$, we obtain

$$\log\Gamma(s) = (s - \tfrac{1}{2})\log s - s + C - \int_0^\infty \frac{\psi(x)}{x+s}\,\mathrm{d}x \tag{5.4}$$

with some absolute constant C. To evaluate C, note first that the integral in (5.4) is $\ll 1/|s|$. This follows on integrating it by parts, since $\psi(x)$ is periodic with period 1, hence $\int_0^x \psi(y)\,\mathrm{d}y \ll 1$. Then in (5.4) we take $s = N, N+1/2, 2N$, use the duplication formula

$$\Gamma(s)\Gamma(s+1/2) = 2\sqrt{\pi}2^{-2s}\Gamma(2s)$$

and let $N \to \infty$ to deduce that $C = \log\sqrt{2\pi}$. Thus we have shown that

$$\log\Gamma(s) = (s-1/2)\log s - s + \log\sqrt{2\pi} + H(s)$$
$$H(s) = -\int_0^\infty \frac{\psi(u)}{u+s}\,\mathrm{d}u \ll \frac{1}{|s|}. \tag{5.5}$$

Write (1.19) as

$$2i\theta(t) = -it\log\pi + \log\Gamma(\tfrac{1}{4} + \tfrac{1}{2}it) - \log\Gamma(\tfrac{1}{4} - \tfrac{1}{2}it). \tag{5.6}$$

Applying (5.5) we see that

$$\begin{aligned}
&\log\Gamma(\tfrac14+\tfrac12 it)-\log\Gamma(\tfrac14-\tfrac12 it)\\
&\quad=-\tfrac14\log\frac{\frac14+\frac12 it}{\frac14-\frac12 it}+\tfrac12 it\log(\tfrac1{16}+\tfrac14 t^2)-it+it\int_0^\infty\frac{\psi(u)\,\mathrm{d}u}{(u+\frac14)^2+\frac14 t^2}\\
&\quad=-\frac{i}{2}\left(\frac{\pi}{2}-\arctan\frac{1}{2t}\right)+\tfrac12 it\log(\tfrac1{16}+\tfrac14 t^2)-it+it\int_0^\infty\frac{\psi(u)\,\mathrm{d}u}{(u+\frac14)^2+\frac14 t^2}.
\end{aligned}$$

Namely, with

$$f(t):=\frac{1}{2i}\log\frac{\frac14+\frac12 it}{\frac14-\frac12 it},\qquad g(t):=\int_{1/(2t)}^\infty\frac{\mathrm{d}x}{1+x^2}=\int_0^t\frac{\mathrm{d}y}{2y^2+\frac12}$$

we find that

$$f'(t)=g'(t)=\frac{1}{2t^2+\frac12}\quad(t\ge 0),\ \ f(0)=g(0)=0,$$

hence $f(t)=g(t)$ for $t\ge 0$, implying

$$f(t)=\frac{1}{2i}\log\frac{\frac14+\frac12 it}{\frac14-\frac12 it}=g(t)=\frac{\pi}{2}-\arctan\frac{1}{2t}\quad(t\ge 0).$$

Therefore when we substitute the above expression for

$$\log\Gamma(\tfrac14+\tfrac12 it)-\log\Gamma(\tfrac14-\tfrac12 it)$$

in (5.6), we obtain (5.1)-(5.2) after some cancelation.

5.2 The formula for the derivatives

Theorem 5.2 *For any integer $k\ge 0$ and $t\ge 2\pi$ we have*

$$\begin{aligned}
Z^{(k)}(t)=2&\sum_{n\le\sqrt{t/(2\pi)}}n^{-1/2}\left(\log\sqrt{\frac{t}{2\pi}}-\log n\right)^k\\
&\times\cos\left(t\log\sqrt{\frac{t}{2\pi}}-t\log n-\frac{t}{2}-\frac{\pi}{8}+\frac{\pi k}{2}\right)+O_k\left(t^{-1/4}\log^{k+1}t\right).
\end{aligned}\tag{5.7}$$

For the proof of Theorem 5.2 we shall need the following lemma.

Lemma 5.3 *For any integer $m \geq 0$ and $t \geq 2\pi$ we have*

$$\begin{aligned}
&\frac{\mathrm{d}^m}{\mathrm{d}t^m}\zeta(\tfrac{1}{2}+it)\\
&= (-i)^m\Big(\sum_{n\leq\sqrt{t/(2\pi)}} n^{-1/2-it} + \mathrm{e}^{i\theta_1(t)}\sum_{n\leq\sqrt{t/(2\pi)}}\log^m\big(\frac{t}{2\pi n}\big)n^{-1/2-it}\Big)\\
&\quad + O_m\left(t^{-1/4}\log^{m+1} t\right), \qquad (5.8)
\end{aligned}$$

where

$$\theta_1(t) := -t\log t + t\log 2\pi + t + \pi/4. \qquad (5.9)$$

Proof of Lemma 5.3 We start from (4.4) with $s = \frac{1}{2}+it$, $\frac{1}{2}T \leq t \leq T$, which we differentiate m times over t. It follows that

$$\begin{aligned}
\frac{\mathrm{d}^m}{\mathrm{d}t^m}\zeta(\tfrac{1}{2}+it) &= (-i)^m\sum_{n\leq N} n^{-1/2-it}\log^m n\\
&\quad + \frac{\mathrm{d}^m}{\mathrm{d}t^m}\left(\frac{N^{1/2-it}}{it-1/2}\right) + O(tN^{-1/2}\log^m N).
\end{aligned}$$

Suppose that $N > N_0 := T/\pi$. The sum $\sum_{N_0<n\leq N} n^{-1/2-it}\log^m n$ is transformed by partial summation, and then by the first and the second derivative test (Lemma 2.2 and Lemma 2.3). We then obtain

$$\begin{aligned}
\frac{\mathrm{d}^m}{\mathrm{d}t^m}\zeta(\tfrac{1}{2}+it) &= (-i)^m\sum_{n\leq\frac{T}{\pi}} n^{-1/2-it}\log^m n\\
&\quad + O_m(t^{-1/2}\log^m t) + O_m(tN^{-1/2}\log^m N). \qquad (5.10)
\end{aligned}$$

Letting $N\to\infty$ in (5.10) we obtain

$$\frac{\mathrm{d}^m}{\mathrm{d}t^m}\zeta(\tfrac{1}{2}+it) = (-i)^m\sum_{n\leq\frac{T}{\pi}} n^{-1/2-it}\log^m n + O_m(t^{-1/2}\log^m t). \qquad (5.11)$$

Finally the portion of the sum in (5.11) for which $\sqrt{t/(2\pi)} < n \leq T/\pi$ is transformed by the *saddle point technique*. This is standard, but rather technical, and the details of the method can be found, for example, in chapter 2 of [Iv1], in Graham-Kolesnik [GrKo], or in Karatsuba-Voronin [KaVo]. Therefore the details are omitted. After the transformation (5.8)-(5.9), Lemma 5.3 will follow.

Lemma 5.4 *For any integer $r \geq 0$ we have*

$$\frac{\mathrm{d}^r}{\mathrm{d}t^r}\mathrm{e}^{i\theta(t)} = i^r(\theta'(t))^r\mathrm{e}^{i\theta(t)} + O_r(t^{-1}\log^{r-1} t). \qquad (5.12)$$

Proof of Lemma 5.4 We use the well-known formula for the r-the derivative of a composite function. This is the so-called *formula of Faà de Bruno*: if $y = F(\tau)$ and $\tau = f(t)$, then

$$\frac{\mathrm{d}^r}{\mathrm{d}t^r}F(f(t)) = \sum_{k_1+2k_2+\cdots+rk_r=r, k_j\ge 0} \frac{r!}{k_1!\cdots k_r!} \cdot \frac{\mathrm{d}^{k_1+\cdots+k_r}F}{\mathrm{d}\tau^{k_1+\cdots+k_r}}\Big(\frac{f'(t)}{1!}\Big)^{k_1}\cdots\Big(\frac{f^{(r)}(t)}{r!}\Big)^{k_r},$$

provided that all the derivatives in the above formula exist. For Lemma 5.4 we take $F(\tau) = \mathrm{e}^\tau$, $\tau = i\theta(t)$. Therefore

$$\frac{\mathrm{d}^r}{\mathrm{d}t^r}\mathrm{e}^{i\theta(t)} = \mathrm{e}^{i\theta(t))}\sum_{k_1+2k_2+\cdots+rk_r=r, k_j\ge 0} \frac{r!}{k_1!\cdots k_r!}\Big(\frac{i\theta'(t)}{1!}\Big)^{k_1}\cdots\Big(\frac{i\theta^{(r)}(t)}{r!}\Big)^{k_r}.$$

The term $k_1 = r$ (supposing henceforth that $r > 0$) yields

$$i^r(\theta'(t))^r\mathrm{e}^{i\theta(t)}.$$

The remaining terms are $\mathrm{e}^{-i\theta(t)}$ times

$$\sum_{k_1+2k_2+\cdots+rk_r=r, 0\le k_1<r,\cdots,0\le k_r} \frac{r!}{k_1!\cdots k_r!}\Big(\frac{i\theta'(t)}{1!}\Big)^{k_1}\cdots\Big(\frac{i\theta^{(r)}(t)}{r!}\Big)^{k_r}$$
$$= \frac{r!}{t^r}\sum_{m=1}^{r-1}\frac{1}{m!}\sum_{k_1+2k_2+\cdots+rk_r=r, k_1+\cdots+k_r=m, 0\le k_1<r,\cdots,0\le k_r}\frac{m!}{k_1!\cdots k_r!}$$
$$\times\Big(\frac{i\theta'(t)t}{1!}\Big)^{k_1}\cdots\Big(\frac{i\theta^{(r)}(t)}{r!}\Big)^{k_r}.$$

This gives

$$\Big|\frac{\mathrm{d}^r}{\mathrm{d}t^r}\mathrm{e}^{i\theta(t)} - i^r(\theta'(t))^r\mathrm{e}^{i\theta(t)}\Big|$$
$$\le \frac{r!}{t^r}\sum_{m=1}^{r-1}\frac{1}{m!}\sum_{k_1+2k_2+\cdots+rk_r, k_1+\cdots+k_r=m, 0\le k_1<r,\cdots,0\le k_r}\frac{m!}{k_1!\cdots k_r!}$$
$$\times\Big|\frac{\theta'(t)t}{1!}\Big|^{k_1}\cdots\Big|\frac{\theta^{(r)}(t)t^r}{r!}\Big|^{k_r}$$
$$\le \frac{r!}{t^r}\sum_{m=1}^{r-1}\frac{1}{m!}\Big(\sum_{\nu=1}^{r}\Big|\frac{\theta^{(\nu)}(t)t^\nu}{\nu!}\Big|\Big)^m. \tag{5.13}$$

But from (5.1)-(5.2) of Lemma 5.1 it follows that

$$\theta'(t) \ll \log t, \qquad \theta^{(\nu)}(t) \ll_\nu t^{1-\nu} \quad (\nu\ge 2).$$

We therefore obtain

$$\frac{r!}{t^r}\sum_{m=1}^{r-1}\frac{1}{m!}\left(\sum_{\nu=1}^{r}\left|\frac{\theta^{(\nu)}(t)t^\nu}{\nu!}\right|\right)^m \ll_r \frac{\log^{r-1}t}{t}. \tag{5.14}$$

The formula (5.12) of Lemma 5.4 then follows from (5.13) and (5.14).

Proof of Theorem 5.2 If the right-hand side of (5.8) is estimated trivially, one obtains

$$\frac{\mathrm{d}^m}{\mathrm{d}t^m}\zeta(\tfrac{1}{2}+it) \ll_m t^{1/4}\log^m t. \tag{5.15}$$

From (1.20) we obtain

$$Z^{(k)}(t) = \sum_{r=0}^{k}\binom{k}{r}\left(\frac{\mathrm{d}^r}{\mathrm{d}t^r}\mathrm{e}^{i\theta(t)}\right)\left(\frac{\mathrm{d}^{k-r}}{\mathrm{d}t^{k-r}}\zeta(\tfrac{1}{2}+it)\right). \tag{5.16}$$

We now substitute (5.8)-(5.9) in (5.16) and use Lemma 5.4. This will yield

$$\begin{aligned}Z^{(k)}(t) &= \sum_{r=0}^{k}\binom{k}{r}i^r(\theta'(t))^r\mathrm{e}^{i\theta(t)}(-i)^{k-r}\Big\{\sum_{n\le\sqrt{t/(2\pi)}} n^{-1/2-it}\log^{k-r}n \\ &\quad + \mathrm{e}^{i\theta_1(t)}\sum_{n\le\sqrt{t/(2\pi)}} n^{-1/2+it}\log^{k-r}(t/(2\pi n)) + O_k(t^{-1/4}\log^{k-r+1}t)\Big\} \\ &= Z_1(t) + Z_2(t) + O_k(t^{-1/4}\log^{k+1}t),\end{aligned} \tag{5.17}$$

say, where

$$\begin{aligned}Z_1(t) &:= (-i)^k\mathrm{e}^{i\theta(t)}\sum_{n\le\sqrt{t/(2\pi)}} n^{-1/2-it}\sum_{r=0}^{k}\binom{k}{r}(-\theta'(t))^r\log^{k-r}n \\ &= i^k\mathrm{e}^{i(\theta(t)}\sum_{n\le\sqrt{t/(2\pi)}} n^{-1/2-it}\Big(\theta'(t)-\log n\Big)^k,\end{aligned}$$

and

$$Z_2(t) := i^k\mathrm{e}^{i(\theta(t)+\theta_1(t))}\sum_{n\le\sqrt{t/(2\pi)}} n^{-1/2+it}\Big(\theta'(t)-\log(t/(2\pi n))\Big)^k.$$

By using the relation (see (1.27))

$$\theta'(t) = \frac{1}{2}\log\frac{t}{2\pi} + O\Big(\frac{1}{t}\Big)$$

we have

$$\Big(\theta'(t) - \log n\Big)^k = \Big(\frac{1}{2}\log\frac{t}{2\pi} - \log n\Big)^k + O_k(t^{-1}\log^{k-1} t),$$
$$\Big(\theta'(t) - \log(t/(2\pi n))\Big)^k = (-1)^k\Big(\frac{1}{2}\log\frac{t}{2\pi} - \log n\Big)^k + O_k(t^{-1}\log^{k-1} t)$$
$$= (-1)^k(\theta'(t) - \log n)^k + O_k(t^{-1}\log^{k-1} t).$$

If we substitute these expressions in the formulas defining $Z_1(t)$ and $Z_2(t)$ we arrive at

$$Z_1(t) + Z_2(t) = \sum_{n\le\sqrt{t/(2\pi)}} n^{-1/2}(\theta'(t) - \log n)^k(\mathrm{e}^{i\theta_2(t)} + \mathrm{e}^{i\theta_3(t)}) + O_k(t^{-3/4}\log t), \tag{5.18}$$

where

$$\theta_2(t) = \theta_2(t;n) = \theta(t) + \tfrac{1}{2}\pi k - t\log n$$
$$\theta_3(t) = \theta_3(t;n) = \theta(t) - \tfrac{1}{2}\pi k + t\log n + \theta_1(t).$$

Since we have

$$\theta(t) = \tfrac{1}{2}t\log t - \tfrac{1}{2}t\log 2\pi - \tfrac{1}{2}t - \tfrac{\pi}{8} + O(1/t) \tag{5.19}$$

and

$$\theta_1(t) = -t\log t + t\log 2\pi + t + \tfrac{1}{4}\pi,$$

we obtain

$$\theta_2(t) = \tfrac{1}{2}t\log t + \tfrac{1}{2}\pi k - t\log n - \tfrac{1}{2}t\log 2\pi - \tfrac{1}{2}t - \tfrac{1}{8}\pi + O(1/t),$$
$$\theta_3(t) = -\tfrac{1}{2}t\log t - \tfrac{1}{2}\pi k + t\log n + \tfrac{1}{2}t\log 2\pi + \tfrac{1}{2}t + \tfrac{1}{8}\pi + O(1/t).$$

Thus

$$\mathrm{e}^{i\theta_2(t)} + \mathrm{e}^{i\theta_3(t)} = \mathrm{e}^{i\theta_2(t)} + \mathrm{e}^{-i\theta_2(t)} + O\Big(\frac{1}{t}\Big)$$
$$= 2\cos\Big(\theta(t) + \tfrac{1}{2}\pi k - t\log n\Big) + O\Big(\frac{1}{t}\Big).$$

Using (5.17) and (5.18) we infer that

$$Z^{(k)}(t) = 2\sum_{n\le\sqrt{t/(2\pi)}} n^{-1/2}(\theta'(t) - \log n)^k\cos\Big(\theta(t) + \tfrac{1}{2}\pi k - t\log n\Big) + O_k(t^{-1/4}\log^{k+1} t).$$

This implies, in view of (5.19), formula (5.7) of Theorem 5.2. In the special case when $k = 0$, equation (5.7) reduces to the weakened form (2.3) of the Riemann-Siegel formula.

Theorem 5.2 was sharpened by A. A. Lavrik [Lav1]. She proved that, for any integer $k \ge 0$ and $t \ge \max(t_0, \mathrm{e}^{2k})$ we have, uniformly in k,

$$Z^{(k)}(t) = 2 \sum_{n \le \sqrt{t/(2\pi)}} n^{-1/2} \left(\theta'(t) - \log n\right)^k \times$$
$$\times \cos\left(\theta(t) - t\log n + \frac{\pi k}{2}\right) + O\left(t^{-1/4}(\tfrac{3}{2}\log t)^{k+1}\right). \quad (5.20)$$

The chief ingredient in the proof of (5.20) is the formula

$$\frac{\mathrm{d}^r}{\mathrm{d}t^r}\mathrm{e}^{i\theta(t)} = i^r(\theta'(t))^r \mathrm{e}^{i\theta(t)} + O(rt^{-1}\log^{r-1} t),$$

which improves (5.12) of Lemma 5.4. This holds uniformly in r for $1 \le r \le t$, $t \ge t_0$. The proof is similar to the proof of (5.12), only more precise estimates are used at relevant places. As a consequence of (5.20), A. A. Lavrik deduces that there exists a constant $T_0 > 0$ such that, for $T \ge T_0 > 0$ and

$$H \ \ge\ \max\left(9\pi\log T,\ T^{1/(6k+6)}\log^{2/(k+1)} T\right),$$

every interval $[T,\ T + H]$ contains a zero of odd order of $Z^{(k)}(t)$. This improves a result of A. A. Karatsuba [Kar1], who obtained the same assertion, but with

$$H \ \ge\ c(k)T^{1/(6k+6)}\log^{2/(k+1)} T$$

and an unspecified constant $c(k)$.

Notes

For a proof of the classical formula of Faà de Bruno (Francesco Fa di Bruno, March 29, 1825-March 27, 1888, an Italian mathematician and priest) for the derivatives of composite functions, see, for example, chapter 1 of E. Goursat [Gours].

K. Matsumoto and Y. Tanigawa [MaTa] prove that

$$N_{0,k}(T) \le \frac{T}{2\pi}\log\frac{T}{2\pi} - \frac{T}{2\pi} + O_k(\log T), \quad (5.21)$$

where $N_{0,k}(T)$ denotes the number of zeros of $Z^{(k)}(t)$ in $(0, T]$. Moreover, under the RH they show that the inequality in (5.21) may be replaced by equality.

R. J. Anderson [And] constructed and studied a meromorphic function $\eta(s)$ (this is different from the function in (1.18)), whose zeros on the line $\sigma = \frac{1}{2}$ coincide with the zeros of $Z'(t)$ (see Notes to Chapter 2). This is the function

$$\eta(s) := \zeta(s) - \frac{2\zeta'(s)}{\omega(s)}, \quad \omega(s) = \frac{\chi'(s)}{\chi(s)}. \quad (5.22)$$

From (1.20) we have, on using $(\chi(\frac{1}{2}+it))^{-1/2} = \mathrm{e}^{i\theta(t)}$,

$$\begin{aligned}
Z'(t) &= i\zeta'(\tfrac{1}{2}+it)\mathrm{e}^{i\theta(t)} + \zeta(\tfrac{1}{2}+it)\Big(\mathrm{e}^{i\theta(t)}\Big)' \\
&= \Big(\mathrm{e}^{i\theta(t)}\Big)'\left\{\zeta(\tfrac{1}{2}+it) + i\zeta'(\tfrac{1}{2}+it)\frac{\mathrm{e}^{i\theta(t)}}{\Big(\mathrm{e}^{i\theta(t)}\Big)'}\right\} \\
&= \Big(\mathrm{e}^{i\theta(t)}\Big)'\left\{\zeta(\tfrac{1}{2}+it) + \frac{\zeta'(\frac{1}{2}+it)}{\frac{-\frac{1}{2}\chi'(\frac{1}{2}+it)}{\chi(\frac{1}{2}+it)}}\right\} \\
&= \Big(\mathrm{e}^{i\theta(t)}\Big)'\eta(\tfrac{1}{2}+it).
\end{aligned}$$

Anderson proved that the number of zeros of $\eta(s)$ (counted with multiplicities) in the rectangle

$$\left\{s = \sigma + it \;\middle|\; -7 < \sigma < 8,\ 0 < t < T\right\}$$

equals

$$\frac{T}{2\pi}\log\frac{T}{2\pi} - \frac{T}{2\pi} + O(\log T).$$

K. Matsumoto and Y. Tanigawa [MaTa] generalize Anderson's result. They introduce the function $(\eta_1(s) \equiv \eta(s),\ k \in \mathbb{N})$

$$\eta_{k+1}(s) = \lambda(s)\eta_k(s) + \eta_k'(s), \quad \lambda(s) = \frac{\omega'(s)}{\omega(s)} - \tfrac{1}{2}\omega(s),$$

where $\omega(s)$ is given by (5.22). They derive several properties of $\eta_k(s)$, such as

$$Z^{(k)}(t) = i^{k-1}\Big(\mathrm{e}^{i\theta(t)}\Big)'\eta_k(\tfrac{1}{2}+it) \quad (k \in \mathbb{N}),$$

and the functional equation

$$h(1-s)\eta_k(s) = (-1)^k h(s)\eta_k(s) \quad (h(s) = \pi^{-s/2}\Gamma(s/2),\ k \in \mathbb{N}).$$

If $N(T;\eta_k)$ is the number of zeros (counted with multiplicities) of $\eta_k(s)$ in the rectangle

$$\left\{s = \sigma + it \;\middle|\; 1-2m < \sigma < 2m,\ 0 < t < T\right\},$$

where $m = m(k)$ which is explicitly constructed, then their main result is the asymptotic formula

$$N(T;\eta_k) = \frac{T}{2\pi}\log\frac{T}{2\pi} - \frac{T}{2\pi} + O_k(\log T). \tag{5.23}$$

This result generalizes and sharpens (5.21).

R. R. Hall [Hal1] proves that, for fixed $k \in \mathbb{N}\cup\{0\}$,

$$\int_0^T \Big(Z^{(k)}(t)\Big)^2\,\mathrm{d}t = \frac{1}{4^k(2k+1)}T P_{2k+1}\Big(\log\frac{T}{2\pi}\Big) + O(T^{3/4}(\log T)^{2k+1/2}),$$

where P_{2k+1} is an explicitly given monic polynomial of degree $2k+1$, whose coefficients depend on k. This is deduced from his AFE for $Z^{(k)}(t)$, which is slightly sharper than (5.7) of Theorem 5.2, and says that

$$Z^{(k)}(t) = i^k \left\{ e^{i\theta(t)} \sum_{n\le x} (\log \tau/n)^k n^{-1/2-it} + e^{-i\theta(t)} \sum_{n\le y} (\log \tau/n)^k n^{-1/2+it} \right\}$$
$$+ O_k\left((x^{-1/2} + y^{-1/2}) \log^k t \right),$$

where $2\pi xy = t, x \ge 1, y \ge 1, \tau = \sqrt{t/(2\pi)}$. When $x = y = \tau$, we obtain (5.7) with the error term $O_k(t^{-1/4} \log^k t)$, which is better by a log-factor.

The author [Iv3] showed that every interval $[T, T-H]$ contains a zero of odd order of $Z^{(k)}(t)$ if $T \ge T_0$ and $H \gg kT^{a(k)} \log T$, where $a(1) = \min(3/37, \mu(1/2)+\varepsilon)$ and $a(k) = \min(27/(164k+168), \mu(1/2)/k+\varepsilon)$ when $k \ge 2$, which improves the bound of A. A. Karatsuba [Kar1].

6

Gram points

6.1 Definition and order of Gram points

As already mentioned in Chapter 1, the function $\theta(t)$ (see (1.19)) is monotonic increasing for $t \geq 7$. For $n \geq -1$, we define the nth *Gram point* g_n to be the unique solution > 7 of the equation

$$\theta(g_n) \;=\; \pi n, \tag{6.1}$$

or equivalently $\sin\theta(g_n) = 0$. Numerical calculations (see, for example, C. B. Haselgrove [Has]) reveal that

$$g_{-1} = 9.666\,908\ldots, \qquad g_0 = 17.845\,600\ldots,$$
$$g_1 = 23.170\,283\ldots, \qquad g_2 = 27.670\,182\ldots,$$

etc. The significance of Gram points is that (1.20) shows that

$$\zeta(\tfrac{1}{2} + ig_n) = (-1)^n Z(g_n), \tag{6.2}$$

hence $\zeta(\frac{1}{2} + ig_n)$ is real. Therefore if $\zeta(\frac{1}{2} + it)$ is positive at successive Gram points g_n and g_{n+1}, then (6.2) shows that there must be a zero of $Z(t)$ for some $t \in (g_n, g_{n+1})$. J. P. Gram [Gra] calculated that $\zeta(\frac{1}{2} + ig_n)$ was positive for $-1 \leq n \leq 15$, and he supposed that this holds for many large values of n. Namely from (2.3), (1.27) and (6.1) it follows that

$$\begin{aligned} Z(g_n) &= 2 \sum_{m \leq \sqrt{g_n/(2\pi)}} m^{-1/2} \cos\Big(\theta(g_n) - g_n \log m + O(g_n^{-1})\Big) + O(g_n^{-1/4}) \\ &= 2(-1)^n \sum_{m \leq \sqrt{g_n/(2\pi)}} m^{-1/2} \cos(g_n \log m) + O(g_n^{-1/4}). \end{aligned} \tag{6.3}$$

Further note that on the right-hand side of (6.3) the first summand is $+1$, after which the summands are oscillatory and decreasing in absolute value. This, heuristically, makes one expect that the initial term $+1$ will dominate in size the whole sum. Therefore one may expect that $(-1)^n Z(g_n) \sim 2$ "often." For the order of g_n we have the following result.

Theorem 6.1

$$\begin{aligned} g_n &= \frac{2\pi n}{\log n}\left\{1 + \frac{1+\log\log n}{\log n} + O\Big(\Big(\frac{\log\log n}{\log n}\Big)^2\Big)\right\}, \\ n &= \frac{g_n}{2\pi}\log g_n\left\{1 - \frac{\log 2\pi \mathrm{e}}{\log g_n} - \frac{\pi}{4g_n\log g_n} + O\Big(\frac{1}{g_n^2\log g_n}\Big)\right\}. \end{aligned} \tag{6.4}$$

Corollary 6.2 *If $N_g(T)$ denotes the number of Gram points g_n not exceeding T, then the second formula in* (6.4) *implies that*

$$N_g(T) = \frac{T}{2\pi}\log T\left\{1 - \frac{\log 2\pi \mathrm{e}}{\log T} - \frac{\pi}{4T\log T} + O\Big(\frac{1}{T^2\log T}\Big)\right\}. \tag{6.5}$$

Proof of Theorem 6.1 We start from (1.21), which we rewrite as

$$\theta(t) = \tfrac{1}{2}t\log\frac{t}{2\pi\mathrm{e}} - \tfrac{1}{8}\pi + \Delta(t).$$

Setting $x = g_n$ and using (6.1) this becomes

$$\pi n = \tfrac{1}{2}g_n\log\Big(\frac{g_n}{2\pi\mathrm{e}}\Big) - \tfrac{1}{8}\pi + \Delta(g_n),$$

and then with

$$X := \log\Big(\frac{g_n}{2\pi\mathrm{e}}\Big), \quad y := \frac{1}{\mathrm{e}}\left(n + \frac{1}{8} - \frac{1}{\pi}\Delta(g_n)\right)$$

we obtain

$$X\mathrm{e}^X = y. \tag{6.6}$$

Since $\frac{\partial y}{\partial n}$ is monotonically increasing (see (1.26)), for $n \ge n_0$, the equation (6.6) has a unique solution $X = X(n)$. From (6.6) we have

$$X + \log X = \log y,$$

giving first $\log X = O(\log\log y)$, hence

$$X = \log y + O(\log\log y).$$

Since $\Delta(t) \ll 1/t$ (see (5.2)), iteration yields that

$$X = \log y - \log\log y + \frac{\log\log y}{\log y} + \frac{(\log\log y)^2}{2\log^2 y} + O\left(\frac{\log\log y}{\log^2 y}\right),$$

which gives

$$\frac{g_n}{2\pi \mathrm{e}} = \frac{y}{\log y} \exp\left\{\frac{\log\log y}{\log y} + \frac{(\log\log y)^2}{2\log^2 y} + O\left(\frac{\log\log y}{\log^2 y}\right)\right\},$$

and the first relation in (6.4) follows. The second one is proved similarly. The process can be, of course, carried further and more terms in the asymptotic formulas obtained.

We shall now give a formula for $g_{n+1} - g_n$, the difference between two consecutive Gram points. It will show that these differences are evenly distributed and tend to zero as $n \to \infty$. This is contained in the following theorem.

Theorem 6.3 *We have*

$$g_{n+1} - g_n = \frac{2\pi}{\log \frac{g_n}{2\pi} + O\left(\frac{1}{g_n \log g_n}\right)}. \tag{6.7}$$

Proof of Theorem 6.3 From (6.1) we have

$$\theta(g_{n+1}) - \theta(g_n) = \pi(n+1) - \pi n = \pi.$$

But the mean value theorem and (1.27) give

$$\theta(g_{n+1}) - \theta(g_n) \gg (g_{n+1} - g_n)\log g_n,$$

which implies that

$$g_{n+1} - g_n \ll 1/\log g_n. \tag{6.8}$$

On using Taylor's formula, (1.27) and (6.8), we obtain

$$\begin{aligned}\theta(g_{n+1}) - \theta(g_n) &= (g_{n+1} - g_n)\theta'(g_n) + O\Big((g_{n+1} - g_n)^2/g_n\Big)\\ &= (g_{n+1} - g_n)\Big(\tfrac{1}{2}\log(g_n/(2\pi)) + O(1/g_n^2)\\ &\quad + O(1/(g_n \log g_n))\Big).\end{aligned}$$

Formula (6.7) now easily follows from the last expression. Clearly, by refining the above argument, one could sharpen (6.7).

Remark 6.4 More generally, if $g_m, g_n \in [T, 2T]$, then one obtains

$$g_m - g_n \sim \frac{2\pi(m-n)}{\log m} \sim \frac{2\pi(m-n)}{\log T} \qquad (T \to \infty).$$

6.2 Gram's law

The so-called *Gram law* was proposed by J. I. Hutchinson [Hut]. Given Gram points g_n and g_{n+1}, Gram's law is said to hold if there is exactly one zero of $\zeta(\frac{1}{2} + it)$ for some t in the interval $(g_n, g_{n+1}]$. Hutchinson's original statement is couched in the zeros of $Z(t)$: "Gram calculated the first fifteen roots γ (of $Z(t)$) and called attention to the fact that the γ's and g_n's separate each other. I will refer to this property as Gram's law." It appears that zeros of multiplicity greater than one were not considered in this definition. In verifying the Riemann hypothesis in [Hut] to a certain height, zeros of multiplicity k are included as k simple zeros. It is therefore natural to suppose that Hutchinson wrote "zero" when he meant "simple zero". Furthermore the above definition is not concerned with zeros off the critical line, namely the presence of these zeros does not contradict Gram's law.

One also defines *the weak Gram law* as the statement that

$$(-1)^n Z(g_n) > 0, \quad (-1)^{n+1} Z(g_{n+1}) > 0. \tag{6.9}$$

Note that this statement and "Gram's law" are not equivalent, because the alternate definition (6.9) guarantees the presence of a zero of odd order in $(g_n, g_{n+1}]$, which may not be necessarily simple, but of order three, five, etc. It is for this reason that the property (6.9) is called the "weak Gram law." It is known today that "Gram's law" fails infinitely often. Hutchinson knew that "Gram's law" was not a "law" in the strict sense when he proposed the name; indeed "Gram's phenomenon" would have been a more appropriate terminology. What is correct is that, on average, there is exactly one zero of $Z(t)$ between two consecutive Gram points. In the calculations of [LRW] it turned out that among the first billion and a half Gram intervals, 72.6% have one zero, 13.8% have no zeros, 13.4% have two zeros, 0.18% have three zeros, and that there are only 33 Gram intervals with four zeros.

Recently, T. S. Trudgian [Tru1], [Tru2], [Tru3] investigated problems involving Gram's law. In [Tru1], he showed that, for $T \geq T_0$, there is a positive proportion of failures the weak Gram law, and therefore of Gram's law, between T and $2T$. He also showed that there exists a positive proportion of Gram intervals between T and $2T$ which contain at least one zero of $Z(t)$. In particular, this implies that the weak Gram law is a true positive proportion of time. As for the distribution of $Z(g_{2n})$ and $Z(g_{2n+1})$ on the average, we have the following result.

Theorem 6.5 *We have*

$$\sum_{n\le N} Z(g_{2n}) = 2N + O(N^{3/4}\log^{1/4} N),$$
$$\sum_{n\le N} Z(g_{2n+1}) = -2N + O(N^{3/4}\log^{1/4} N). \tag{6.10}$$

It immediately follows that $Z(g_{2n})$ is positive for an infinity of values of n and that $Z(g_{2n+1})$ is negative for an infinity of values of n. This yields once again that there are infinitely many zeros of $Z(t)$. Combining the formulas in (6.10) we also have

$$\sum_{n\le N} Z(g_n) = O(N^{3/4}\log^{1/4} N),$$
$$\sum_{n\le N} (-1)^n Z(g_n) = 2N + O(N^{3/4}\log^{1/4} N).$$

Proof of Theorem 6.5 Both formulas in (6.10) are proved in a similar way, so only the first one will be treated in detail. The starting point is (6.3). In view of (6.5) it follows that

$$\begin{aligned}\sum_{n\le N} Z(g_{2n}) &= 2\sum_{n\le N}\sum_{m\le\sqrt{g_{2n}/(2\pi)}} m^{-1/2}\cos(g_{2n}\log m) + O(N^{3/4}\log^{1/4} N)\\ &= 2N + 2\sum_{n\le N}\sum_{2\le m\le\sqrt{g_{2n}/(2\pi)}} m^{-1/2}\cos(g_{2n}\log m)\\ &\quad + O(N^{3/4}\log^{1/4} N)\\ &= 2N + O(N^{3/4}\log^{1/4} N) + 2\sum_{2\le m\le\sqrt{g_{2N}/(2\pi)}} m^{-1/2}\sum\nolimits_m,\end{aligned}$$

say, where

$$\sum\nolimits_m := \sum_{2\pi m^2\le g_{2n}\le g_{2N}} \cos(g_{2n}\log m) = 2\Re\Big\{\sum_{2\pi m^2\le g_{2n}\le g_{2N}} \exp(2\pi i\varphi(n))\Big\}$$

with

$$\varphi(n) = \varphi(n;m) := \frac{g_{2n}\log m}{2\pi i}. \tag{6.11}$$

From (6.11) and $\theta(g_n) = \pi n$ it follows that

$$\varphi'(n) = \frac{\log m}{2\pi}\frac{\mathrm{d}g_{2n}}{\mathrm{d}n}, \quad \theta'(g_{2n})\frac{\mathrm{d}g_{2n}}{\mathrm{d}n} = 2\pi, \quad \varphi'(n) = \frac{\log m}{\theta'(g_{2n})}.$$

Hence $\varphi'(n)$ is positive and steadily increasing, and

$$\varphi''(n) = -2\pi \log m \,\frac{\theta''(g_{2n})}{(\theta'(g_n))^3} \sim -\frac{8\pi \log m}{g_{2n}\log^3 g_{2n}}. \tag{6.12}$$

To complete the proof, we need

Lemma 6.6 *If $f(x)$ is real-valued and twice differentiable, and*

$$0 < \lambda_2 \le f''(x) \le h\lambda_2 \quad (\text{or } \lambda_2 \le -f''(x) \le h\lambda_2) \quad (x \in [a,b],\ b-a \ge 1),$$

then

$$\sum_{a<n\le b} e^{2\pi i f(n)} \ll h(b-a)\lambda_2^{1/2} + \lambda_2^{-1/2}. \tag{6.13}$$

Proof of Lemma 6.6 If $\lambda_2 \ge 1$ the result is trivial, since the sum is $\ll b-a$. Otherwise we use Lemma 2.3 and the well-known transformation formula

$$\sum_{a<n\le b} e^{2\pi i f(n)} = \sum_{\alpha-\eta<\nu<\beta+\eta} \int_a^b e^{2\pi i (f(x)-\nu x)}\,dx + O(\log(\beta-\alpha+2)). \tag{6.14}$$

In (6.14) we suppose that $f(x)$ is a real-valued function with a continuous and decreasing derivative in (a,b), and let $f'(b)=\alpha$, $f'(a)=\beta$, while $0<\eta<1$ is a constant. It follows that the sum in (6.13) is

$$\ll (\beta-\alpha+1)\lambda_2^{-1/2} + \log(\beta-\alpha+2), \tag{6.15}$$

where

$$\beta-\alpha = f'(a)-f'(b) \ll (b-a)h\lambda_2.$$

Since

$$\log(\beta-\alpha+2) \ll \beta-\alpha+2 \ll (b-a)h\lambda_2+1 \ll (b-a)h\lambda_2^{1/2}+1,$$

the bound in (6.13) follows from (6.15).

In view of (6.12) we apply Lemma 6.6 with

$$\lambda_2 \asymp \frac{\log m}{M\log^{3/2} M}$$

to obtain, uniformly in m,

$$\sum_{M<g_{2n}<M'\le 2M} \exp\Big(2\pi i\varphi(n)\Big) \ll M^{1/2}\left(\frac{\log^{1/2} m}{\log^{3/2} M} + \frac{\log^{3/2} M}{\log^{1/2} m}\right). \tag{6.16}$$

Taking in (6.16) $M = 2^{-j} g_{2N}$ $(j = 1, 2, \ldots)$ and summing the resulting bounds we have

$$\sum_{2 \le m \le \sqrt{\frac{g_{2N}}{2\pi}}} m^{-1/2} \sum\nolimits_m \ll \sum_{2 \le m \le \sqrt{\frac{g_{2N}}{2\pi}}} m^{-1/2} g_{2N}^{1/2} \left(\frac{\log^{1/2} m}{\log^{3/2} g_{2N}} + \frac{\log^{3/2} g_{2N}}{\log^{1/2} m} \right)$$

$$\ll g_{2N}^{1/2} \left(g_{2N}^{1/4} \frac{\log^{1/2} g_{2N}}{\log^{3/2} g_{2N}} + \frac{g_{2N}^{1/4}}{\log^{1/2} g_{2N}} \log^{3/2} g_{2N} \right)$$

$$\ll g_{2N}^{3/4} \log g_{2N} \ll N^{3/4} \log^{1/4} N$$

as $g_n \ll n/\log n$ by (6.4). This yields the first formula in (6.10), and the second one is proved similarly.

6.3 A mean value result

We now follow F. V. Atkinson's work [Atk3] and define, for a fixed $\alpha > 0$, t_α as a continuous function of t, by the relation

$$\theta(t_\alpha) - \theta(t) = \alpha.$$

This defining relation makes t_α somewhat related to Gram points. To evaluate it explicitly note that by (1.27) we have

$$(t_\alpha - t)\theta'(t) + O((t_\alpha - t)^2 t^{-1}) = \alpha,$$

hence

$$t_\alpha = t + \frac{2\alpha}{\log(t/2\pi)} + O\left(\frac{1}{t \log^3 t}\right), \tag{6.17}$$

which is a good approximation to t_α. We shall prove the following result.

Theorem 6.7 *We have, for a fixed $\alpha > 0$,*

$$\int_{T/2}^{T} Z(t)Z(t_\alpha)\mathrm{d}t \ = \ \frac{\sin\alpha}{2\alpha} T \log T + O(T). \tag{6.18}$$

Proof of Theorem 6.7 We use (1.20), $\overline{Z(t)} = Z(t)$ and (6.17) to deduce that

$$\int_{T/2}^{T} Z(t)Z(t_\alpha)\,\mathrm{d}t = \int_{T/2}^{T} Z(t)Z\Big(t + \frac{2\alpha}{\log(t/2\pi)}\Big)\,\mathrm{d}t + O(T^{1/2})$$

$$= \mathrm{e}^{-i\alpha} \int_{T/2}^{T} \zeta(\tfrac{1}{2} + it)\zeta\Big(\tfrac{1}{2} - it - \frac{2i\alpha}{\log(t/2\pi)}\Big)\,\mathrm{d}t + O(T^{1/2}). \tag{6.19}$$

For the zeta terms in the last integral we use (4.5) of Theorem 4.3, namely

$$\zeta(\tfrac12+it)=\sum_{m\le\sqrt{t/(2\pi)}} m^{-1/2-it}+\chi(\tfrac12+it)\sum_{m\le\sqrt{t/(2\pi)}} m^{-1/2+it}+O(t^{-1/4}),$$

$$\zeta\Big(\tfrac12-it-\frac{2i\alpha}{\log(t/2\pi)}\Big)=\sum_{n\le\sqrt{t/(2\pi)}} n^{-1/2+it+\frac{2i\alpha}{\log(t/2\pi)}}$$

$$+\chi\Big(\tfrac12-it-\frac{2i\alpha}{\log(t/2\pi)}\Big)\sum_{n\le\sqrt{t/(2\pi)}} n^{-1/2-it-\frac{2i\alpha}{\log(t/2\pi)}}+O(t^{-1/4}).\tag{6.20}$$

We multiply together the expressions in (6.20) and integrate. Note that each of the terms $O(t^{-1/4})$ will make contributions that will be (δ equals either 0 or 1)

$$\ll T^{-1/4}\int_{T/2}^{T}\Bigg|\sum_{n\le\sqrt{t/(2\pi)}} n^{-1/2-it-\delta\frac{2i\alpha}{\log(t/2\pi)}}\Bigg|\,\mathrm{d}t\ \ll\ T^{3/4}\log^{1/2}T,$$

on using the Cauchy-Schwarz inequality for integrals and developing the square. The contribution of the terms with one χ-term will be, by the second derivative test,

$$\ll T^{1/2}\sum_{m\le\sqrt{T/(2\pi)}} m^{-1/2}\sum_{n\le\sqrt{T/(2\pi)}} n^{-1/2}\ll T,$$

since (see (2.17))

$$\chi(s)=\Big(\frac{2\pi}{t}\Big)^{\sigma+it-1/2}\mathrm{e}^{i(t+\pi/4)}\Big(1+O\Big(\frac1t\Big)\Big).$$

With the aid of this formula it is found that for the term containing two χ-terms we shall have

$$\chi(\tfrac12+it)\chi\Big(\tfrac12-it-\frac{2i\alpha}{\log(t/2\pi)}\Big)=\mathrm{e}^{-2i\alpha/\log(t/2\pi)}\Big(1+O\Big(\frac1t\Big)\Big).$$

Therefore the left-hand side of (6.19) is $O(T)$ plus

$$\int_{T/2}^{T}\big(\mathrm{e}^{-i\alpha}\Sigma(t)S(t)+\mathrm{e}^{i\alpha}\overline{\Sigma(t)S(t)}\big)\,\mathrm{d}t=2\Re\Big\{\mathrm{e}^{-i\alpha}\int_{T/2}^{T}\Sigma(t)S(t)\,\mathrm{d}t\Big\},\tag{6.21}$$

where

$$\Sigma(t):=\sum_{m\le\sqrt{t/(2\pi)}} m^{-1/2-it},\quad S(t):=\sum_{n\le\sqrt{t/(2\pi)}} n^{-1/2+it+\frac{2i\alpha}{\log(t/2\pi)}}.$$

We write

$$\int_{T/2}^{T} \Sigma(t)S(t)\,\mathrm{d}t = J_1 + J_2, \tag{6.22}$$

say, where in J_1 summation is over the terms for which $m = n$, and in J_2 over the terms for which $m \neq n$. Thus

$$\begin{aligned} J_1 &:= \int_{T/2}^{T} \sum_{n \le \sqrt{t/(2\pi)}} n^{-1+\frac{2i\alpha}{\log(t/2\pi)}}\,\mathrm{d}t \\ &= \int_{T/2}^{T} \sum_{n \le \sqrt{T}} n^{-1+\frac{2i\alpha}{\log(t/2\pi)}}\,\mathrm{d}t + O(T) \\ &= \sum_{n \le \sqrt{T}} n^{-1} I(n,T) + O(T), \end{aligned}$$

say, with

$$I(n,T) := \int_{T/2}^{T} \mathrm{e}^{\frac{2i\alpha\log n}{\log(t/2\pi)}}\,\mathrm{d}t.$$

Integration by parts yields

$$\begin{aligned} I(n,T) &= t\mathrm{e}^{\frac{2i\alpha\log n}{\log(t/2\pi)}}\Big|_{T/2}^{T} + \int_{T/2}^{T} \frac{2i\alpha\log n}{\log^2(t/2\pi)}\mathrm{e}^{\frac{2i\alpha\log n}{\log(t/2\pi)}}\,\mathrm{d}t \\ &= t\mathrm{e}^{\frac{2i\alpha\log n}{\log(t/2\pi)}}\Big|_{T/2}^{T} + O\left(T\frac{\log n}{\log^2 T}\right). \end{aligned}$$

It follows that

$$J_1 = \left(t \sum_{n \le \sqrt{T}} n^{-1+\frac{2i\alpha}{\log(t/2\pi)}}\right)\Bigg|_{T/2}^{T} + O(T).$$

Setting $u := 2\alpha/\log(t/2\pi)$ we obtain, by the Euler-Maclaurin summation formula, that the last sum over n equals

$$\sum_{1/2 < n \le \sqrt{T}} n^{-1+iu} = \int_{1/2}^{\sqrt{T}} x^{-1+iu}\,\mathrm{d}x + O(1).$$

Further we have

$$\begin{aligned} \int_{1/2}^{\sqrt{T}} x^{-1+iu}\,\mathrm{d}x &= \frac{1}{iu}\left(\mathrm{e}^{\frac{1}{2}iu\log T} - \mathrm{e}^{iu\log\frac{1}{2}}\right) \\ &= \frac{\log(t/2\pi)}{2i\alpha}\left(\mathrm{e}^{\frac{\alpha i\log T}{\log(t/(2\pi))}} - \mathrm{e}^{\frac{2i\alpha\log\frac{1}{2}}{\log(t/(2\pi))}}\right). \end{aligned}$$

But since

$$\frac{\log T}{\log(t/(2\pi))} = \frac{\log \frac{t}{2\pi} + \log \frac{T}{t/(2\pi)}}{\log(t/(2\pi))} = 1 + O\left(\frac{1}{\log T}\right),$$

we obtain

$$\int_{1/2}^{\sqrt{T}} x^{-1+iu}\,\mathrm{d}x = \frac{\log(t/2\pi)}{2i\alpha}\left(\mathrm{e}^{\alpha i} - 1 + O\left(\frac{1}{\log T}\right)\right).$$

This gives

$$\begin{aligned} 2\mathbb{R}\mathrm{e}\,\{J_1\} &= 2\mathbb{R}\mathrm{e}\left\{\frac{\mathrm{e}^{i\alpha}-1}{2i\alpha}\right\} t\log\frac{t}{2\pi}\Big|_{T/2}^{T} + O(T) \\ &= \frac{\sin\alpha}{\alpha} t\log\frac{t}{2\pi}\Big|_{T/2}^{T} + O(T) \\ &= \frac{\sin\alpha}{2\alpha} T\log T + O(T). \end{aligned} \tag{6.23}$$

It remains yet to deal with J_2, where we have $m \neq n$, and the function in the exponential is $ig(t; m, n)$ with

$$g(t;m,n) := t\log\frac{n}{m} + \frac{2\alpha\log n}{\log(t/(2\pi))}, \quad \frac{\mathrm{d}g(t;m,n)}{\mathrm{d}t} = \log\frac{n}{m} - \frac{2\alpha\log n}{t\log^2(t/(2\pi))}.$$

For $n \le m/2$ or $n \ge 2m$ we have

$$\frac{\mathrm{d}g(t;m,n)}{\mathrm{d}t} \gg \left|\log\frac{n}{m}\right| \gg 1,$$

hence by the first derivative test the contribution of such m, n is clearly $\ll \sqrt{T}$. In the remaining range

$$\log\frac{n}{m} = \log\left(1 + \frac{n-m}{m}\right) \gg \left|\frac{n-m}{m}\right| \gg \frac{1}{m} \ge \frac{1}{\sqrt{T}} > \frac{4\alpha\log n}{t\log^2(t/(2\pi))},$$

hence

$$\frac{\mathrm{d}g(t;m,n)}{\mathrm{d}t} \gg \log\frac{n}{m} \gg \frac{|n-m|}{m}.$$

Thus, again by the first derivative test, the contribution of such m, n is

$$\sum_{m\neq n\le\sqrt{T},\,2m>n>m/2} (mn)^{-1/2}\frac{m}{|n-m|} \ll \sqrt{T}\log T \qquad (T \ge T_0),$$

and then

$$J_2 \ll \sqrt{T}\log T. \tag{6.24}$$

The assertion (6.18) of Theorem 6.7 follows then from (6.21)-(6.24).

The interest in a result like Theorem 6.7 is that provides a fairly good lower bound for $N_0(T)$, the number of zeros of $Z(t)$ in $(0, T]$, which is the same as the number of zeros of $\zeta(s)$ on the critical line $\mathbb{R}\mathrm{e}\, s = 1/2$ whose imaginary parts lie in $(0, T]$. We have the following

Corollary 6.8

$$N_0(T) \gg \frac{T}{\log T}. \tag{6.25}$$

To obtain (6.25) from (6.18), let first

$$\mathcal{S}(T) := \Big\{ t \in (T/2, T) \ : \ Z(t)Z(t_\alpha) < 0 \Big\}.$$

Choose henceforth $\alpha = 3\pi/2$. Then note that from (6.18) with this particular value one obtains

$$\int_{T/2}^{T} Z(t)Z(t_\alpha)\,\mathrm{d}t = -\frac{1}{3\pi} T \log T + O(T).$$

Thus we have

$$\int_{\mathcal{S}(T)} |Z(t)Z(t_\alpha)|\,\mathrm{d}t \gg T \log T. \tag{6.26}$$

Namely, suppose that this lower bound does not hold, and that

$$\int_{\mathcal{S}(T)} |Z(t)Z(t_\alpha)|\,\mathrm{d}t < \varepsilon T \log T \qquad (T \geq T_o(\varepsilon))$$

for every $\varepsilon > 0$, so that

$$\int_{\mathcal{S}(T)} Z(t)Z(t_\alpha)\,\mathrm{d}t > -\varepsilon T \log T.$$

Then with $R(T) := [T/2, T] \setminus \mathcal{S}(T)$ we have

$$\begin{aligned} -\frac{1}{3\pi} T \log T &\sim \int_{\mathcal{S}(T)} Z(t)Z(t_\alpha)\,\mathrm{d}t + \int_{R(T)} Z(t)Z(t_\alpha)\,\mathrm{d}t \\ &\geq \int_{\mathcal{S}(T)} Z(t)Z(t_\alpha)\,\mathrm{d}t > -\varepsilon T \log T, \end{aligned}$$

which is a contradiction for sufficiently large T. It follows from (6.26) by the Cauchy-Schwarz inequality for integrals that

$$\mu(\mathcal{S}(T)) \int_{\mathcal{S}(T)} \Big|Z(t)Z(t_\alpha)\Big|^2 \mathrm{d}t \gg T^2 \log^2 T,$$

where $\mu(\cdot)$ denotes measure. But we have

$$\int_{\mathcal{S}(T)} |Z(t)Z(t_\alpha)|^2\,\mathrm{d}t \le \left\{\int_{T/2}^{T} |Z(t)|^4\,\mathrm{d}t \int_{T/2}^{T} |Z(t_\alpha)|^4\,\mathrm{d}t\right\}^{1/2}.$$

Note that

$$\int_{T/2}^{T} |Z(t)|^4\,\mathrm{d}t = \int_{T/2}^{T} \left|\zeta(\tfrac{1}{2}+it)\right|^4\,\mathrm{d}t \ll T\log^4 T,$$

and a similar upper bound is readily derived for the integral with t_α; for example, one can use the method of proof of theorem 5.2 of [Iv1]. Hence

$$\mu(\mathcal{S}(T)) \gg \frac{T}{\log^2 T}.$$

If $t \in \mathcal{S}(T)$, then there is at least one zero of $Z(t)$ in $[t, t + C/\log T]$ for suitable $C > 0$. Therefore each zero contributes an interval of length $\ll 1/\log T$ to $\mathcal{S}(T)$, and consequently

$$\mu(\mathcal{S}(T)) \ll \Big(N_0(T) - N_0(T/2)\Big)(\log T)^{-1},$$

and we obtain

$$N_0(T) - N_0(T/2) \gg \frac{T}{\log T}.$$

Replacing T by $T2^{-j}$, summing over $j = 1, 2, \ldots$ and adding the resulting estimates, we obtain (6.25).

Notes

A calculation of the first 15 sign changes of $Z(t)$ was published by J. P. Gram [Gra] in 1903 (Jørgen Pedersen Gram, June 27, 1850-April 29, 1916, a Danish actuary and mathematician). He showed that these first 15 zeros are the only non-trivial zeros of $\zeta(s)$ up to height $t = 50$, in other words they satisfy the Riemann hypothesis (RH). Each of these zeros was found to lie between successive Gram points. The work of Gram was continued by R. J. Backlund [Bac] in 1912, who calculated the first 79 zeros of $Z(t)$ and verified the RH and Gram's law up to height $t = 200$. This was extended in 1925 by Hutchinson [Hut], who found the first 138 zeros of $Z(t)$. They all verify the RH, but Gram's law failed with four exceptions. He also established that the interval (g_{125}, g_{126}) contains no zeros, while (g_{126}, g_{127}) contains two zeros. Today the RH has been verified for billions of zeros. For example, J. van de Lune *et al.* [LRW] showed in 1986 that the first 1.5 billion zeros of $\zeta(s)$ up to height $t = 2.5 \times 10^8$ are simple and satisfy the RH, while X. Gourdon [Gourd] extended this in 2004 to the first 10^{13} zeta-zeros up to height 2.5×10^{12}. No doubt numerical data will continue to accrue.

"Gram's law" actually has different senses in different papers. A detailed historical survey of this subject, as well as some new results are extensively discussed in the forthcoming paper of M. A. Korolev [Kor5].

E. C. Titchmarsh [Tit1] proved Theorem 6.1. He also proved that

$$\sum_{n\le N} Z(g_n)Z(g_{n+1}) = -2(1+\gamma)N + o(N) \qquad (N\to\infty),$$

where γ is Euler's constant. Titchmarsh also conjectured (*op. cit.*) that

$$\sum_{n\le N} Z^2(g_n)Z^2(g_{n+1}) \ll N\log^A N \tag{6.27}$$

for some constant $A > 0$. This was proved by J. Moser [Mos1]. In a series of papers he investigated various sums of Hardy's function over Gram points, extending and sharpening Titchmarsh's results. In particular, in [Mos1] he proved that

$$\sum_{n\le N} Z^4(g_n) \ \ll\ N\log^4 N, \tag{6.28}$$

hence (6.27) follows from (6.28) by the Cauchy-Schwarz inequality. Later in [Mos2] he sharpened (6.27) to

$$\sum_{T\le g_\nu\le 2T} Z^2(g_\nu)Z^2(g_{\nu+1}) = \frac{3}{4\pi^5}T\ln^5 T + O(T\ln^4 T).$$

The upper bound in (6.28) was sharpened by A. A. Lavrik [Lav2] (Alla Aleksandrovna Lavrik, 1964-2003, Russian mathematician) to the asymptotic formula

$$\sum_{n\le N} Z^4(g_n) \ =\ \frac{1}{2\pi^2}N\log^4 N\Big(1 + O(\log^{-1/2} N)\Big). \tag{6.29}$$

Thus (6.29) shows that the bound in (6.28) is of the correct order of magnitude.

F. V. Atkinson [Atk3] (Frederick Valentine Atkinson, January 25, 1916-November 13, 2002, Canadian mathematician of British origin), proved that, for fixed $T_0 > 0$,

$$\int_{T_0}^{T} Z(t)Z(t_\alpha)\,\mathrm{d}t = \frac{\sin\alpha}{\alpha}T\log T + O(T\log^{3/4} T), \tag{6.30}$$

hence Theorem 6.7 is a new result. The deduction of (6.25) from (6.30) (hence it follows a priori from (6.18)) is due to Atkinson [Atk3]. The result is in fact by a factor of $\log^2 T$ poorer than the true order of magnitude of $N_0(T)$, as shown by A. Selberg's (1.49). However, the proof of Selberg's bound is much more involved than the proof of (6.25).

As stated in the Notes to Chapter 2, (6.18) is similar to R. R. Hall's (2.60), but the above integral is more complicated to handle, since t_α depends on t whereas Hall's a $(=\alpha/\log T)$ in (2.60) does not. If one wanted to sharpen (6.18) and obtain a result analogous (2.60), one would have to estimate

$$\int_{T/2}^{T} |Z(t)|\left|Z\Big(t+\frac{2\alpha}{\log T}\Big) - Z\Big(t+\frac{2\alpha}{\log t}\Big)\right|\mathrm{d}t,$$

but this does not seem to be easy. In a recent paper H. Kösters [Kos] proved several results connected to Atkinson's (6.18). In particular he showed that, for real constants μ, ν,

$$\lim_{T\to\infty}\frac{1}{T\log T}\int_2^T \zeta\Big(\frac12 + i\Big(t+\frac{2\pi\mu}{\log t}\Big)\Big)\zeta\Big(\frac12 - i\Big(t+\frac{2\pi\nu}{\log t}\Big)\Big)\,\mathrm{d}t = \mathrm{e}^{-\pi(\mu-\nu)}\mathcal{S}(\pi(\mu-\nu)),$$

where

$$\mathcal{S}(x) = \begin{cases} \frac{\sin x}{x} & \text{if } x\ne 0,\\ 1 & \text{if } x = 0.\end{cases}$$

He also showed that

$$\lim_{T\to\infty}\frac{1}{T\log^4 T}\int_2^T\left|\zeta\left(\frac{1}{2}+i\left(t+\frac{2\pi\mu}{\log t}\right)\right)\right|^2\left|\zeta\left(\frac{1}{2}-i\left(t+\frac{2\pi\nu}{\log t}\right)\right)\right|^2 \mathrm{d}t=\frac{3}{2\pi^2}\mathcal{T}(\pi(\mu-\nu)),$$

where

$$\mathcal{T}(x)=\begin{cases}\frac{1}{x^2}\left(1-\left(\frac{\sin x}{x}\right)\right)^2 & \text{if } x\neq 0,\\ 1/3 & \text{if } x=0.\end{cases}$$

Kösters also makes a conjecture, based on random matrix theory (this is connected to the works of J. B. Conrey *et al.* [CFKRS1], [CFKRS2]), for the integral of product of M zeta-shifts with the plus sign, times M shifts with the minus sign, of which the first formula above is a special case. In the case when $\mu=\nu=0$, the formulas reduce to the classical results on the mean square and mean fourth moment of $|\zeta(\frac{1}{2}+it)|^2$, respectively.

Another related result is due to S. Bettin [Bet], who obtains an asymptotic formula for

$$\int_0^T \zeta(\tfrac{1}{2}+a+it)\zeta(\tfrac{1}{2}-b-it)\,\mathrm{d}t$$
$$\left(T\geq 2, a=a(T), b=b(T), \mathbb{R}\mathrm{e}\, a\ll 1/\log T, \mathbb{R}\mathrm{e}\, b\ll 1/\log T\right). \tag{6.31}$$

He begins the proof by noting that the left-hand side of (6.31) equals

$$-i\int_{1/2}^{1/2+iT}\zeta(s+a)\zeta(1-s-b)\,\mathrm{d}s=-i\int_{1/2}^{1/2+iT}\zeta(s+a)\chi(1-s-b)\zeta(s+b)\,\mathrm{d}s,$$

changing then the segment of integration, so that the assumption that $a=a(T), b=b(T)$ is essential. Therefore, as was the case with Köster's result, Bettin's formula cannot be applied directly, since in (6.19) we need $a=0$, $b=2\alpha/\log(t/2\pi)$, so that b depends on t, and not on T.

Theorem 6.5 is stronger than theorem 10.10 of [Tit3], which says that, as $N\to\infty$,

$$\sum_{n\leq N} Z(g_{2n}) = 2N+o(N),$$
$$\sum_{n\leq N} Z(g_{2n+1}) = -2N+o(N).$$

The transformation formula (6.14) is theorem 10.14 of [Tit3] or lemma 2.11 of [Iv1]. Its proof follows by using the familiar *Euler-Maclaurin summation formula* (see (A.23) of [Iv1]). We obtain that the left-hand side of (6.14) is equal to

$$\int_a^b \mathrm{e}(f(x))\,\mathrm{d}x+2\pi i\int_a^b \psi(x)f'(x)\mathrm{e}(f(x))\,\mathrm{d}x+O(1), \tag{6.32}$$

where. as usual, $\mathrm{e}(z)=\mathrm{e}^{2\pi iz}$ and

$$\psi(x)=x-[x]-\tfrac{1}{2}=-\sum_{m=1}^{\infty}\frac{1}{\pi m}\sin(2\pi mx)\qquad (x\notin\mathbb{Z}). \tag{6.33}$$

We use the Fourier expansion (6.33) in (6.32), integrate by parts and apply the second derivative test (Lemma 2.3) to the portion of the sum over m for which $m>\beta$ to arrive eventually at (6.14).

7

The moments of Hardy's function

7.1 The asymptotic formula for the moments

In this chapter we shall deal with the moments

$$F_k(T) := \int_1^T Z^k(t)\,\mathrm{d}t \qquad (k \in \mathbb{N},\ T \geq 1), \tag{7.1}$$

so that, in particular, $F(T) \equiv F_1(T)$ is a primitive of Hardy's function $Z(t)$. The study of a primitive of a highly oscillating function, such as $Z(t)$, certainly is of interest. If we set

$$\mathcal{M}_k(s) = \int_1^\infty Z^k(x)x^{-s}\,\mathrm{d}x \quad (k \in \mathbb{N},\, \sigma \geq \sigma_0(k)), \tag{7.2}$$

then on integrating by parts we find that

$$\mathcal{M}_k(s) = s\int_1^\infty F_k(x)x^{-s-1}\,\mathrm{d}x, \tag{7.3}$$

so that the properties of $F_k(x)$ are reflected on $\mathcal{M}_k(s)$. Conversely, the Mellin inversion formula gives

$$Z^k(x) = \frac{1}{2\pi i}\int_{(c)} \mathcal{M}_k(s)x^{s-1}\,\mathrm{d}s \qquad (x \geq 1) \tag{7.4}$$

for suitable $c = c(k)$ (> 0). Hence from (7.4) we obtain by integration

$$F_k(x) = \frac{1}{2\pi i}\int_{(c)} \mathcal{M}_k(s)\frac{x^s}{s}\,\mathrm{d}s.$$

Our main result concerning $F_k(T)$ is the following

Theorem 7.1 *For fixed $k = 1, 2, 3, 4$ we have*

$$\int_T^{2T} Z^k(t)\,dt = 2\pi\sqrt{\frac{2}{k}} \sum_{(\frac{T}{2\pi})^{k/2} \le n \le (\frac{T}{\pi})^{k/2}} d_k(n) n^{-\frac{1}{2}+\frac{1}{k}} \cos\left(k\pi n^{\frac{2}{k}} + \tfrac{1}{8}(k-2)\pi\right) + \cdots + O_\varepsilon(T^{k/4+\varepsilon}), \tag{7.5}$$

where $+\cdots$ denotes terms similar to the one on the right-hand side of (7.5), *with the similar cosine term, but of a lower order of magnitude.*

Proof of Theorem 7.1 To prove (7.5) we shall make use of the approximate functional equation (4.81), in which we take $b = 2$. Recall that $\rho(x)$ is a non-negative, smooth function supported in $[0, 2]$, such that $\rho(x) = 1$ for $0 \le x \le 1/2$, and $\rho(x) + \rho(1/x) = 1$ for all x. It follows that

$$\int_T^{2T} Z^k(t)\,\mathrm{d}t = 2\int_T^{2T} \sum_{n \le 2\sqrt{\tau}} \rho\Big(\frac{n}{\sqrt{\tau}}\Big) d_k(n) n^{-1/2} \cos \mathcal{F}_k(t)\,\mathrm{d}t + O\left(T^{k/4}\log^{k-1} T\right), \tag{7.6}$$

where $\tau = \tau(t,k) = (t/(2\pi))^k$ and

$$\mathcal{F}_k(t) := t\log\left\{\frac{\left(\frac{t}{2\pi}\right)^{k/2}}{n}\right\} - \frac{kt}{2} - \frac{k\pi}{8}. \tag{7.7}$$

To evaluate the right-hand side of (7.6) we write first

$$2\int_T^{2T} \sum_{n \le 2\sqrt{\tau}} \rho\Big(\frac{n}{\sqrt{\tau}}\Big) d_k(n) n^{-1/2} \cos \mathcal{F}_k(t)\,\mathrm{d}t = 2\sum_{n \le T_0} d_k(n) n^{-1/2} \mathbb{R}\mathrm{e}\left\{\int_{T_1}^{2T} \rho\Big(\frac{n}{\sqrt{\tau}}\Big) \mathrm{e}^{i\mathcal{F}_k(t)}\,\mathrm{d}t\right\}. \tag{7.8}$$

Here

$$T_0 = 2\sqrt{\tau(2T,k)} = 2(T/\pi)^k, \qquad T_1 = \max\Big(T,\ 2\pi(n/2)^{2/k}\Big).$$

Now we split the range of summation over n on the right-hand side of (7.8) as follows. Let

$$I_1 := \left[1,\ \left(\frac{T}{2\pi}\right)^{k/2} - T^{k/2-1/2+\varepsilon}\right),$$

$$I_2 := \left[\left(\frac{T}{2\pi}\right)^{k/2} - T^{k/2-1/2+\varepsilon},\ \left(\frac{T}{2\pi}\right)^{k/2} + T^{k/2-1/2+\varepsilon}\right),$$

$$I_3 := \left[\left(\frac{T}{2\pi}\right)^{k/2} + T^{k/2-1/2+\varepsilon}, \left(\frac{T}{\pi}\right)^{k/2} - T^{k/2-1/2+\varepsilon}\right],$$
$$I_4 := \left(\left(\frac{T}{\pi}\right)^{k/2} - T^{k/2-1/2+\varepsilon}, \left(\frac{T}{\pi}\right)^{k/2} + T^{k/2-1/2+\varepsilon}\right],$$
$$I_5 := \left(\left(\frac{T}{\pi}\right)^{k/2} + T^{k/2-1/2+\varepsilon}, T_0\right]. \tag{7.9}$$

In the integrals in (7.8) where $n \in I_1$ and $n \in I_5$ we integrate by parts, writing

$$\int \rho\Big(\frac{n}{\sqrt{\tau}}\Big) e^{i\mathcal{F}_k(t)}\, dt = \int \frac{\rho\Big(\frac{n}{\sqrt{\tau}}\Big)}{i\log\{(t/2\pi)^{k/2}/n\}}\, de^{i\mathcal{F}_k(t)}. \tag{7.10}$$

Note that the derivatives of $\rho\Big(n/\sqrt{\tau}\Big)$, considered as functions of t, decrease after each integration by parts by a factor of t, while in $\sum_{n\in I_1}\int$ we have

$$\left(\frac{1}{\log\{(t/2\pi)^{k/2}/n\}}\right)' = -\frac{2}{kt\log^2\{(t/2\pi)^{k/2}/n\}}$$
$$\ll_\varepsilon \frac{1}{T\log^2\left\{\frac{CT^{k/2}}{T^{k/2}+O(T^{k/2-1/2+\varepsilon})}\right\}} \ll_\varepsilon T^{-2\varepsilon}. \tag{7.11}$$

Therefore if we integrate by parts sufficiently many times, the contribution will be negligible. The sums over the integrated terms are essentially partial sums of $\zeta^k(\frac{1}{2}+iu), u \asymp T$, when we remove the monotonic coefficients ρ from the sums over n by partial summation. The resulting sums are bounded by Perron's inversion formula (see for example, the appendix of [Iv1]). Since $\zeta(\frac{1}{2}+it) \ll t^c$ for some $c < 1/6$ (*ibid.*, chapter 7), we see that

$$\sum_{n\in I_1} + \sum_{n\in I_5} \ll T^{k/6}. \tag{7.12}$$

Note that (cf. (7.7))

$$\mathcal{F}_k'(t) = \log\{(t/2\pi)^{k/2}/n\}, \quad \mathcal{F}_k''(t) = k/(2t). \tag{7.13}$$

The integrals when $n \in I_2 \cup I_4$ are estimated as $\ll T^{1/2}$ by the second derivative test (see Lemma 2.3), and then trivial estimation gives

$$\sum_{n\in I_2} + \sum_{n\in I_4} \ll_\varepsilon T^{1/2}T^{k/2-1/2+\varepsilon}T^{-k/4} = T^{k/4+\varepsilon}. \tag{7.14}$$

Finally when in (7.8) we have $n \in I_3$, then the saddle point (root of $\mathcal{F}_k'(t) = 0$), namely

$$t_0 \equiv c_n := 2\pi n^{2/k} \tag{7.15}$$

lies in $[T_1, 2T]$. For $\int_{T_1}^{2T}$ we could use a general result on exponential integrals, such as the following (see [KaVo], lemma III.2).

Lemma 7.2 *We have*

$$\begin{aligned}\int_a^b \varphi(x)\exp(2\pi i f(x))\,\mathrm{d}x = {} & \frac{\varphi(c)}{\sqrt{f''(c)}}\mathrm{e}^{2\pi i f(c)+\pi i/4} + O(HAU^{-1}) \\ & + O\big(H\min(|f'(a)|^{-1}, \sqrt{A}\big) \\ & + O\big(H\min(|f'(b)|^{-1}, \sqrt{A}\big),\end{aligned} \tag{7.16}$$

if $f'(c) = 0$, $a \le c \le b$, *and the following conditions hold:* $f(x) \in C^4[a,b]$, $\varphi(x) \in C^2[a,b]$, $f''(x) > 0$ *in* $[a,b]$, $f''(x) \asymp A^{-1}$, $f^{(3)}(x) \ll A^{-1}U^{-1}$, $f^{(4)}(x) \ll A^{-1}U^{-2}$, $\varphi^{(r)}(x) \ll HU^{-r}$ $(r = 0, 1, 2)$ *in* $[a,b]$, $0 < H, A < U$, $0 < b - a \le U$.

In our case $f(x) = \frac{1}{2\pi}\mathcal{F}_k(x)$, $c = c_n$, so that $f''(c) = k/(4\pi c)$, and

$$\frac{\varphi(c_n)}{\sqrt{f''(c_n)}}\mathrm{e}^{2\pi i f(c_n)+\pi i/4} = \pi\sqrt{\frac{2}{k}}\,n^{\frac{1}{k}}\exp\Big(-k\pi i n^{\frac{2}{k}} + \frac{(2-k)\pi i}{8}\Big)\Big\{1 + O\big(\frac{1}{T^2}\big)\Big\}. \tag{7.17}$$

But in our case the last two error terms in (7.16) are large, and thus it is more expedient to carry out the evaluation by the saddle point technique directly, that is, by using a suitable contour in the complex plane, although the main terms will of course be identical.

To this end, if $T_1 = T$ (the other case is similar) let $\mathcal{L}_1$ be the segment $T - iv$ $(0 \le v \le \frac{1}{\sqrt{2}}T^{1-\varepsilon})$, $\mathcal{L}_2$ is the segment $x - i\frac{1}{\sqrt{2}}T^{1-\varepsilon}$ $(0 \le x \le c_n - \frac{1}{\sqrt{2}}T^{1-\varepsilon})$, $\mathcal{L}_3$ is the segment $c_n + v\mathrm{e}^{\frac{1}{4}\pi i}$, $-\frac{1}{\sqrt{2}}T^{1-\varepsilon} \le v \le \frac{1}{\sqrt{2}}T^{1-\varepsilon}$, $\mathcal{L}_4$ is the segment $x + i\frac{1}{\sqrt{2}}T^{1-\varepsilon}$ $(c_n + \frac{1}{\sqrt{2}}T^{1-\varepsilon} \le x \le 2T)$, and finally $\mathcal{L}_4$ is the segment joining the points $2T + i\frac{1}{\sqrt{2}}T^{1-\varepsilon}$ and $2T$.

As a simplification we develop $\rho(n/\sqrt{\tau})$ by Taylor's formula at the point $t_0 = c_n$ when $t \in [c_n - T^{1-\varepsilon}, c_n + T^{1-\varepsilon}]$, and at other appropriate points for other values of t. An alternative approach is to use the Mellin inversion formula:

$$\rho(x) = \frac{1}{2\pi}\int_{d-i\infty}^{d+i\infty} R(s)x^{-s}\,\mathrm{d}s \quad (d > 0), \qquad R(s) = \int_0^\infty \rho(x)x^{s-1}\,\mathrm{d}x.$$

The function $R(s)$ is odd, and of fast decay; see (4.73).

As already noted the derivatives of $\rho(n/\sqrt{\tau})$, considered as a function of t, decrease each time by a factor of t. Since the length of the interval is $2T^{1-\varepsilon}$, it is possible to take finitely many terms in Taylor's formula so that the total contribution of the error term is negligible, namely $\ll 1$. The remaining integrals will be all of the same type (this is where the terms $+\cdots+$ in (7.5) come from), with the same exponential factor, and the largest one will be the first one, namely the one with ($c_n = 2\pi n^{2/k}$)

$$\rho\left(\frac{n}{\sqrt{\tau(k,c_n)}}\right) = \rho\left(\frac{n}{n(1+O(T^{-2}))}\right) = \rho(1) + O(T^{-2}) = \frac{1}{2} + O(T^{-2}),$$

since $\rho(x)+\rho(1/x)=1$. Here the O-term above has an asymptotic expansion. After that we replace the integral over $[T_1, 2T]$, using Cauchy's theorem, by $\cup_{j=1}^{5}\int_{\mathcal{L}_j}$. Therefore

$$\begin{aligned}&2\sum_{n\in I_3} d_k(n)n^{-1/2}\mathbb{R}\mathrm{e}\left\{\int_{T_1}^{2T}\rho\Big(\frac{n}{\sqrt{\tau}}\Big)\mathrm{e}^{i\mathcal{F}_k(t)}\,\mathrm{d}t\right\}\\ &= 2\sum_{n\in I_3} d_k(n)n^{-1/2}\mathbb{R}\mathrm{e}\left\{\bigcup_{j=1}^{5}\int_{\mathcal{L}_j}\mathrm{e}^{i\mathcal{F}_k(z)}\,\mathrm{d}z\right\}+\cdots, \qquad (7.18)\end{aligned}$$

where $+\cdots$ has the same meaning as before. On $\mathcal{L}_3$ we have (since $\mathcal{F}_k'(c_n)=0$)

$$\begin{aligned}i\mathcal{F}_k(z) &= i\mathcal{F}_k(c_n) + i\frac{v^2}{2!}\mathrm{e}^{\frac{1}{2}\pi i}\mathcal{F}_k''(c_n) + i\frac{v^3}{3!}\mathrm{e}^{\frac{3}{4}\pi i}\mathcal{F}_k'''(c_n)\\ &\quad + i\frac{v^4}{4!}\mathrm{e}^{\pi i}\mathcal{F}_k^{(4)}(c_n)+\cdots. \qquad (7.19)\end{aligned}$$

Note that, since $v \ll T^{1-\varepsilon}$,

$$v^m\mathcal{F}_k^{(m)}(c_n) \ll_{m,\varepsilon} T^{m(1-\varepsilon)}T^{1-m} = T^{1-m\varepsilon} \qquad (m>1). \qquad (7.20)$$

Hence if we choose $M = M(k,\varepsilon)$ sufficiently large, then (7.20) shows that the terms of the series in (7.19) for $m > M$, on using $\exp z = 1 + O(|z|)$ for $|z|\le 1$, will make a negligible contribution. We have

$$\exp(i\mathcal{F}_k(z)) = \exp\big(i\mathcal{F}_k(c_n)\big)\exp\Big(-\tfrac{1}{2}v^2F''(c_n)\Big)\exp\left(\sum_{m=3}^{M} d_m v^m F^{(m)}(c_n)\right)$$

plus a small error term, where $d_m = \exp((m+2)\frac{\pi i}{4})/m!$. The last exponential factor is expanded by Taylor's series, and again the terms of the series (with v^m) for large m will make a negligible contribution. In the remaining terms we restore integration over v to the whole real line, making a very small error.

Then we use the classical integral (see (1.13))

$$\int_{-\infty}^{\infty} \exp(Ax - Bx^2)\,\mathrm{d}x \;=\; \sqrt{\frac{\pi}{B}}\exp\left(\frac{A^2}{4B}\right) \qquad (\mathbb{R}\mathrm{e}\, B > 0). \tag{7.21}$$

By differentiating (7.21) as a function of A we may explicitly evaluate integrals of the type

$$\int_{-\infty}^{\infty} x^{2m}\exp(-Bx^2)\,\mathrm{d}x \qquad (\mathbb{R}\mathrm{e}\, B > 0,\ m = 0, 1, 2, \dots).$$

It transpires that the largest contribution ($= \sqrt{\pi}$) will come from the integral with $m = 0$, which will coincide with the contribution of the main term in (7.16) (or (7.17)).

It remains for us to deal with the remaining integrals over $\mathcal{L}_j$. The integrals over $\mathcal{L}_1$ and $\mathcal{L}_5$, and likewise the integrals over $\mathcal{L}_2$ and $\mathcal{L}_4$ are estimated analogously. On $\mathcal{L}_4$ we have

$$z = x + iH,\ c_n + \frac{H}{\sqrt{2}} \le x \le 2T\ ,\ H = T^{1-\varepsilon}.$$

On using Taylor's formula we obtain

$$\begin{aligned}\exp(i\mathcal{F}_k(z)) &= \exp\Big(i\mathcal{F}_k(x) - i\frac{H^2}{2!}\mathcal{F}_k''(x) + \cdots\Big)\\ &\quad\times \exp\Big(-H\mathcal{F}_k'(x) + \frac{H^3}{3!}\mathcal{F}_k'''(x) - \cdots\Big).\end{aligned}$$

Similarly, as in (7.20), it follows that we may truncate the series after a finite number of terms with a negligible error. Observe that the real-valued term in the exponential is negative, and that the derivative of the imaginary part is dominated by

$$\begin{aligned}\mathcal{F}_k'(x) &= \log\frac{\left(\frac{x}{2\pi}\right)^{k/2}}{n} \ge \log\frac{\left(n^{2/k} + H/\sqrt{2}\right)^{k/2}}{n}\\ &= \log\Big(1 + \frac{H}{\sqrt{8\pi}\,n^{2/k}}\Big)^{k/2} \ge A_k H T^{-1} = A_k T^{-\varepsilon}\end{aligned}$$

for some constant $A_k > 0$. Hence by the first derivative test the total contribution of such terms is

$$\ll_\varepsilon\ T^{k/4+\varepsilon}. \tag{7.22}$$

On $\mathcal{L}_5$ we have $z = 2T + iy, 0 \le y \le H, H = T^{1-\varepsilon}$. This gives

$$i\mathcal{F}_k(z) = i\mathcal{F}_k(2T) - y\mathcal{F}_k'(2T) - i\frac{y^2}{2!}\mathcal{F}_k''(2T) + \frac{y^3}{3!}\mathcal{F}_k'''(T) - \cdots,$$

where, as before, we may truncate the series after a finite number of terms with a negligible error. Therefore the integral over $\mathcal{L}_5$ becomes

$$i\mathrm{e}^{i\mathcal{F}_k(2T)}\int_0^H \mathrm{e}^{f(y)}\mathrm{e}^{ig(y)}\,\mathrm{d}y,$$

say, with real-valued

$$f(y) := -y\mathcal{F}_k'(2T) + \frac{y^3}{3!}\mathcal{F}_k'''(T)\ldots,$$
$$g(y) := -\frac{y^2}{2!}\mathcal{F}_k''(2T) + \frac{y^4}{4!}\mathcal{F}_k^{(4)}(2T) + \cdots.$$

Then we have

$$\int_0^H = \int_0^{\sqrt{T}} + \int_{\sqrt{T}}^H = J_1 + J_2,$$

say. We write J_1 as

$$J_1 = -\frac{1}{\mathcal{F}_k'(2T) + \frac{y^2}{2!}\mathcal{F}_k'''(2T) + \cdots}\int_0^{\sqrt{T}} \mathrm{e}^{ig(y)}\,\mathrm{d}\Big(\mathrm{e}^{f(y)}\Big)$$

and integrate by parts. We obtain the same type of exponential integral, only smaller by a factor of

$$\ll y\frac{\mathcal{F}_k''(2T)}{\mathcal{F}_k'(2T)} \ll T^{1/2}\cdot\frac{1}{T}\cdot T^{1/2-\varepsilon} = T^{-\varepsilon},$$

since

$$\mathcal{F}_k'(2T) \ge \log\frac{(T/\pi)^{k/2}}{n} \ge \log\frac{(T/\pi)^{k/2}}{(T/\pi)^{k/2} - T^{k/2-1/2+\varepsilon}} \ge T^{\varepsilon-1/2}.$$

This means that, after sufficiently many integrations by parts, the ensuing integral will be negligible, while the integrated terms will be $\ll T^{k/4+\varepsilon}$. Finally in J_2

$$y\mathcal{F}_k'(2T) - \frac{y^3}{3!}\mathcal{F}_k'''(T)\cdots \ \ge\ CT^{1/2}\cdot T^{\varepsilon-1/2} = CT^{\varepsilon},$$

so that $\mathrm{e}^{f(y)}$ is negligibly small. The net result of our considerations is that in the evaluation of the left-hand side of (7.5) the main terms, arising from the saddle point terms, are given by (7.17), while all the error terms are $\ll_\varepsilon T^{k/4+\varepsilon}$.

7.2 Remarks

Remark 7.3 With a more careful analysis one can get rid of the terms implied by $+\cdots$ in (7.5). The same also follows if one uses an idea of Professor Matti Jutila, who kindly informed me that the above proof may be simplified as follows. The method may be traced back to E. C. Titchmarsh [Tit3, p. 261], and a sketch is as follows. Note that

$$\begin{aligned}\int_T^{2T} Z^k(t)\,\mathrm{d}t &= -i\int_{\frac12+iT}^{\frac12+2iT} \chi^{-k/2}(s)\zeta^k(s)\,\mathrm{d}s\\ &= -i\left(\int_{1+\varepsilon+iT}^{1+\varepsilon+2iT} + \int_{\frac12+iT}^{1+\varepsilon+iT} - \int_{\frac12+2iT}^{1+\varepsilon+2iT}\right)\chi^{-k/2}(s)\zeta^k(s)\,\mathrm{d}s.\end{aligned}$$

On $\sigma = 1+\varepsilon$ we have

$$\zeta^k(s) = \sum_{n=1}^{\infty} d_k(n) n^{-s},$$

so that the above expression is seen to be

$$\sum_{n=1}^{\infty} d_k(n) n^{-1-\varepsilon}\int_T^{2T}\left(\frac{t}{2\pi}\right)^{\frac{k}{4}+\frac{k\varepsilon}{2}} \mathrm{e}^{i\mathcal{F}_k(t)}\,\mathrm{d}t + O_{\varepsilon,k}(T^{k/4+\varepsilon})$$

for $k \le 4$. The exponential integral is evaluated by (7.16), and Theorem 7.1 will follow without the terms implied by $+\cdots$. I am grateful to Professor Jutila for pointing this out to me.

Remark 7.4 In the formulation of Theorem 7.1 we assumed that $k \le 4$, primarily because of the exponent $k/4$ in (7.5), which is greater than unity when $k > 4$. It is also clear that, unfortunately, the practical value of Theorem 7.1 is only when $k = 3$. Namely for $k = 1$ we obtain

$$\int_T^{2T} Z(t)\,\mathrm{d}t = 2\sqrt{2}\pi \sum_{\sqrt{T/(2\pi)}\le n\le\sqrt{T/\pi}} n^{1/2}\cos(\pi n^2 - \pi/8) + O_\varepsilon(T^{1/4+\varepsilon}). \tag{7.23}$$

But as

$$\cos(\pi n^2 - \pi/8) = \begin{cases} \cos(\pi/8) & \text{if } n = 2m,\\ -\cos(\pi/8) & \text{if } n = 2m-1,\end{cases}$$

it is seen that the sum in (7.23) is a multiple of

$$\sum_{\sqrt{T/(2\pi)}\le n\le\sqrt{T/\pi}} (-1)^n n^{1/2} \ll T^{1/4}.$$

Hence, replacing T by $T2^{-j}$ $(j = 1, 2, \dots)$ and adding up all the estimates, we obtain from (7.23)

$$F(T) = \int_1^T Z(t)\,\mathrm{d}t = O_\varepsilon(T^{1/4+\varepsilon}). \tag{7.24}$$

This is the result of [Iv16], sharpened independently by M. Korolev [Kor3], [Kor4] and M. Jutila [Jut5], [Jut8], both of whom obtained the bound $O(T^{1/4})$ (see (7.34) and (8.18)).

When $k = 2$ note that

$$\int_T^{2T} Z^2(t)\,\mathrm{d}t = \int_T^{2T} |\zeta(\tfrac{1}{2} + it)|^2\,\mathrm{d}t. \tag{7.25}$$

But we have (see chapter 15 of [Iv1] or chapter 2 of [Iv4], or (4.54))

$$\int_0^T |\zeta(\tfrac{1}{2} + it)|^2\,\mathrm{d}t = T\log\frac{T}{2\pi} + (2\gamma - 1)T + E(T),$$

where $\gamma = -\Gamma'(1)$ is Euler's constant, and the error term function $E(T)$ satisfies

$$E(T) = O(T^{1/3}). \tag{7.26}$$

Therefore

$$\int_T^{2T} |\zeta(\tfrac{1}{2} + it)|^2\,\mathrm{d}t = T\log\frac{2T}{\pi} + (2\gamma - 1)T + O(T^{1/3}). \tag{7.27}$$

On the other hand we have (see (4.86) and chapter 13 of [Iv1] for more details)

$$\sum_{n\le x} d(n) = x(\log x + 2\gamma - 1) + O(x^{1/3}). \tag{7.28}$$

From (7.5), (7.25), (7.27) and (7.28) it follows that, as $t \to \infty$,

$$\frac{1}{2\pi}\int_T^{2T} |\zeta(\tfrac{1}{2} + it)|^2\,\mathrm{d}t \sim \sum_{T/(2\pi)<n\le T/\pi} d(n)$$
$$\sim \frac{T}{2\pi}\log\frac{2T}{\pi} + (2\gamma - 1)\frac{T}{2\pi},$$

which coincides with the main term in (7.27). A more careful analysis of the proof of Theorem 7.1 when $k = 2$ would lead to

$$\int_T^{2T} |\zeta(\tfrac{1}{2} + it)|^2\,\mathrm{d}t = T\log\frac{2T}{\pi} + (2\gamma - 1)T + O_\varepsilon(T^{1/2+\varepsilon}), \tag{7.29}$$

which would be the limit of the method. The exponent 1/2 in (7.29) is much weaker than the one in (7.26), which is not even the sharpest one known; the best one is currently $131/416 = 0.314\,903\ldots$, due to N. Watt [Watt] (see also [HuIv]).

When $k = 4$ the error term in Theorem 7.1 is already $O_\varepsilon(T^{1+\varepsilon})$. We know however (see chapter 4 of [Iv1]) that

$$I_4(T) = \int_T^{2T} |\zeta(\tfrac{1}{2}+it)|^4\,\mathrm{d}t = \frac{1}{2\pi^2}T\log^4 T + O(T\log^3 T), \tag{7.30}$$

and even a full asymptotic formula for $I_4(T)$ is known (see (4.55)). However, getting even the trivial bound $I_4(T) \ll_\varepsilon T^{1+\varepsilon}$ from (7.5) does not appear easy.

There remains the case $k = 3$ of Theorem 7.1 which, as already mentioned, is the interesting one. When $k = 3$ trivial estimation gives

$$I_3(T) \ll \sum_{(T/2\pi)^{3/2} < n \le (T/\pi)^{3/2}} d_3(n)n^{-1/6} + T^{3/4+\varepsilon} \ll T^{5/4}\log^2 T, \tag{7.31}$$

on using the elementary bound $\sum_{n\le x} d_3(n) \ll x\log^2 x$. Unfortunately this is very poor, as the bound that is trivially deduced by the Cauchy-Schwarz inequality, namely

$$I_3^2(T) \le I_2(T)I_4(T),$$

gives, with the aid of (7.29) and (7.30),

$$I_3(T) = \int_T^{2T} Z^3(t)\,\mathrm{d}t \ll T\log^{5/2} T. \tag{7.32}$$

Notes

Theorem 7.1 was proved in [Iv15]. The bound in (7.24), which precedes this result and significantly improves on (2.4), was obtained in [Iv12]. It was asked there: is it true that perhaps

$$\int_0^T Z(t)\,\mathrm{d}t = \Omega(T^{1/4}) \quad (= \Omega_\pm(T^{1/4}))? \tag{7.33}$$

If yes, then (7.24) would be (up to the factor "ε") the best possible. The reason why one expects (7.33) to hold is that $T^{1/4}$ is the order of the terms coming from the saddle points (the main term in (7.5) when $k = 1$), and in the evaluation of exponential integrals one usually expects the saddle points to produce the largest contribution. The $\Omega_\pm$-result in (7.33) was proved later by M. Korolev [Kor3] and [Kor4]. Korolev also proved a sharpening of (7.24), namely

$$\int_0^T Z(t)\,\mathrm{d}t = O(T^{1/4}), \tag{7.34}$$

which combined with his $\Omega_\pm$-result completely solves the order of the primitive of $Z(t)$, up to the numerical constants which are involved. He obtained an asymptotic formula for $F(T)$, and even proved that, for T sufficiently large,

$$\left| \int_{2\pi}^{T} Z(t)\,\mathrm{d}t \right| < 18.2T^{1/4}, \tag{7.35}$$

which is quite explicit. The chief ingredient in his proofs is a very sharp version of Lemma 7.2.

Independently of M. Korolev, and by using different methods, M. Jutila [Jut6], [Jut8] and [Jut9], also obtained the above mentioned results. His method is in some aspects less difficult than Korolev's and more susceptible to generalizations. Jutila's results will be analyzed in detail in Chapter 8. However, the merit of Theorem 7.1 is that it provides an explicit formula for the kth moment of $Z(t)$ when $k = 1, 2, 3, 4$, which is interesting when $k = 3$.

There are many evaluations of saddle point integrals in the literature. For example, one may use general results contained in [Iv1], [Iv4], [Kra]. Lemma 7.2 in the text is taken from [KaVo], lemma III.2, where the reader may find a detailed proof. In results such as Lemma 7.2 one makes various assumptions on $f(x)$; the quality of the error terms clearly depends on such conditions. It is often more convenient to work directly with the saddle method principle, as was actually done in the text above. For a systematic account of results on exponential sums and integrals the reader is referred to the monograph [GrKo] of S. W. Graham and G. Kolesnik.

To deal with the exponential sum appearing in (7.5) when $k = 3$, namely

$$\overline{S} = \overline{S}(N, N') := \sum_{N<n\le N'\le 2N} d_3(n)n^{-1/6}\exp(3\pi i n^{2/3}), \tag{7.36}$$

there are several ways to proceed. First, by partial summation one can easily remove the monotonic coefficients $n^{-1/6}$ from the sum in (7.36), so that in fact it is sufficient to treat

$$S = S(N, N') := \sum_{N<n\le N'\le 2N} d_3(n)\exp(3\pi i n^{2/3}). \tag{7.37}$$

A noteworthy feature of the exponential sum in (7.37) is that the summands do not depend on T. A natural way to transform the sum S in (7.37) is to use the variant of the Voronoï summation formula for $d_3(n)$ (Georgy Feodosevich Voronoï, April 28, 1868-November 20, 1908, Ukrainian mathematician), as expounded by the present author in [Iv7]. However, after we introduce smooth weights in the sum in question, we shall see that (up to the error terms) we shall wind up with the original sum, i.e. the Voronoï formula is essentially (in this case, and even for the general divisor function $d_k(n)$) an involution. So, at least in the present situation, the Voronoï formula is of no great use.

Writing $n = abc$ in (7.37) it is seen that S becomes a three-dimensional exponential sum over the variables $a, b, c \in \mathbb{N}$. Such sums can be treated, for example, by the method of three-dimensional exponent pairs. For this theory one can consult B. R. Srinivasan [Sri]. If (ℓ_0, ℓ_1) is a three-dimensional exponent pair, then we get

$$S \ll N^{1+\ell_0-\ell_1}.$$

But as noticed in [Sri], one has the inequalities

$$0.107\,520\,466 < \ell_1 - \ell_0 < 0.107\,520\,47$$

for any three-dimensional exponent (ℓ_0, ℓ_1). This leads to

$$\overline{S} \ll N^{0.725\,818\,67}, \tag{7.38}$$

giving $I_3(T) \ll T^{1.088\,719\,301}$. However, to get a non-trivial bound one should have the exponent in (7.38) $< 2/3$. Hence Srinivasan's theory alone does not suffice for our purpose, but an improved variant is to be found in the work of G. Kolesnik [Kol], which is more complicated to apply.

Nevertheless, it seems quite difficult to get a bound

$$\int_0^T Z^3(t)\,\mathrm{d}t \ll_\varepsilon T^{c+\varepsilon} \tag{7.39}$$

with any $c < 1$, which can be justly called "non-trivial." We shall deal again with the estimation of $S(N, N')$ in Chapter 9. Proving (7.39) with some $0 < c < 1$ is still an open problem.

8

The primitive of Hardy's function

8.1 Introduction

The integral

$$F(T) \equiv F_1(T) = \int_1^T Z(t)\mathrm{d}t \qquad (T \geq 1) \tag{8.1}$$

measures the oscillatory nature of the function $Z(t)$ and thus indicates the occurrence of its sign changes. Recall that from Theorem 7.1 for $k = 1$ we obtain

$$F(2T) - F(T) = 2\sqrt{2\pi} \sum_{\sqrt{T/(2\pi)} \leq n \leq \sqrt{T/\pi}} n^{1/2} \cos(\pi n^2 - \pi/8) + O_\varepsilon(T^{1/4+\varepsilon}),$$

which gives

$$F(T) = 2\sqrt{2\pi} \sum_{n \leq \sqrt{T/\pi}} n^{1/2} \cos(\pi n^2 - \pi/8) + O_\varepsilon(T^{1/4+\varepsilon}). \tag{8.2}$$

This in turn implies the bound

$$F(T) = O_\varepsilon(T^{1/4+\varepsilon}), \tag{8.3}$$

but as noted in Chapter 7, M. Korolev [Kor3], [Kor4] and M. Jutila [Jut6] independently showed that one actually has

$$F(T) = O(T^{1/4}), \quad F(T) = \Omega_\pm(T^{1/4}). \tag{8.4}$$

In this chapter we shall follow Jutila's method [Jut6], and present an explicit formula for $F(T)$. This is different from (8.2). Not only that it has a sharper error term (the error term $O_\varepsilon(T^{1/4+\varepsilon})$ in (8.2) sets the limit to the bound in (8.3)), but the main term in Jutila's formula is analogous to the famous *Atkinson formula* for $E(T)$, the error term in the mean square formula for $|\zeta(\frac{1}{2}+it)|$. As already noted in Chapter 7 after eq. (7.25), this is

$$\int_0^T |\zeta(\tfrac{1}{2}+it)|^2\,\mathrm{d}t = T\log\frac{T}{2\pi} + (2\gamma-1)T + E(T). \tag{8.5}$$

M. Jutila [Jut3] used the Laplace transform of $Z^2(t)$, with the goal to furnish a unified method for proving both the original Atkinson formula and its analog for cusp form L-functions. This dispenses with the use of the Riemann-Siegel formula for $Z(t)$, which appeared M. Korolev's work.

We state here, without proof, Atkinson's formula for $E(T)$ as follows.

Lemma 8.1 *Let* $0 < A < A'$ *be any two fixed constants such that* $AT < N < A'T$, *and let* $N' = N'(T) = T/(2\pi) + N/2 - (N^2/4 + NT/(2\pi))^{1/2}$. *Then*

$$E(T) = \Sigma_1(T) + \Sigma_2(T) + O(\log^2 T), \tag{8.6}$$

where

$$\Sigma_1(T) = 2^{1/2}(T/(2\pi))^{1/4}\sum_{n\le N}(-1)^n d(n)n^{-3/4}e(T,n)\cos(f(T,n)), \tag{8.7}$$

$$\Sigma_2(T) = -2\sum_{n\le N'} d(n)n^{-1/2}(\log T/(2\pi n))^{-1}\cos(g(T,n)), \tag{8.8}$$

with

$$\begin{aligned} f(T,n) &= 2T\,\mathrm{arsinh}\left(\sqrt{\pi n/(2T)}\right) + \sqrt{2\pi nT + \pi^2n^2} - \pi/4 \\ &= -\tfrac{1}{4}\pi + 2\sqrt{2\pi nT} + \tfrac{1}{6}\sqrt{2\pi^3}n^{3/2}T^{-1/2} + a_5n^{5/2}T^{-3/2} \\ &\quad + a_7n^{7/2}T^{-5/2} + \cdots, \\ g(T,n) &= T\log\left(\frac{T}{2\pi n}\right) - T + \pi/4, \end{aligned} \tag{8.9}$$

where

$$\begin{aligned} e(T,n) &= (1+\pi n/(2T))^{-1/4}\left\{(2T/\pi n)^{1/2}\mathrm{arsinh}\,(\sqrt{\pi n/(2T)})\right\}^{-1} \\ &= 1 + O(n/T) \qquad (1\le n<T), \end{aligned} \tag{8.10}$$

and $\mathrm{arsinh}\,x = \log(x+\sqrt{1+x^2}\,)$.

The new formula for $F(T)$ is contained in

Theorem 8.2 *Let T be a large positive number, $N \asymp T$, and $e(T,n)$, $f(T,n)$, $g(T,n)$ and N' as in Atkinson's formula. Then*

$$F(T) = S_1(T) + S_2(T) + O\left((\log T)^{5/4}\right), \tag{8.11}$$

where

$$S_1(T) = 2\sqrt{2}(T/(2\pi))^{1/4} \sum_{0\le n\le\sqrt{N}} (-1)^{n(n+1)/2} e\left(T, \left(n+\tfrac{1}{2}\right)^2\right)\left(n+\tfrac{1}{2}\right)^{-1}$$
$$\times \cos\left(\tfrac{1}{2} f\left(T, \left(n+\tfrac{1}{2}\right)^2\right) - 3\pi/8\right) \tag{8.12}$$

and

$$S_2(T) = -4 \sum_{1\le n\le\sqrt{N'}} n^{-1/2}\left(\log(T/2\pi n^2)\right)^{-1} \cos\left(\tfrac{1}{2} g\left(T, n^2\right) + \pi/4\right). \tag{8.13}$$

The analogies between Atkinson's formula and Theorem 8.2 are obvious. The notation $S_1(T)$, $S_2(T)$ in (8.12)-(8.13) was chosen deliberately by M. Jutila to correspond to $\Sigma_1(T)$, $\Sigma_2(T)$ in (8.7) and (8.8), respectively.

One of the first applications of Atkinson's formula was the mean value formula

$$\int_0^T E^2(t)\,\mathrm{d}t \;\sim\; cT^{3/2} \qquad (T\to\infty), \tag{8.14}$$

where c is a positive constant. A corollary of (8.14) is the omega result

$$E(T) \;=\; \Omega(T^{1/4}).$$

By analogy it follows that

$$\int_0^T F^2(t)\,\mathrm{d}t \;\sim\; dT^{3/2} \qquad (T\to\infty) \tag{8.15}$$

with some constant $d > 0$. Therefore (8.15) is the analog of (8.14), and in turn it gives the omega result

$$F(T) \;=\; \Omega(T^{1/4}). \tag{8.16}$$

From Theorem 8.2 one can actually deduce a sharper omega result than (8.16), namely Korolev's

$$F(T) \;=\; \Omega_{\pm}(T^{1/4}). \tag{8.17}$$

One obtains (8.17), as well as the pointwise bound

$$F(T) \;=\; O(T^{1/4}) \tag{8.18}$$

from the following theorem, which follows by careful analysis from Theorem 8.2. Let

$$K(x) = \begin{cases} 0 & \text{if } 0 \le x < 1/4 \text{ and } 3/4 < x \le 1, \\ 2\pi & \text{if } 1/4 < x < 3/4. \end{cases}$$

Then we have the following.

Theorem 8.3 *Let T be a large positive number and write $\sqrt{T/2\pi} = L + \vartheta$ with $L \in \mathbb{N}$ and $0 \le \vartheta < 1$. Define $\vartheta_0 = \min(|\vartheta - 1/4|, |\vartheta - 3/4|)$. Then, for $\vartheta_0 \ne 0$, we have*

$$\begin{aligned} F(T) = (T/2\pi)^{1/4}(-1)^L K(\vartheta) + O(T^{1/6}\log T) \\ + O\left(\min\left(T^{1/4}, T^{1/8}\vartheta_0^{-3/4}\right)\right), \end{aligned} \tag{8.19}$$

and further

$$\begin{aligned} F(T) &= (-1)^L \frac{4\pi}{3}(T/2\pi)^{1/4} + O(T^{1/6}\log T) \quad \text{for } \vartheta = 1/4, \\ F(T) &= (-1)^L \frac{2\pi}{3}(T/2\pi)^{1/4} + O(T^{1/6}\log T) \quad \text{for } \vartheta = 3/4. \end{aligned} \tag{8.20}$$

Remark 8.4 We have $L = \left[\sqrt{T/(2\pi)}\right]$ and $\theta = \left\{\sqrt{T/(2\pi)}\right\}$, where $\{y\}$ denotes the fractional part of y. Hence taking $T = 2\pi(2n + 3/2)^2$, $T = 2\pi(2n + 1/2)^2$ with $n \in \mathbb{N}$, we see that the omega result in (8.17) follows. The bound (8.18) trivially follows from (8.19).

8.2 The Laplace transform of Hardy's function

The method of proof of Theorem 8.2 rests on the use of the Laplace transform of Hardy's function. If p denotes a complex variable, this is

$$L(p) = \int_0^\infty Z(t)\mathrm{e}^{-pt}\,\mathrm{d}t,$$

which is a holomorphic function in the half-plane $\mathbb{R}\mathrm{e}\, p > 0$. However, in order to get a neat arithmetic formula for our Laplace transform, we insert into the integrand the auxiliary factor $H(\frac{1}{2} + it)$, where

$$H(s) := \sqrt{2}\pi^{-1/4-s/2}\cos(\pi s/4)\big(\chi(s)\big)^{1/2}\Gamma(s/2 + 1/4).$$

Its role is to lead to a nice arithmetic formula for the corresponding Laplace transform, and it might be pointed out that the purpose is to produce a suitable gamma-factor in the integrand as a preparation for an application of Mellin's classical formula.

By Stirling's formula (2.15), we have, in any fixed vertical strip, the asymptotic expansion

$$\log H(s) \sim \frac{a}{s} + \cdots$$

in negative powers of s, where a is a constant. Therefore $H(s) = 1 + O(|s|^{-1})$ in such a strip as $|\mathbb{I}\mathrm{m}\, s| \to \infty$. We apply Cauchy's integral formula for the derivative of a holomorphic function to $\log H(s) - a/s$. This function is $\ll |s|^{-2}$, and then the derivative is of the same order if Cauchy's formula is applied to circle of radius 1, say. On the other hand, the derivative is $H'(s)/H(s) + as^{-2}$, thus

$$H'(s)/H(s) \ll |s|^{-2}.$$

Hence

$$H(\tfrac{1}{2} + it) = 1 + O\left((|t| + 1)^{-1}\right), \quad H'(\tfrac{1}{2} + it) \ll (|t| + 1)^{-2}, \tag{8.21}$$

and thus the transform

$$\tilde{L}(p) := \int_0^\infty Z(t)H(\tfrac{1}{2} + it)\mathrm{e}^{-pt}\,\mathrm{d}t$$

should be a good approximation to $L(p)$. We have the following lemma.

Lemma 8.5 *For* $0 < \mathbb{R}\mathrm{e}\, p < \pi/2$, *we have*

$$\begin{aligned}\tilde{L}(p) = &-\sqrt{2}\pi^{1/4}\mathrm{e}^{-\pi i/4}\mathrm{e}^{pi/2}\Gamma(3/4)\\ &+ 2\sqrt{2}\pi\mathrm{e}^{\pi i/8}\mathrm{e}^{-ip}\sum_{n=1}^{\infty} n^{1/2}\exp\left(-\pi i n^2\mathrm{e}^{-2ip}\right) + \lambda(p),\end{aligned} \tag{8.22}$$

where $\lambda(p)$ *can be analytically continued to the wider strip* $|\mathbb{R}\mathrm{e}\, p| < \pi/2$. *Moreover, in any fixed strip* $|\mathbb{R}\mathrm{e}\, p| \le \theta$ *with* $0 < \theta < \pi/2$, *we have*

$$\lambda(p) \ll (|p| + 1)^{-1}. \tag{8.23}$$

Proof of Lemma 8.5 With p as in the lemma, we shall deal with the function

$$J = J(p) := -i\mathrm{e}^{-ip/2}\int_{(1/2)} \frac{H(s)\zeta(s)\chi^{1/2}(1-s)\mathrm{e}^{is(p-\pi/4)}}{2\cos(\pi s/4)}\,\mathrm{d}s$$

in two ways.

First, putting $s = \frac{1}{2} + it$, we have

$$J = e^{-\pi i/8} \int_{-\infty}^{\infty} \frac{H(\frac{1}{2}+it)Z(t)e^{-pt}e^{\pi t/4}}{e^{\pi i/8-\pi t/4} + e^{-\pi i/8+\pi t/4}}\, dt.$$

The positive values of t give a contribution $\tilde{L}(p)$ up to a correction function that is holomorphic in the wider strip $|\mathbb{R}\mathrm{e}\, p| < \pi/2$, and the negative values of t give a similar correction function. Thus $\tilde{L}(p) = J + \lambda(p)$, where the function $\lambda(p)$ is as in the lemma. The estimate (8.23) can be verified using integration by parts with respect to the factor e^{-pt}, noting that the integrals defining $\lambda(p)$ converge exponentially.

On the other hand, we evaluate J explicitly upon moving the integration to the right to some line $\mathbb{R}\mathrm{e}\, s = a > 1$. Then, taking into account the pole at $s = 1$, we see, by the theorem of residues, that

$$\begin{aligned} J = &-\sqrt{2}\pi^{1/4}e^{-\pi i/4}e^{pi/2}\Gamma(3/4) \\ &- \frac{ie^{-ip/2}}{\sqrt{2}\pi^{1/4}} \int_{(a)} \pi^{-s/2}\zeta(s)e^{is(p-\pi/4)}\Gamma\Big(\frac{s}{2}+\frac{1}{4}\Big)\, ds. \end{aligned}$$

The new integral decays exponentially in the strip $0 < \mathbb{R}\mathrm{e}\, p < \pi/2$, and the zeta-factor in the integrand can be written as a series. We integrate termwise and finally apply the inversion formula (4.58) to obtain

$$\begin{aligned} &\int_{(a)} \pi^{-s/2}\zeta(s)e^{is(p-\pi/4)}\Gamma\Big(\frac{s}{2}+\frac{1}{4}\Big)\, ds \\ &= \sum_{n=1}^{\infty} \int_{(a)} \left(\pi i n^2 e^{-2ip}\right)^{-s/2} \Gamma\Big(\frac{s}{2}+\frac{1}{4}\Big)\, ds \\ &= 2\sum_{n=1}^{\infty} \left(\pi i n^2 e^{-2ip}\right)^{1/4} \int_{(a/2+1/4)} \left(\pi i n^2 e^{-2ip}\right)^{-w} \Gamma(w)\, dw \\ &= 4\pi^{5/4} i e^{\pi i/8} e^{-pi/2} \sum_{n=1}^{\infty} n^{1/2} \exp\left(-\pi i n^2 e^{-2ip}\right). \end{aligned}$$

The assertion of Lemma 8.5 now follows on comparing the expressions for J.

Lemma 8.6 *Let $p = a + iu$, where $a = 1/T$ and T is sufficiently large. Then, for $0 \le u \ll (T^{1/2}\log T)^{-1}$, we have*

$$\tilde{L}(p) \ll 1. \tag{8.24}$$

Proof of Lemma 8.6 In view of Lemma 8.5 it is sufficient to prove the bound

$$\sum_{n=1}^{\infty} n^{1/2} \exp\left(-\pi i n^2 \mathrm{e}^{-2ip}\right) \ll 1. \tag{8.25}$$

Setting $p = a + iu$ $(a, u \in \mathbb{R})$, we obtain

$$\begin{aligned}\exp\left(-\pi i n^2 \mathrm{e}^{-2ip}\right) &= (-1)^n \exp\left(-\pi i n^2 (\mathrm{e}^{-2ip} - 1)\right)\\ &= (-1)^n \exp\left\{\left(1 - \mathrm{e}^{2u}\cos(2a)\right)\pi i n^2 - \pi n^2 \mathrm{e}^{2u}\sin(2a)\right\}.\end{aligned}$$

Hence the series (8.25) can be truncated to $n \ll (T \log T)^{1/2}$. Let us write the general term of this sum as $(-1)^n \psi(n)$. We split up the sum into weighted dyadic parts, introducing smooth weight functions $w(x)$ which have their support on some interval $[N, 2N]$ with $1 \ll N \ll (T \log T)^{1/2}$, and which moreover satisfy $w^{(j)}(x) \ll_j N^{-j}$ for sufficiently many derivatives. We are going to show that

$$\sum_{n=1}^{\infty} w(n)(-1)^n \psi(n) \ll N^{-1/10}, \tag{8.26}$$

which clearly implies (8.25).

First, let us assume that u lies in the range $0 \le u \le N^{-9/5}$. By the quadratic Taylor formula for the function $w(x)\psi(x)$ at $x = 2n$, we see that

$$\begin{aligned}&-\tfrac{1}{2}w(2n-1)\psi(2n-1) + w(2n)\psi(2n) - \tfrac{1}{2}w(2n+1)\psi(2n+1)\\ &\quad \ll N^{1/2}(N^2u^2 + N^2T^{-2} + N^{-2} + u + T^{-1}) \ll N^{-11/10}.\end{aligned}$$

Summing the above over n, we obtain the sum in (8.26), and hence the latter is $\ll N^{-1/10}$.

Now let $N^{-9/5} \ll u \ll (T^{1/2} \log T)^{-1}$. We apply Poisson's summation formula (see, e.g., A.6 of [Iv1]) separately for even and odd n. Note that

$$\frac{\mathrm{d}}{\mathrm{d}n}\left((1 - \mathrm{e}^{2u}\cos(2a))\pi i n^2\right) \asymp uN.$$

Since $N^{-4/5} \ll uN \ll (\log T)^{-1/2}$, all of the terms in the summation formula will be oscillating exponential integrals, and repeated integration by parts shows that their contribution is small, in any case $\ll N^{-1/10}$.

Lemma 8.7 *For* $0 < \operatorname{Re} p \le \theta < \pi/2$, *we have*

$$L(p) - \tilde{L}(p) \ll \left(\log(1/\operatorname{Re} p)\right)^{9/4} |p|^{-1}. \tag{8.27}$$

Proof of Lemma 8.7 The bound (8.27) follows if we integrate by parts with respect to the factor e^{-pt} in the integral

$$\tilde{L}(p) - L(p) = \int_0^\infty Z(t)\Big(H(\tfrac{1}{2} + it) - 1\Big)e^{-pt}\,dt$$

and use the properties (8.21) of the H-function together with the estimates

$$\int_0^T |Z^{(j)}(t)|\,dt \ll T(\log T)^{1/4+j} \qquad (j = 0, 1). \tag{8.28}$$

8.3 Proof of Theorem 8.2

The Laplace transform of $F(t)$ is $L(p)/p$, and hence

$$F(T) = \frac{1}{2\pi i}\int_{(a)} L(p)p^{-1}e^{pT}\,dp \qquad (a > 0)$$

by the Laplace inversion formula, where we shall take $a = 1/T$. We may work in the upper half-plane $\mathbb{Im}\, p \geq 0$ since the integrals over $\mathbb{Im}\, p \geq 0$ and $\mathbb{Im}\, p \leq 0$ give complex conjugate contributions to $F(T)$, and therefore

$$F(T) = 2\mathbb{Re}\left\{\frac{1}{2\pi i}\int_a^{a+i\infty} L(p)p^{-1}e^{pT}\,dp\right\}.$$

We replace $L(p)$ by $\tilde{L}(p)$, which is easier to handle, and write

$$F(T) = 2\mathbb{Re}\left\{\frac{1}{2\pi i}\int_a^{a+i\infty} \tilde{L}(p)p^{-1}e^{pT}\,dp\right\} + F_0(T), \tag{8.29}$$

where

$$F_0(T) := 2\mathbb{Re}\left\{\frac{1}{2\pi i}\int_a^{a+i\infty} (L(p) - \tilde{L}(p))p^{-1}e^{pT}\,dp\right\} \tag{8.30}$$

is a correction term. In fact, we shall show next that

$$F_0(T) \ll (\log T)^{5/4}, \tag{8.31}$$

which is indeed small as far as the proof of Theorem 8.2 is concerned. Set $p = a + ui$ and consider in (8.30) the integrals over the ranges $0 \leq u \leq \log T$ and $u > \log T$ separately. By Lemma 8.7, the latter integral is of the desired order. We write the remaining part of $F_0(T)$ as

$$2\mathbb{Re}\left\{\frac{1}{2\pi i}\int_a^{a+i\log T}\left(\left(\int_0^{T/2} + \int_{T/2}^\infty\right) Z(t)(1 - H(\tfrac{1}{2} + it))e^{-pt}\,dt\right)\right.$$
$$\left.\times\, p^{-1}e^{pT}\,dp\right\}.$$

The integral over $t \geq T/2$ gives a contribution which is $\ll (\log T)^{5/4}$ if we estimate the t-integral directly using (8.21) and (8.28). To deal with the range $0 \leq t \leq T/2$, we integrate by parts over p with respect to the factor $\mathrm{e}^{(T-t)p}$, and obtain a similar contribution. This completes the proof of (8.31).

By Lemma 8.6, we may omit the range $0 \leq u \ll U$ with $U = (T^{1/2}\log T)^{-1}$ in (8.30), producing an error that is $\ll \log T$. For technical reasons, this will be done with the aid of a smooth weight function $w(u)$ such that $w(u) = 1$ for $u \geq 2U$ and $w(u) = 0$ for $u < U$; and moreover for which $w^{(j)}(u) \ll U^{-j}$ for sufficiently many derivatives. Thus we are left with the expression

$$\mathcal{F}(T) := 2\mathrm{e}\mathbb{R}\mathrm{e}\left\{\frac{1}{2\pi}\int_0^\infty w(u)\tilde{L}(a+iu)(a+iu)^{-1}\mathrm{e}^{iuT}\,\mathrm{d}u\right\} \tag{8.32}$$

such that

$$F(T) = \mathcal{F}(T) + O\Big((\log T)^{5/4}\Big). \tag{8.33}$$

We substitute the formula for $\tilde{L}(p)$ from Lemma 8.1 into (8.32), noting that the first and third terms give error terms $\ll \log T$. Hence, substituting the second term, we have

$$\begin{aligned} F(T) = O\Big((\log T)^{5/4}\Big) + 2\mathbb{R}\mathrm{e}\Big\{\sqrt{2}\mathrm{e}^{1-ia+\pi i/8}\sum_{n=1}^{\infty}\sqrt{n}\int_0^\infty w(u) \\ \times \exp\left(u - \pi i n^2 \mathrm{e}^{2u}\mathrm{e}^{-2ai} + iuT\right)(a+iu)^{-1}\mathrm{d}u\Big\}. \end{aligned}$$

Simplifying the resulting expression we arrive at

$$F(T) = 2\mathbb{R}\mathrm{e}\, I + O\Big((\log T)^{5/4}\Big), \tag{8.34}$$

where

$$\begin{aligned} I := \sqrt{2}\mathrm{e}^{1+\pi i/8}\sum_{n=1}^{\infty} n^{1/2}\int_0^\infty w(u) \\ \times \exp\left(u - 2a\pi\mathrm{e}^{2u}n^2 - \pi i n^2\mathrm{e}^{2u} + iuT\right)(a+iu)^{-1}\mathrm{d}u. \end{aligned}$$

Substituting $y = \mathrm{e}^{2u} - 1$, we have

$$\begin{aligned} I &= 2^{-1/2}\mathrm{e}^{1+\pi i/8}\sum_{n=1}^{\infty}(-1)^n n^{1/2} \\ &\quad\times \int_0^\infty \frac{w(\frac{1}{2}\log(1+y))\exp\left(-2\pi(y+1)an^2 - \pi i n^2 y + \frac{1}{2}iT\log(1+y)\right)}{\sqrt{y+1}(a+\frac{1}{2}i\log(1+y))}\mathrm{d}y \\ &= I_1 + I_2 + I_3, \end{aligned} \tag{8.35}$$

say, letting n run, respectively, over the intervals $[\sqrt{N'}, \sqrt{N''}]$, $[1, \sqrt{N'})$, and $(\sqrt{N''}, \infty)$, with N' as in Theorem 8.2 and $N'' = A(T/2\pi)$, where $A > 1$ is a constant.

The terms I_1 and I_2 will produce the terms $S_1(T)$ and $S_2(T)$ in Theorem 8.2, respectively, which motivates the notation; the term I_3 will be negligibly small, as we will see shortly.

The exponential integrals in (8.35) can be treated by a standard method: by the first or second derivative test, or by the saddle-point method, or simply by integration by parts. The value or size of the nth integral depends largely on the saddle point

$$y_0 = \frac{T}{2\pi n^2} - 1, \tag{8.36}$$

which is the zero of the derivative of the function

$$f(y) = -\frac{1}{2}n^2 y + \frac{T}{4\pi}\log(1+y). \tag{8.37}$$

Now $y_0 > 0$ only if $n < \sqrt{T/2\pi}$, and so the integrals in I_3 have no saddle point. Therefore repeated integration by parts with respect to the factor $\mathrm{e}(f(y))$ can be applied to show that

$$I_3 \ll 1. \tag{8.38}$$

To evaluate the saddle-point integrals we shall need the following result, which we state without proof.

Lemma 8.8 *Let $f(z)$ and $g(z)$ be two functions of the complex variable z and $[a, b]$ be a real interval such that:*

(i) *for $a \le x \le b$, the function $f(x)$ is real and $f''(x) > 0$;*
(ii) *for a certain positive continuously differentiable function $\mu(x)$, the functions $f(z)$ and $g(z)$ are holomorphic in the set $|z - x| \le \mu(x)$, $a \le x \le b$;*
(iiii) *there exist positive functions $F(x)$, $G(x)$ such that for $|z - x| \le \mu(x)$, $a \le x \le b$, we have*

$$g(z) \ll G(x),$$
$$f'(z) \ll F(x)\mu(x)^{-1},$$
$$f''(x) \gg F(x)\mu(x)^{-2}.$$

Let k be any real number, and if $f'(x) + k$ has a zero in $[a, b]$, denote it by x_0. Let the values of functions at a, x_0 and b be characterized by the subscripts a,

0 and b, respectively. Then

$$\int_a^b g(x)\mathrm{e}\Big(f(x)+kx\Big)\mathrm{d}x = g_0(f_0'')^{-1/2}\mathrm{e}(f_0+kx_0+1/8)$$
$$+O\bigg(\int_a^b G(x)\exp(-C|k|\mu(x)-CF(x))(1+|\mu'(x)|)\mathrm{d}x\bigg)+O(G_0\mu_0F_0^{-3/2})$$
$$+O\big(G_a(|f_a'+k|+(f_a'')^{1/2})^{-1}\big)+O\big(G_b(|f_b'+k|+(f_b'')^{1/2})^{-1}\big). \tag{8.39}$$

If $f'(x)+k$ has no zero for $a\le x\le b$, then the terms involving x_0 are to be omitted. If the function f satisfies the above conditions except that $f''(x)<0$ and the condition for $f''(x)$ is understood as a lower bound for $|f''(x)|$, then the formula remains valid with the following changes: $1/8$ in the leading term is to be replaced by $-1/8$, and f_0'' is to be replaced by $|f_0''|$.

We begin with the evaluation of I_1. We have

$$I_1 = 2^{-1/2}\mathrm{e}^{1+\pi i/8}\int_0^\infty \frac{w(\frac12\log(1+y))S(y)(1+y)^{iT/2}}{\sqrt{y+1}(a+\frac12 i\log(1+y))}\mathrm{d}y,$$

where

$$S(y) := \sum_{\sqrt{N'}\le n\le\sqrt{N''}} (-1)^n n^{1/2}\exp(-2\pi(y+1)an^2-\pi i n^2 y).$$

The integral in the formula above is truncated to a bounded interval. For this, note that the saddle point (8.36) for n lying in the present range is bounded. Now, if the function $w(\frac12\log(1+y))$ is truncated to a smooth weight function, say $\tilde w(y)$, of support $[\mathrm{e}^{2U}-1, c]$ for a sufficiently large constant c, then repeated integration by parts shows that the new integral deviates from I_1 by a negligible amount, say $O(1/T)$. Therefore

$$I_1 = 2^{-1/2}\mathrm{e}^{1+\pi i/8}\int_0^\infty \frac{\tilde w(y)S(y)(1+y)^{iT/2}}{\sqrt{y+1}\Big(a+\frac12 i\log(1+y)\Big)}\mathrm{d}y + O\Big(\frac1T\Big). \tag{8.40}$$

The sum $S(y)$ is transformed by a modified Poisson summation formula, namely

$$\sum_{A\le n\le B}(-1)^n f(n) = \sum_{n=-\infty}^{\infty}\int_A^B f(x)\mathrm{e}\left(-\left(n+\tfrac12\right)x\right)\mathrm{d}x, \tag{8.41}$$

where $f(x)$ is a continuous function of bounded variation on a finite interval $[A, B]$, with the convention that if A or B is an integer, then the corresponding term is to be halved. One obtains (8.41) from the classical Poisson summation formula

$$\sum_{A\le n\le B} f(n) = \int_A^B f(x)\,\mathrm{d}x + 2\sum_{n=1}^{\infty} f(x)\cos(2\pi nx)\,\mathrm{d}x$$

with $f(n)$ replaced by $(-1)^n f(n) = \mathrm{e}(-n/2)f(n)$. When applying (8.41) to the sum $S(y)$ we may suppose, on changing N' or N'' slightly if necessary, that $\{\sqrt{N'}\} = \{\sqrt{N''}\} = \frac{1}{2}$. Then we obtain

$$S(y) = \sum_{n=-\infty}^{\infty} \int_{\sqrt{N'}}^{\sqrt{N''}} x^{1/2} \exp\left\{-2\pi(y+1)ax^2 - \pi i x^2 y + 2\pi i(n+\tfrac{1}{2})x\right\} \mathrm{d}x. \tag{8.42}$$

Now we analyze formula (8.40) for I_1 as follows: ignoring error terms for a moment, we apply the saddle-point method to (8.42) to transform the sum $S(y)$ into another exponential sum, and this new sum is then substituted into (8.40), where the integrals are again evaluated by the saddle-point method. Thus error terms arise at two stages, and though their estimation is analogous to that in [Jut2], we give some details later for completeness. But first we calculate the main terms of I_1.

To begin with, we note that the series in (8.42) can be truncated to $n \ll \sqrt{T}$. Namely, if n exceeds a suitable multiple of $\sqrt{T}$, then the integral has no saddle point, and if we integrate by parts twice with respect to the oscillating factor, then the integrated terms form an alternating series owing to our choice of N' and N''. The contribution of these values of n to $S(y)$ is $\ll T^{-1/4}$, and thus their contribution to I_1 is $\ll T^{-1/4}\log T$.

The saddle point for the nth term in (8.42) is $x_0 = (n+\frac{1}{2})/y$, and the saddle-point terms of Lemma 8.7 yield the transformed sum

$$\mathrm{e}^{-\pi i/4}/y \sum_{\sqrt{N'}y\le n+\frac{1}{2}\le\sqrt{N''}y} (n+\tfrac{1}{2})^{1/2}$$
$$\times \exp\left\{\pi i(n+\tfrac{1}{2})^2/y - 2\pi a(n+\tfrac{1}{2})^2(y+1)/y^2\right\},$$

up to a certain error term, which will be discussed later. We substitute this into I_1 and obtain the sum

$$2^{-1/2}\mathrm{e}^{1-\pi i/8} \sum_{0\le n\ll\sqrt{T}} (n+\tfrac{1}{2})^{1/2} \int_{(n+\frac{1}{2})/\sqrt{N''}}^{(n+\frac{1}{2})/\sqrt{N'}}$$
$$\times \frac{\tilde{w}(y)\exp\left\{\frac{1}{2}iT\log(1+y) + \pi i(n+\frac{1}{2})^2/y - 2\pi a(n+\frac{1}{2})^2(y+1)/y^2\right\}}{y\sqrt{y+1}(a+\frac{1}{2}i\log(1+y))}\mathrm{d}y.$$

After the substitution $v = 1/y$ the preceding expression becomes

$$2^{-1/2} e^{1-\pi i/8} \sum_{0 \le n \ll \sqrt{T}} (n+\tfrac{1}{2})^{1/2} \int_{\sqrt{N'}/(n+\frac{1}{2})}^{\sqrt{N''}/(n+\frac{1}{2})} \times$$
$$\times \frac{\tilde{w}(1/v) \exp\left\{\frac{1}{2} i T \log(1+\frac{1}{v}) + \pi i (n+\frac{1}{2})^2 v - 2\pi a (n+\frac{1}{2})^2 (v^2+v)\right\}}{\sqrt{v^2+v}\,\left(a + \frac{1}{2} i \log(1+\frac{1}{v})\right)} dv. \tag{8.43}$$

The saddle-point condition is

$$\frac{T}{2\pi(v_0^2+v_0)} = \left(n+\tfrac{1}{2}\right)^2, \tag{8.44}$$

which turns out to be (4.21) in [Jut2] if n is replaced by $(n+\frac{1}{2})^2$. Therefore we may simply obtain the saddle-point terms for I_1 directly from [Jut2], or repeat the calculations. Note that

$$v_0 = -\frac{1}{2} + \sqrt{\frac{1}{4} + \frac{T}{2\pi\left(n+\frac{1}{2}\right)^2}}, \tag{8.45}$$

and that the condition

$$\sqrt{N'}/(n+\tfrac{1}{2}) \le v_0 \le \sqrt{N''}/(n+\tfrac{1}{2})$$

is equivalent to the condition $(n+\frac{1}{2})^2 \le N$ in $S_1(T)$. The calculations in [Jut2, section 4.7], give

$$(v_0^2+v_0)^{-1/2} = (n+\tfrac{1}{2})\sqrt{2\pi/T},$$

$$\pi(n+\tfrac{1}{2})^2 v_0 = -\tfrac{\pi}{2}(n+\tfrac{1}{2})^2 + \tfrac{1}{2}\sqrt{\pi^2(n+\tfrac{1}{2})^4 + 2\pi(n+\tfrac{1}{2})^2 T},$$

$$\log\left(1+\frac{1}{v_0}\right) = 2\operatorname{arsinh}\left((n+\tfrac{1}{2})\sqrt{\pi/2T}\right),$$

$$\frac{(2v_0+1)T}{4\pi(v_0+v_0^2)^2} = 2\pi(n+\tfrac{1}{2})^4 T^{-1}\left(\frac{1}{4} + \frac{T}{2\pi(n+\frac{1}{2})^2}\right)^{1/2},$$

and $\tilde{w}(1/v_0) = 1$. Therefore the main term for I_1 is

$$e^{\pi i/8} \sum_{0<n+\frac{1}{2} \le \sqrt{N}} (n+\tfrac{1}{2})^{-1/2} (-1)^{n(n+1)/2} \exp\left(\tfrac{1}{2} i f(T, (n+\tfrac{1}{2})^2)\right)$$
$$\times \left(\frac{1}{4} + \frac{T}{2\pi(n+\frac{1}{2})^2}\right)^{-1/4} \left(a + i \operatorname{arsinh}\left((n+\tfrac{1}{2})\sqrt{\pi/2T}\right)\right)^{-1}.$$

Here the term a can be omitted with an error $\ll 1$, and so the main term for $2\mathbb{R}\mathrm{e}\, I_1$ coincides with $S_1(T)$. It now remains to estimate the contribution of the error terms of the saddle-point method.

Let us first consider the error terms of Lemma 8.8 for the integrals in formula (8.42) for $S(y)$. As mentioned earlier, the truncations $y \ll 1$ and $n \ll \sqrt{T}$ are admissible. These error terms are treated in a manner similar to the proof of Atkinson's Lemma 8.8.

We observed already that the saddle point for the nth integral in (8.42) is $x_0 = x_0(n, y) = (n + \frac{1}{2})/y$. Suppose first that $\sqrt{N'} < x_0 < \sqrt{N''}$. Let $\rho = e^{-\pi i/4}$ and replace the interval $[\sqrt{N'}, \sqrt{N''}]$ by a path connecting the points $\sqrt{N'}$, $\sqrt{N'} - c\rho\sqrt{T}$, $x_0 - c\rho\sqrt{T}$, $x_0 + c\rho\sqrt{T}$, $\sqrt{N''} + c\rho\sqrt{T}$, and $\sqrt{N''}$ by line segments; here c is a sufficiently small positive constant. The middle part of the contour which goes through x_0 gives the main term of Lemma 8.7 together with the error term involving x_0. The horizontal parts give the first error term, which is very small in our case. The first and last parts are responsible for the last two error terms, and we prefer to keep these terms explicit in the course of the estimation of the y-integral. If $x_0 \le \sqrt{N'}$, then the contour will connect the points $\sqrt{N'}$, $\sqrt{N'} + c\rho\sqrt{T}$, $\sqrt{N''} + c\rho\sqrt{N''}$, and $\sqrt{N''}$, while for $x_0 \ge \sqrt{N''}$ we use an analogous contour in the upper half-plane.

The location of x_0 with respect to $\sqrt{N'}$ can be characterized as follows: putting

$$y(n) = (n + \tfrac{1}{2})/\sqrt{N'},$$

we have

$$x_0(n, y) > \sqrt{N'} \iff y < y(n) \quad \text{and} \quad x_0(n, y) < \sqrt{N'} \iff y > y(n).$$

Consider the error terms which depend on x_0. We have $F(x) = Ty$, $\mu(x) = b\sqrt{T}$ for some constant $b > 0$, and $G(x) = T^{1/4}$, so that

$$G_0 \mu_0 F_0^{-3/2} \ll T^{-3/4} y^{-3/2}.$$

Since $(n + \frac{1}{2})/y \asymp \sqrt{T}$ by the saddle-point condition, we have $y \gg T^{-1/2}$ and $0 \le n \ll y\sqrt{T}$, and so the contribution of these terms to I_1 is

$$\ll \int_{c_1 T^{-1/2}}^{c_2} T^{-3/4} y^{-5/2} \mathrm{d}y \ll 1$$

for some constants c_1 and c_2.

Next we consider the integrals over the first segment starting from $x = \sqrt{N'}$. First suppose that $x_0(n, y) > \sqrt{N'}$ and write $x = x(t) = \sqrt{N'} - t\rho$ for $0 \le t \ll \sqrt{T}$. These x-integrals, combined with the relevant y-integrals over

$y < y(n)$, lead to the following sum of double integrals:

$$-\rho \sum_{n \ll \sqrt{T}} \int_0^{c\sqrt{T}} x^{1/2} e((n + \tfrac{1}{2})x)$$
$$\times \left(\int_0^{y(n)} \frac{\tilde{w}(y) \exp\left(-2\pi(y+1)ax^2 - \pi i x^2 y + \frac{1}{2} i T \log(1+y)\right)}{\sqrt{y+1}\left(a + \frac{1}{2} i \log(1+y)\right)} \mathrm{d}y \right) \mathrm{d}t. \tag{8.46}$$

If $x_0(n, y) \le \sqrt{N'}$, then we let t run over the interval $[-c\sqrt{T}, 0]$ and the range for y is $y \ge y(n)$.

In any case, the absolute value of the integrand in our double integral is

$$\ll T^{1/4} y^{-1} \exp\left(-\pi y t^2 - \sqrt{2}\pi t \left(n + \tfrac{1}{2} - \sqrt{N' y}\right)\right),$$

hence the t-integral can be truncated to

$$t \ll \min\left(y^{-1/2}, \left|n + \tfrac{1}{2} - \sqrt{N' y}\right|^{-1}\right) \log T, \tag{8.47}$$

and we get a contribution of $O(T^{-1/4} \log^2 T)$ for the integral in (8.46).

Turning to other values of y, we note that since t changes sign as y passes the value $y(n)$, the ranges $y < y(n)$ and $y > y(n)$ must be treated separately (the values of $|y - y_1| \le T^{-1/4}$ are excluded from this consideration). Therefore integration by parts with respect to the function $\exp(\varphi(y))$ produces integrated terms at $y_1 \pm T^{-1/2}$ and $y(n)$, except that the latter terms are relevant only if $|y(n) - y_1| > T^{-1/2}$. Since the length of the t-range for $y = y(n)$ is $\ll y(n)^{-1/2} \log T \ll T^{1/4}(n + \frac{1}{2})^{-1/2} \log T$, the contribution of the integrated terms in question is

$$\ll T^{1/4} \sum_{0 \le n \ll \sqrt{T}} \left(\frac{n + \frac{1}{2}}{T^{1/2}}\right)^{-3/2} \frac{\log T}{\sqrt{T}(|n - \sqrt{N}| + 1)} \ll \log T,$$

if one considers separately the cases $n \ge \sqrt{N}/2$ and $n < \sqrt{N}/2$.

If we repeat integration by parts, then the integrated terms at the ends of the support of the function $\tilde{w}(y)$ vanish and the integrated terms at the points considered above are diminished step by step. The same holds for the integrands arising in the course of this argument, so that finally we end up with negligibly small integrals.

The estimation of the integrals from $\sqrt{N''}$ to $\sqrt{N''} \pm c\rho\sqrt{T}$ is similar but easier, since the y-integral has no saddle point in its range. This completes the estimation of the effect of the error terms for the integrals in (8.42).

It remains for us to consider the error terms for the integrals in formula (8.43) for I_1. Now $G(v) = (n + \frac{1}{2})^{1/2}$, $F(v) = T/v$, and $\mu(v) = bv$ for some constant b. The contribution of the error terms depending on $v_0 \asymp T^{1/2}/(n + \frac{1}{2})$ is

$$\ll \sum_{0 \le n \ll \sqrt{T}} G_0 \mu_0 F_0^{-3/2} \ll T^{-1/4} \sum_{0 \le n \ll \sqrt{T}} (n + \tfrac{1}{2})^{-2} \ll T^{-1/4}.$$

The first error term in Lemma 8.8 is negligibly small. As to the last two error terms, we have to estimate the derivative

$$\frac{\mathrm{d}}{\mathrm{d}v} \left(\tfrac{1}{2} T \log(1 + \tfrac{1}{v}) + \pi (n + \tfrac{1}{2})^2 v \right) = -\frac{T}{2(v^2 + v)} + \pi (n + \tfrac{1}{2})^2$$

at $v = \sqrt{N'}/(n + \frac{1}{2})$ and $v = \sqrt{N''}/(n + \frac{1}{2})$. For $v = \sqrt{N'}/(n + \frac{1}{2})$, the above derivative is a function of $n + \frac{1}{2}$ which vanishes at $n + \frac{1}{2} = \sqrt{N}$, since

$$N' + \sqrt{NN'} = \frac{T}{2\pi}.$$

Hence this derivative is $\gg \sqrt{T} \left| \sqrt{N} - (n + \frac{1}{2}) \right|$ for $n \asymp \sqrt{T}$, and for smaller values of n it is $\gg (n + \frac{1}{2})^2$. Therefore the contribution of the error terms related to $v = \sqrt{N'}/(n + \frac{1}{2})$ is

$$\ll \sum_{n \asymp \sqrt{T}} n^{1/2} T^{-1/2} \min \left(\left| \sqrt{N} - (n + \tfrac{1}{2}) \right|^{-1}, 1 \right) + \sum_{0 \le n \ll \sqrt{T}} (n + \tfrac{1}{2})^{-3/2} \ll 1.$$

This settles the third error term of Lemma 8.8. The last error term is similar but easier. Thus all in all, the error term for I_1 is $\ll \log T$, giving

$$2\mathbb{R}\mathrm{e}\, I_1 = S_1(T) + O(\log T).$$

In a similar way we find that

$$2\mathbb{R}\mathrm{e}\, I_2 = S_2(T) + O(T^{-1/4}),$$

so that Theorem 8.2 follows from (8.35).

8.4 Proof of Theorem 8.3

We have to show that the formula for $F(T)$ in Theorem 8.2 can be approximately rewritten in terms of a simple step-function as asserted in Theorem 8.3. First, note that

$$S_2(T) \ll T^{1/6} \log T, \tag{8.48}$$

since summation by parts reduces this sum to ordinary "zeta-sums." Thus we can use classical methods (see [Iv1] and [Tit3]), which yield the bound $\zeta(\frac{1}{2}+it) \ll t^{1/6}\log t$ to obtain (8.48). Hence we have

$$F(T) = S_1(T) + O(T^{1/6}\log T). \tag{8.49}$$

It remains to analyze the sum $S_1(T)$. A key role will be played by the Fourier series of $K(x)$, where $K(x)$ appears in the formulation of Theorem 8.3. We extend this function to an odd periodic function of period 2 defined in the usual way at the discontinuities, i.e. $K(1/4) = K(3/4) = -K(-1/4) = -K(-3/4) = \pi$. The Fourier series in question is

$$K(x) = 2\sqrt{2}\sum_{n=0}^{\infty}(-1)^{n(n+1)/2}\left(n+\tfrac{1}{2}\right)^{-1}\sin(2\pi(n+\tfrac{1}{2})x). \tag{8.50}$$

The connection between $S_1(T)$ and $K(\vartheta)$ with $\vartheta = \{\sqrt{T/2\pi}\}$ becomes apparent if we rewrite $S_1(T)$ using (8.9), according to which

$$\tfrac{1}{2}f(T,(n+\tfrac{1}{2})^2) - 3\pi/8 = -\pi/2 + 2\pi(T/2\pi)^{1/2}(n+\tfrac{1}{2}) + \psi(n),$$

where

$$\psi(n) = \frac{1}{12}\sqrt{2\pi^3}(n+\tfrac{1}{2})^3T^{-1/2} + \cdots$$

is a series in powers of $(n+\frac{1}{2})^2/T$. Hence, with $(T/2\pi)^{1/2} = L + \vartheta$, we have

$$\cos\left(\tfrac{1}{2}f(T,(n+\tfrac{1}{2})^2) - 3\pi/8\right) = (-1)^L\sin\left(2\pi(n+\tfrac{1}{2})\vartheta + \psi(n)\right),$$

and thus

$$S_1(T) = 2\sqrt{2}(T/2\pi)^{1/4}(-1)^L\sum_{0\le n\le\sqrt{N}} \mathrm{e}(T,(n+\tfrac{1}{2})^2)\left(n+\tfrac{1}{2}\right)^{-1}(-1)^{n(n+1)/2}$$
$$\times\sin(2\pi(n+\tfrac{1}{2})\vartheta + \psi(n)).$$

If we omit $\mathrm{e}(T,(n+\frac{1}{2})^2)$ and $\psi(n)$ in the above, and moreover let n run to infinity, then the Fourier series representation of $K(\vartheta)$ emerges as a factor, i.e. heuristically,

$$S_1(T) \approx (T/2\pi)^{1/4}(-1)^LK(\vartheta). \tag{8.51}$$

With this in mind, we write

$$(2\sqrt{2})^{-1}(T/2\pi)^{-1/4}(-1)^LS_1(T) = (2\sqrt{2})^{-1}K(\vartheta) + A - B + C, \tag{8.52}$$

where

$$A := \sum_{0\le n\le N_0} (-1)^{n(n+1)/2}(n+\tfrac{1}{2})^{-1}\Big(\mathrm{e}(T,(n+\tfrac{1}{2})^2)$$
$$\times \sin(2\pi(n+\tfrac{1}{2})\vartheta+\psi(n)) - \sin(2\pi(n+\tfrac{1}{2})\vartheta)\Big), \qquad (8.53)$$

$$B := \sum_{n>N_0}(-1)^{n(n+1)/2}(n+\tfrac{1}{2})^{-1}\sin(2\pi(n+\tfrac{1}{2})\vartheta), \qquad (8.54)$$

$$C := \sum_{N_0<n\le\sqrt{N}} (-1)^{n(n+1)/2}\mathrm{e}(T,(n+\tfrac{1}{2})^2)(n+\tfrac{1}{2})^{-1}\sin(2\pi(n+\tfrac{1}{2})\vartheta+\psi(n)), \qquad (8.55)$$

and $N_0 \in [1,\sqrt{N}]$ will be chosen later in an appropriate manner for the cases $\vartheta_0 \ne 0$ and $\vartheta_0 = 0$.

Note that since $\mathrm{e}(T,n) = 1 + O(nT^{-1})$, we may omit $\mathrm{e}(T,(n+\frac{1}{2})^2)$ in A with an error $\ll N_0^2T^{-1}$. Hence (8.53) becomes

$$A = \sum_{0\le n\le N_0} (-1)^{n(n+1)/2}(n+\tfrac{1}{2})^{-1}$$
$$\times \big\{\sin(2\pi(n+\tfrac{1}{2})\vartheta+\psi(n)) - \sin(2\pi(n+\tfrac{1}{2})\vartheta)\big\} + O(N_0^2T^{-1}). \qquad (8.56)$$

To deal with the sums A, B and C, we need estimates for trigonometric sums related to the Fourier series (8.50) and to the corresponding cosine series.

We need estimates for sums of the form

$$U_s(N_1,N_2;\vartheta,\alpha) := \sum_{N_1\le n\le N_2} (-1)^{n(n+1)/2}(n+\tfrac{1}{2})^{-\alpha}\sin(2\pi(n+\tfrac{1}{2})\vartheta)$$

for $\alpha = 0, 1$, and for analogous sums $U_c(N_1,N_2;\vartheta,\alpha)$ with the cosine replacing the sine. These sums depend on the location of ϑ with respect to $1/4$ and $3/4$. More precisely, with ϑ_0 as in Theorem 8.3,

$$U_s(N_1,N_2;\vartheta,1) \ll \min\big(N_2\vartheta_0,(N_1\vartheta_0)^{-1}\big) \quad (\vartheta_0\ne 0),$$
$$U_c(N_1,N_2;\vartheta,1) \ll (N_1\vartheta_0)^{-1} \quad (\vartheta_0\ne 0),$$

and moreover we have

$$U_s(N_1,N_2;j/4,1) \ll N_1^{-1} \qquad (0\le j\le 4).$$

These bounds follow, by partial summation, from the estimates

$$U_s(N_1,N_2;\vartheta,0) \ll \min\big(N_2^2\vartheta_0,\vartheta_0^{-1}\big) \quad (\vartheta_0\ne 0),$$
$$U_c(N_1,N_2;\vartheta,0) \ll \vartheta_0^{-1} \quad (\vartheta_0\ne 0),$$
$$U_s(N_1,N_2;j/4,0) \ll 1. \qquad (8.57)$$

The sign in the expression for $U_s(N_1, N_2; \vartheta, \alpha)$ depends on n modulo 4, and hence we write $n = 4m + b$ with $0 \le b \le 3$. Also, we write $\vartheta = j/4 + \delta$ with $0 \le j \le 4$ and $|\delta| \le 1/8$. Thus $\vartheta_0 = |\delta|$ for $j = 1, 3$, and $\vartheta_0 \ge 1/8$ otherwise. Then we obtain

$$U_s(N_1, N_2; \vartheta, 0) = \sum_{b=0}^{3}(-1)^{b(b+1)/2}\Big(\cos\left(2\pi\left(b + \tfrac{1}{2}\right)\vartheta\right) \sum_{N_1/4 \le m \le N_2/4} \sin(8\pi m\delta)$$
$$+ \sin\left(2\pi\left(b + \tfrac{1}{2}\right)\vartheta\right) \sum_{N_1/4 \le m \le N_2/4} \cos(8\pi\delta m)\Big) + O(1).$$

Note that

$$\sum_{N_1/4 \le m \le N_2/4} \sin(8\pi m\delta) \ll \min\left(N_2^2|\delta|, |\delta|^{-1}\right),$$
$$\sum_{N_1/4 \le m \le N_2/4} \cos(8\pi m\delta) \ll \min\left(N_2, |\delta|^{-1}\right),$$

and that

$$\sum_{b=0}^{3}(-1)^{b(b+1)/2} \sin(2\pi(b + \tfrac{1}{2})j/4) = 0 \quad (j \in \mathbb{Z}),$$
$$\sum_{b=0}^{3}(-1)^{b(b+1)/2} \cos(2\pi(b + \tfrac{1}{2})j/4) = 0 \quad (j = 0, 2, 4).$$

Hence

$$\sum_{b=0}^{3}(-1)^{b(b+1)/2} \sin(2\pi(b + \tfrac{1}{2})\vartheta) \ll |\delta|,$$
$$\sum_{b=0}^{3}(-1)^{b(b+1)/2} \cos(2\pi(b + \tfrac{1}{2})\vartheta) \ll 1 - 4\vartheta_0.$$

These bounds furnish (8.57)

The cases $\vartheta \ne 0$ and $\vartheta = 0$ are similar, but technically involved, and for this reason will not be dealt here in detail. The net result follows from (8.49) and the estimates for all the exponential sums that appear above. This completes the discussion of Theorem 8.3.

Notes

F. V. Atkinson [Atk4] proved his fundamental formula for $E(T)$ in 1949, where a proof of Lemma 8.8 is also to be found (see also [Iv1], [Iv4]). Proofs that are different from Atkinson's original proof were given by Y. Motohashi [Mot1], M. Jutila [Jut2] and M. Lukkarinen [Luk]. Motohashi obtained a slightly better error term, namely $O(\log T)$, and his proof is based on his approximate functional equation for $\zeta^2(\frac{1}{2}+it)$. The proofs of M. Jutila and M. Lukkarinen, on the other hand, are based on the use of Laplace transforms of $|\zeta(\frac{1}{2}+ix)|^2$ (Pierre-Simon, marquis de Laplace, March 23, 1749-March 5, 1827, great French mathematician and astronomer).

Atkinson's formula remained unnoticed for almost 30 years, until D. R. Heath-Brown [Hea1], [Hea2] made the first important applications of it. Further authors used this formula extensively, which is one of the basic tools in the theory of mean values of $\zeta(\frac{1}{2}+it)$. The paper [Hea1] contains the proof of

$$\int_0^T |\zeta(\tfrac{1}{2}+it)|^{12}\,\mathrm{d}t \ll T^2\log^{17} T,$$

which is essentially still the best result on higher moments of $|\zeta(\frac{1}{2}+it)|$. In [Hea2] Heath-Brown proved the mean square formula

$$\int_0^T E^2(t)\,\mathrm{d}t = CT^{3/2} + O(T^{5/4}\log^2 T),$$

where

$$C = \frac{2}{3}(2\pi)^{-1/2}\sum_{n=1}^{\infty} d^2(n)n^{-3/2},$$

and the series clearly converges since $d(n) \ll_\varepsilon n^\varepsilon$. The above error term was improved to $O(T\log^4 T)$ by E. Preissmann [Pre]. The mean square formula yields at once $E(T)=\Omega(T^{1/4})$, and improvements were obtained by J. L. Hafner and the present author [HaIv1], [HaIv2]. They showed that there exist positive constants C and D such that

$$E(T) = \Omega_+\left\{(T\log T)^{1/4}(\log\log T)^{(3+\log 4)/4}\exp(-C\sqrt{\log\log\log T})\right\}$$

and

$$E(T) = \Omega_-\left\{T^{1/4}\exp\left(D(\log\log T)^{1/4}(\log\log\log T)^{-3/4}\right)\right\}.$$

Y.-K. Lau and K.-M. Tsang [LaTs] used the method of K. Soundararajan [Sou1], who proved the analogous result for $\Delta(x)$ (the error term in the divisor problem), namely they proved

$$E(T) = \Omega\left\{(T\log T)^{\frac{1}{4}}(\log\log T)^{\frac{3}{4}(2^{\frac{4}{3}}-1)}(\log\log\log T)^{-\frac{5}{8}}\right\}.$$

Soundararajan's method does not yield an Ω_+ or Ω_--result, but just an Ω-result.

Theorem 8.2 and Theorem 8.3 are proved by M. Jutila [Jut6]. Theorem 8.3 is analogous to M. Korolev's theorem 1 in [Kor3], [Kor4]. Indeed, if we let T run over values such that the fractional part of $\sqrt{T/(2\pi)}$ equals a fixed number ϑ, then Korolev's theorem follows from Theorem 8.3. The main novelty of Jutila's result is uniformity in T and ϑ. The idea of the use of the Laplace transforms in a similar context, with the aim to develop a unified method for proving both the original Atkinson formula and its analog for cusp form L-functions, was presented by M. Jutila [Jut2].

An alternate version (see, e.g., appendix D of [MoVa]) of the Poisson summation formula (Siméon Denis Poisson, June 21, 1781-April 25, 1840, French mathematician, geometer and physicist) is

$$\sum_{n=-\infty}^{\infty} f(n) = \sum_{k=-\infty}^{\infty} \hat{f}(k),$$

which certainly holds if $f(x)\in L^1(-\infty,\infty)$, x is a real variable and $\hat{f}$ is the Fourier transform of f.

Lemma 8.8 on exponential integrals with a saddle point is a version of F. V. Atkinson's classical result [Atk4] (see also chapter 2 of [Iv1]), used in his proof of the formula for $E(T)$; even the notation follows that introduced by Atkinson. Lemma 8.8 is obtained by the same method as Atkinson's original result.

The bounds in (8.28) are to be found in K. Ramachandra [Ram]. The use of the Cauchy-Schwarz inequality and standard mean square bounds would result in somewhat poorer log-factors in (8.27).

In [Jut9] M. Jutila refined his results on $F(T)$, the primitive of Hardy's function. Let

$$\beta(u) = \frac{|2\pi u|^{1/2}}{3\pi A^{1/2}} K_{1/3}\left(\frac{2|2\pi u|^{3/2}}{3\sqrt{3A}}\right) \qquad (u < 0)$$

and

$$\beta(u) = \frac{(2\pi u)^{1/2}}{3\sqrt{3}A^{1/2}}\left(J_{1/3}\left(\frac{2(2\pi u)^{3/2}}{3\sqrt{3A}}\right) + J_{-1/3}\left(\frac{2(2\pi u)^{3/2}}{3\sqrt{3A}}\right)\right) \qquad (u > 0),$$

where $K_{1/3}, J_{\pm 1/3}$ are the classical *Bessel functions* (named after the German mathematician Friedrich Wilhelm Bessel, July 22, 1784-March 17, 1846) in standard notation (see G. N. Watson [Wats] for a comprehensive account). George Neville Watson (January 31, 1886-February 2, 1965) was an English mathematician, noted for the application of complex analysis to the theory of special functions. The Bessel functions are defined as

$$J_p(z) = \sum_{k=0}^{\infty} \frac{(-1)^k (z/2)^{p+2k}}{k!\Gamma(p+k+1)} \qquad (p \in \mathbb{R}),$$

$$K_p(z) = \frac{\pi}{2}\,\frac{I_{-p}(z) - I_p(z)}{\sin(\pi z)}, \qquad I_p(z) := i^{-p} J_p(iz).$$

Further, $\beta(u)$ is also *the Airy function* in the form

$$\beta(u) = \frac{1}{\pi}\int_0^{\infty} \cos(Ay^3 - 2\pi u y)\,\mathrm{d}y, \tag{8.58}$$

where $A > 0$ is a parameter which (in [Jut9]) depends on T. The Airy functions, named after the British astronomer George Biddell Airy (1801-1892),

$$\mathrm{Ai}(x) := \frac{1}{\pi}\int_0^{\infty} \cos(\tfrac{1}{3}t^3 + xt)\,\mathrm{d}t, \quad \mathrm{Bi}(x) := \frac{1}{\pi}\int_0^{\infty}\left\{\exp(-\tfrac{1}{3}t^3 + xt) + \sin(\tfrac{1}{3}t^3 + xt)\right\}\mathrm{d}t$$

are linearly independent solutions of the differential equation

$$\frac{\mathrm{d}^2 y}{\mathrm{d}^2 x} - xy = 0.$$

The significance of $\beta(u)$ in the present context is that (8.58) is a very useful transformation formula for the term Ay^3 in the argument of the cosine.

Let T be a large positive integer and write $\sqrt{T/2\pi} = L + \vartheta$ with $L \in \mathbb{N}$ and $0 \le \vartheta < 1$. Let further $A = \frac{1}{12}\sqrt{2\pi^3}T^{-1/2}$. Then Jutila [Jut9] proved that

$$F(T) = (T/2\pi)^{1/4}(-1)^L \tilde{K}(\vartheta) + O(T^{1/6}\log T),$$

where

$$\tilde{K}(\vartheta) = K(\vartheta) + 2\pi\int_{-1/2}^{1/2} w(u)\beta(u)\{(\vartheta + u) - K(\vartheta)\}\,\mathrm{d}u.$$

Here $w(u)$ is a smooth weight function such that

$$w(u) = \begin{cases} 1 & \text{if } |u| \le \frac{1}{4}, \\ 0 & \text{if } |u| \ge \frac{1}{2}, \end{cases}$$

and $w^{(\nu)}(u) \ll_\nu 1$ for sufficiently many derivatives. Further, define $\vartheta_0 = \min(|\vartheta - \frac{1}{4}|, |\vartheta - \frac{3}{4}|)$. Then for $\vartheta_0 \neq 0$ we have

$$K(\vartheta) - \tilde{K}(\vartheta) \ll \min\left(1, T^{-1/8}\vartheta_0^{-3/4}\right).$$

Also one has

$$\tilde{K}\left(\frac{1}{4}\right) = \frac{4\pi}{3} + O(T^{-1/8})$$

and

$$\tilde{K}\left(\frac{3}{4}\right) = \frac{2\pi}{3} + O(T^{-1/8}).$$

This result tells that, in a neighborhood of length about $T^{1/3}$ of any point $T = 2\pi(n + j/4)^2$ with n a natural number and $j = 1$ or 3, the function $F(t)$ jumps an amount $\asymp T^{1/4}$ upwards or downwards and the theorem moreover indicates how these jumps take place, at least approximately. The existence of the jumps can be illustrated as follows: the center of gravity of the curve $Z(t)$ over any of the critical intervals mentioned above lies at a distance $\asymp T^{-1/12}$ above or below the axis.

9

The Mellin transforms of powers of $Z(t)$

9.1 Introduction

Integral transforms (Laplace, Fourier and Mellin, among others) play an important rôle in analytic number theory. Of special interest in the theory of the Riemann zeta-function $\zeta(s)$ are the Laplace transforms

$$L_k(s) := \int_0^\infty |\zeta(\tfrac{1}{2}+ix)|^{2k} \mathrm{e}^{-sx}\,\mathrm{d}x \qquad (k \in \mathbb{N},\ \sigma = \mathrm{Re}\, s > 0) \tag{9.1}$$

and the (modified) Mellin transforms

$$\mathcal{Z}_k(s) := \int_1^\infty |\zeta(\tfrac{1}{2}+ix)|^{2k} x^{-s}\,\mathrm{d}x \qquad (k \in \mathbb{N},\ \sigma = \mathrm{Re}\, s \ge c(k) > 1), \tag{9.2}$$

where $c(k)$ is such a constant for which the integral in (9.2) converges absolutely. The term "modified" Mellin transform seems appropriate, since customarily the Mellin transform of $f(x)$ is defined as

$$\mathcal{M}[f(x)] = F(s) := \int_0^\infty f(x)x^{s-1}\,\mathrm{d}x \qquad (s = \sigma + it \in \mathbb{C}). \tag{9.3}$$

Note that the lower bound of integration in (9.2) is not zero, as it is in (9.3). The choice of unity as the lower bound of integration dispenses with convergence problems at that point, while the appearance of the factor x^{-s} instead of the customary x^{s-1} is technically more convenient. Also $\mathcal{Z}_k(s)$ may be compared with the discrete representation

$$\zeta^{2k}(s) = \sum_{n=1}^\infty d_{2k}(n)n^{-s} \qquad (\mathrm{Re}\, s > 1,\ k \in \mathbb{N}).$$

Since we have (see [Iv1], chapter 8)

$$\int_0^T |\zeta(\tfrac{1}{2}+it)|^{2k}\,\mathrm{d}t \ll T^{(k+2)/4}\log^{C(k)} T \qquad (2 \le k \le 6), \tag{9.4}$$

it follows that the integral defining $\mathcal{Z}_k(s)$ is absolutely convergent for $\sigma > 1$ if $0 \le k \le 2$ and for $\sigma > (k+2)/4$ if $2 \le k \le 6$.

By a change of variable Mellin transforms can be viewed as special cases of Fourier transforms, and their theory built by using the theory of Fourier transforms. Namely, if $F(s)$ is the Mellin transform of $f(x)$, then ($\xi = \mathrm{e}^x$)

$$F(\sigma+it) = \int_0^\infty \xi^{\sigma+it-1} f(\xi)\,\mathrm{d}\xi = \int_{-\infty}^\infty \mathrm{e}^{ixt} f(\mathrm{e}^x)\mathrm{e}^{\sigma x}\,\mathrm{d}x$$

is the Fourier transform of $f(\mathrm{e}^x)\mathrm{e}^{\sigma x}$.

One of the basic properties of Mellin transforms is *the inversion formula*. It states that if $F(s) = \mathcal{M}[f(x)]$, $y^{\sigma-1}f(y) \in L^1(0, \infty)$ and $f(y)$ is of bounded variation in a neighborhood of $y = x$, then

$$\frac{f(x+0)+f(x-0)}{2} = \frac{1}{2\pi i}\int_{(\sigma)} F(s)x^{-s}\,\mathrm{d}s := \frac{1}{2\pi i}\lim_{T\to\infty}\int_{\sigma-iT}^{\sigma+iT} F(s)x^{-s}\,\mathrm{d}s. \tag{9.5}$$

We recall that if $f(x)$ denotes measurable functions, then

$$L^p(a,b) := \left\{ f(x) \Big| \int_a^b |f(x)|^p\,\mathrm{d}x < +\infty \right\},$$

where a and b are not necessarily finite. The Mellin inversion formula (9.5) is usually derived from the inversion formula for Fourier transforms. Namely, if

$$\widehat{g}(y) = \int_{-\infty}^\infty \mathrm{e}^{ixy} g(x)\,\mathrm{d}x$$

is the Fourier transform of $g(x)$, then under suitable conditions this is equivalent to

$$g(x) = \frac{1}{2\pi}\int_{-\infty}^\infty \mathrm{e}^{-ixy}\widehat{g}(y)\,\mathrm{d}y. \tag{9.6}$$

For example if $g, \hat{g} \in L^1(-\infty, \infty)$, then (9.6) holds for almost all $x \in (-\infty, \infty)$. If additionally g is continuous in $(-\infty, \infty)$, then (9.6) holds for all $x \in (-\infty, \infty)$. A variant of *Parseval's formula* for Fourier transforms is the identity

$$\frac{1}{2\pi}\int_{-\infty}^\infty |\hat{f}(x)|^2\,\mathrm{d}x = \int_{-\infty}^\infty |f(x)|^2\,\mathrm{d}x, \tag{9.7}$$

and it can be used to derive Parseval's formula for Mellin transforms. Namely if $F(s), G(s)$ denote the Mellin transform of $f(x), g(x)$, respectively, then assuming $f(x)$ and $g(x)$ to be real-valued, we formally have

$$\begin{aligned}\frac{1}{2\pi i}\int_{(\sigma)} F(s)\overline{G(s)}\,\mathrm{d}s &= \int_0^\infty g(x)\left(\frac{1}{2\pi i}\int_{(\sigma)} F(s)x^{\sigma-it-1}\,\mathrm{d}s\right)\mathrm{d}x\\ &= \int_0^\infty g(x)x^{2\sigma-1}\left(\frac{1}{2\pi i}\int_{(\sigma)} F(s)x^{-s}\,\mathrm{d}s\right)\mathrm{d}x\\ &= \int_0^\infty f(x)g(x)x^{2\sigma-1}\,\mathrm{d}x. \qquad (9.8)\end{aligned}$$

The relation (9.8) is a form of Parseval's formula for Mellin transforms, and it offers various possibilities for mean square bounds. A condition under which (9.8) surely holds is that $x^\sigma f(x)$ and $x^\sigma g(x)$ belong to $L^2((0,\infty),\ \mathrm{d}x/x)$. A variant of (9.8) is

$$\frac{1}{2\pi i}\int_{(c)} F(w)G(s-w)\,\mathrm{d}w \;=\; \int_0^\infty f(x)g(x)x^{s-1}\,\mathrm{d}x, \qquad (9.9)$$

which holds if $x^c f(x)$ and $x^{\sigma-c}g(x)$ belong to $L^2((0,\infty),\ \mathrm{d}x/x)$.

9.2 Some properties of the modified Mellin transforms

The main object of study in this chapter is the modified Mellin transform function

$$\mathcal{M}_k(s) := \int_1^\infty Z^k(x)x^{-s}\,\mathrm{d}x, \qquad (9.10)$$

which is a regular function of s for any given $k \ge 0$ and $\sigma \ge \sigma_0(k)\,(> 1)$. However, from the viewpoint of possible applications it is expedient to assume that $k \in \mathbb{N}$.

The general modified Mellin transform $m[f(x)]$, of which $\mathcal{M}_k(s)$ is a special case, is defined as

$$F^*(s) \;=\; m[f(x)] \;=\; \int_1^\infty f(x)x^{-s}\,\mathrm{d}x \qquad (s = \sigma + it), \qquad (9.11)$$

which is often more convenient to have in applications than the ordinary Mellin transform, since now there are no convergence problems at $x = 0$. If

$$\bar{f}(x) = \begin{cases} f(1/x) & \text{if } 0 < x < 1,\\ 0 & \text{if } x \ge 1,\end{cases}$$

then

$$m[f(x)] = \mathcal{M}\left[\frac{1}{x}\bar{f}(x)\right], \tag{9.12}$$

so that the properties of $m[f(x)]$ can be deduced from the properties of the ordinary Mellin transform $\mathcal{M}[f(x)]$ by the use of (9.12). In what follows five lemmas concerning the properties of the modified Mellin transform will be proved.

Lemma 9.1 *If* $x^{-\sigma} f(x) \in L^1(1, \infty)$ *and* $f(x)$ *is continuous for* $x > 1$, *then*

$$f(x) = \frac{1}{2\pi i}\int_{(\sigma)} F^*(s)x^{s-1}\mathrm{d}s, \qquad F^*(s) = m[f(x)]. \tag{9.13}$$

Proof of Lemma 9.1 If $F^*(s) = m[f(x)]$, then from (9.12) and (9.5) we have

$$\frac{1}{x}\bar{f}(x) = \frac{1}{2\pi i}\int_{(\sigma)} F^*(s)x^{-s}\,\mathrm{d}s \tag{9.14}$$

provided that $y^{\sigma-1}\frac{1}{y}\bar{f}(y) \in L^1(0, \infty)$. This means that

$$\int_0^1 y^{\sigma-1}\frac{1}{y}\left|f\left(\frac{1}{y}\right)\right|\,\mathrm{d}y < +\infty,$$

and making the change of variable $y = 1/x$, we obtain that the last condition is equivalent to $x^{-\sigma} f(x) \in L^1(1, \infty)$. Changing x to $1/x$ in (9.14) we then obtain (9.13).

Lemma 9.2 *If* $F^*(s) = m[f(x)]$, $G^*(s) = m[g(x)]$, *and* $f(x)$, $g(x)$ *are real-valued, continuous functions for* $x > 1$, *such that*

$$x^{\frac{1}{2}-c} f(x) \in L^2(1, \infty), \quad x^{c-\frac{1}{2}-\sigma} g(x) \in L^2(1, \infty),$$

then

$$m[f(x)g(x)] = \frac{1}{2\pi i}\int_{(c)} F^*(w)G^*(s + 1 - w)\mathrm{d}w. \tag{9.15}$$

Proof of Lemma 9.2 We start from

$$\int_0^\infty f(x)g(x)x^{s-1}\,\mathrm{d}x = \frac{1}{2\pi i}\int_{(c)} F(w)G(s - w)\,\mathrm{d}w, \tag{9.16}$$

which certainly holds if

$$F(s) = \mathcal{M}[f(x)], \ G(s) = \mathcal{M}[g(x)],$$
$$x^{c-\frac{1}{2}} f(x) \in L^2(0, \infty), \ x^{\sigma-c-\frac{1}{2}} g(x) \in L^2(0, \infty).$$

In place of $f(x)$ and $g(x)$ in (9.16) we shall take $\frac{1}{x}\bar{f}(x)$ and $\bar{g}(x)$, respectively. By (9.12) we have $\mathcal{M}[\frac{1}{x}\bar{f}(x)] = F^*(s)$ and

$$\mathcal{M}[\bar{g}(x)] = \int_0^1 g\left(\frac{1}{x}\right) x^{s-1}\,\mathrm{d}x = G^*(s+1).$$

Consequently (9.16) gives

$$\int_0^\infty \frac{1}{x}\bar{f}(x)\bar{g}(x)x^{s-1}\,\mathrm{d}x = \frac{1}{2\pi i}\int_{(c)} F^*(w)G^*(s+1-w)\,\mathrm{d}w.$$

After the change of variable $y = 1/x$ this reduces to (9.15) if

$$x^{c-\frac{1}{2}} \cdot \frac{1}{x}\bar{f}(x) \in L^2(0,\,\infty), \quad x^{\sigma-c-\frac{1}{2}} \cdot \frac{1}{x}\bar{g}(x) \in L^2(0,\,\infty),$$

and these conditions are easily seen to be equivalent to

$$x^{\frac{1}{2}-c} f(x) \in L^2(1,\infty), \qquad x^{c-\frac{1}{2}-\sigma} g(x) \in L^2(1,\infty).$$

Lemma 9.3 *If $F^*(s) = m[f(x)]$, $G^*(s) = m[g(x)]$, and $f(x)$, $g(x)$ are real-valued, continuous functions for $x > 1$, such that*

$$x^{\frac{1}{2}-\sigma} f(x) \in L^2(1,\infty), \quad x^{\frac{1}{2}-\sigma} g(x) \in L^2(1,\infty),$$

then

$$\int_1^\infty f(x)g(x)x^{1-2\sigma}\,\mathrm{d}x = \frac{1}{2\pi i}\int_{(\sigma)} F^*(s)\overline{G^*(s)}\mathrm{d}s. \tag{9.17}$$

Proof of Lemma 9.3 Follows similarly to the preceding proof from Parseval's formula for Mellin transforms in the form

$$\int_0^\infty f(x)g(x)x^{2\sigma-1}\,\mathrm{d}x = \frac{1}{2\pi i}\int_{(\sigma)} F(s)\overline{G(s)}\,\mathrm{d}s$$

if

$$F(s) = \mathcal{M}[f(x)], \; G(s) = \mathcal{M}[g(x)],$$
$$x^{\sigma-\frac{1}{2}} f(x) \in L^2(0,\,\infty), \; x^{\sigma-\frac{1}{2}} g(x) \in L^2(0,\,\infty).$$

The last relation follows, for example, from [Tit2, theorem 72] by a change of variable.

Lemma 9.3, in the special case $f(x) = g(x)$, is a natural tool for investigating mean square formulas connected with the modified Mellin transform. In particular, it offers the possibility to obtain mean square estimates of $f(x)$ from the mean square estimates of $m[f(x)]$, provided we have adequate analytic information about the latter. A result in this direction, which is useful for the applications that we have mind, will be given now as

Lemma 9.4 *Suppose that $g(x)$ is a real-valued, integrable function on $[a, b]$, a subinterval of $[2, \infty)$, which is not necessarily finite. Then*

$$\int_0^T \left|\int_a^b g(x)x^{-s}\mathrm{d}x\right|^2 \mathrm{d}t \le 2\pi \int_a^b g^2(x)x^{1-2\sigma}\mathrm{d}x \quad (s = \sigma + it,\, T > 0,\ a < b). \tag{9.18}$$

Proof of Lemma 9.4 Let

$$I := \int_0^T \left|\int_a^b g(x)x^{-s}\,\mathrm{d}x\right|^2 \mathrm{d}t \qquad (s = \sigma + it,\ T > 0). \tag{9.19}$$

In Lemma 9.3 set

$$f(x) = \begin{cases} g(x) & \text{if } x \in [a, b], \\ 0 & \text{if } x \notin [a, b]. \end{cases}$$

Then

$$F^*(s) = m[f(x)] = \int_a^b g(x)x^{-s}\,\mathrm{d}x, \quad F^*(s) = G^*(s),$$

and $x^{\frac{1}{2}-\sigma} f(x) \in L^2(1, \infty)$ for any σ. Consequently (9.17) of Lemma 9.3 (with $f \equiv g$) gives

$$\frac{I}{2\pi} \le \frac{1}{2\pi}\int_{-\infty}^{\infty} |F^*(\sigma + it)|^2\,\mathrm{d}t = \int_a^b g^2(x)x^{1-2\sigma}\,\mathrm{d}x, \tag{9.20}$$

and (9.18) follows from (9.19) and (9.20).

In cases where there is sufficient information on f and F^*, we can obtain an explicit bound for the mean square estimate of $f(x)$. Such a result is furnished by the next lemma.

Lemma 9.5 *Let $f(x) \in C^\infty[2, \infty)$ be a real-valued function for which*

(a) $\int_1^X |f^{(r)}(x)|\mathrm{d}x \ll_{\varepsilon,r} X^{1+\varepsilon}$ $(r = 0, 1, 2, \dots)$ *and*
(b) $F^*(s) = m[f(x)]$ *has a pole at $s = 1$ of order ℓ, but otherwise can be analytically continued to the region $\mathbb{R}\mathrm{e}\, s > \frac{1}{2}$, where it is of polynomial growth in $|\mathbb{I}\mathrm{m}\, s|$. Then for $\frac{1}{2} < \sigma < 1$ and any given $\varepsilon > 0$ we have*

$$\int_T^{2T} f^2(x)\mathrm{d}x \ll_\varepsilon \log^{\ell-1} T \cdot \int_{T/2}^{5T/2} |f(x)|\mathrm{d}x + T^{2\sigma-1}\int_0^{T^{1+\varepsilon}} |F^*(\sigma + it)|^2\mathrm{d}t. \tag{9.21}$$

Proof of Lemma 9.5 Let $\varphi(x) \in C^\infty(0,\infty)$ be a smooth function such that $\varphi(x) \ge 0$, $\varphi(x) = 1$ for $T \le x \le 2T$, $\varphi(x) = 0$ for $x < \frac{1}{2}T$ or $x > \frac{5}{2}T$ ($T \ge T_0 > 0$), $\varphi(x)$ is increasing in $[\frac{1}{2}T, T]$ and decreasing in $[2T, \frac{5}{2}T]$. We have

$$\varphi^{(r)}(x) \ll_r T^{-r} \qquad (r = 0, 1, 2, \dots).$$

Then obviously

$$I_1 \le I_2,$$

where

$$I_1 := \int_T^{2T} f^2(x)\,\mathrm{d}x, \qquad I_2 := \int_{T/2}^{5T/2} \varphi(x) f^2(x)\,\mathrm{d}x.$$

From the assumption a) we have $x^{-\sigma} f(x) \in L^1(1,\infty)$ if $\sigma > 1$. Therefore Lemma 9.1 gives

$$f(x) = \frac{1}{2\pi i}\int_{(1+\varepsilon)} F^*(w)x^{w-1}\,\mathrm{d}w = \frac{1}{2\pi i}\int_{\mathcal{L}} F^*(w)x^{w-1}\,\mathrm{d}w + Q_{\ell-1}(\log x), \tag{9.22}$$

where $Q_{\ell-1}(\log x)$ is a polynomial in $\log x$ of degree $\ell - 1$, and $\mathcal{L}$ is the line $\mathbb{R}\mathrm{e}\, w = 1 + \varepsilon$ with a small indentation to the left at the pole $w = 1$ of $F^*(w)$ of order ℓ, so that by the residue theorem we pick a contribution equal to $Q_{\ell-1}(\log x)$. Thus (9.22) yields

$$I_2 = \frac{1}{2\pi i}\int_{\mathcal{L}} F^*(w)\left(\int_{T/2}^{5T/2} \varphi(x) f(x) x^{w-1}\,\mathrm{d}x\right)\mathrm{d}w + O\left(\log^{\ell-1} T \int_{T/2}^{5T/2} \varphi(x)|f(x)|\,\mathrm{d}x\right). \tag{9.23}$$

We integrate r times by parts to obtain

$$\int_{T/2}^{5T/2} \varphi(x) f(x) x^{w-1}\,\mathrm{d}x = \frac{(-1)^r}{w\cdots(w+r-1)}\int_{T/2}^{5T/2} (\varphi(x)f(x))^{(r)} x^{w+r-1}\,\mathrm{d}x. \tag{9.24}$$

Since $\varphi^{(r)}(x) \ll_r T^{-r}$ and (a) of the lemma holds, then by (9.24) it follows that the portion of the integral over w in (9.23) for which $|\Im w| = |v| > T^{1+\varepsilon}$ makes a negligible contribution, namely $\ll T^{-A}$ for any given $A > 0$, provided that we choose $r = r(A, \varepsilon)$ sufficiently large. This remains true even if we move the contour of integration to the left. Hence if $\frac{1}{2} < u < 1$, $w = u + iv$, then by (2.16) and the Cauchy-Schwarz inequality for integrals we infer that

$$\begin{aligned}
I_2 &= \frac{1}{2\pi}\int_{-T^{1+\varepsilon}}^{T^{1+\varepsilon}} F^*(u+iv)\int_{T/2}^{5T/2}\varphi(x)f(x)x^{u+iv-1}\,\mathrm{d}x\,\mathrm{d}v \\
&\quad + O\left(\log^{\ell-1}T\cdot\int_{T/2}^{5T/2}\varphi(x)|f(x)|\,\mathrm{d}x\right) \\
&\ll \left\{\int_0^{T^{1+\varepsilon}}|F^*(u+iv)|^2\,\mathrm{d}v\int_0^{T^{1+\varepsilon}}\left|\int_{T/2}^{5T/2}\varphi(x)f(x)x^{u+iv-1}\,\mathrm{d}x\right|^2\mathrm{d}v\right\}^{1/2} \\
&\quad + \log^{\ell-1}T\cdot\int_{T/2}^{5T/2}\varphi(x)|f(x)|\,\mathrm{d}x.
\end{aligned}$$

Now we apply Lemma 9.4 (with $g(x)=\varphi(x)f(x),\ s=1-u+iv, T$ replaced by $T^{1+\varepsilon}$) to deduce that

$$\begin{aligned}
&\int_0^{T^{1+\varepsilon}}\left|\int_{T/2}^{5T/2}\varphi(x)f(x)x^{u+iv-1}\,\mathrm{d}x\right|^2\mathrm{d}v \\
&\ll \int_{T/2}^{5T/2}\varphi^2(x)f^2(x)x^{2u-1}\,\mathrm{d}x \ll T^{2u-1}\int_{T/2}^{5T/2}\varphi(x)f^2(x)\,\mathrm{d}x = T^{2u-1}I_2,
\end{aligned}$$

since $0\le\varphi(x)\le 1$. It follows that

$$I_2 \ll \log^{\ell-1}T\cdot\int\limits_{T/2}^{5T/2}\varphi(x)|f(x)|\,\mathrm{d}x + \left(\int\limits_0^{T^{1+\varepsilon}}|F^*(u+iv)|^2\,\mathrm{d}v\cdot T^{2u-1}I_2\right)^{1/2},$$

which easily gives (9.21) with $\sigma=u$.

9.3 Analytic continuation of $\mathcal{M}_k(s)$

In this section we shall discuss the analytic continuation of $\mathcal{M}_k(s)$. This is primarily of interest when $1\le k\le 4$, since our knowledge of the kth moment of $Z(t)$ (or $|\zeta(\frac{1}{2}+it)|$) when $k\ge 5$ is imperfect, as no bound of the form

$$\int_0^T Z^k(t)\,\mathrm{d}t \ll_{\varepsilon,k} T^{1+\varepsilon}$$

seems to be known when $k\ge 5$. We begin with $\mathcal{M}_1(s)$, and we shall prove the following theorem.

Theorem 9.6 *The function $\mathcal{M}_1(s)$ has analytic continuation to the region $\sigma>0$, where it is regular.*

Proof of Theorem 9.6 Let

$$\bar{L}(s) := \int_1^\infty Z(y)\mathrm{e}^{-ys}\,\mathrm{d}y, \quad L(s) := \int_0^\infty Z(y)\mathrm{e}^{-ys}\,\mathrm{d}y \quad (\sigma = \Re\mathrm{e}\, s > 0).$$

Then we have, by absolute convergence, taking initially σ to be sufficiently large and making the change of variable $xy = t$,

$$\begin{aligned}\int_0^\infty \bar{L}(x)x^{s-1}\,\mathrm{d}x &= \int_0^\infty \left(\int_1^\infty Z(y)\mathrm{e}^{-xy}\,\mathrm{d}y\right) x^{s-1}\,\mathrm{d}x\\ &= \int_1^\infty Z(y)\left(\int_0^\infty x^{s-1}\mathrm{e}^{-xy}\,\mathrm{d}x\right)\mathrm{d}y\\ &= \int_1^\infty Z(y)y^{-s}\,\mathrm{d}y \int_0^\infty \mathrm{e}^{-t}t^{s-1}\,\mathrm{d}t = \mathcal{M}_1(s)\Gamma(s). \qquad (9.28)\end{aligned}$$

Since $\Gamma(s)$ has no zeros, it suffices to prove the assertion about the analytic continuation of $\mathcal{M}_1(s)$ for the function

$$\begin{aligned}\int_0^\infty \bar{L}(x)x^{s-1}\,\mathrm{d}x &= \int_0^1 \bar{L}(x)x^{s-1}\,\mathrm{d}x + \int_1^\infty \bar{L}(x)x^{s-1}\,\mathrm{d}x\\ &= \int_1^\infty \bar{L}(1/x)x^{-1-s}\,\mathrm{d}x + A(s) \qquad (\sigma > 1),\end{aligned}$$

say, where

$$A(s) := \int_1^\infty \bar{L}(x)x^{s-1}\,\mathrm{d}x$$

is an entire function. Since

$$\bar{L}(1/x) = L(1/x) - \int_0^1 Z(y)\mathrm{e}^{-y/x}\,\mathrm{d}y \qquad (x \ge 1),$$

it remains for us to consider

$$\begin{aligned}\int_1^\infty \bar{L}(1/x)x^{-s-1}\,\mathrm{d}x &= \int_1^\infty L(1/x)x^{-s-1}\,\mathrm{d}x - \int_1^\infty\left(\int_0^1 Z(y)\mathrm{e}^{-y/x}\,\mathrm{d}y\right)x^{-s-1}\,\mathrm{d}x\\ &= I_1(s) - I_2(s),\end{aligned}$$

say. Note that in $I_2(s)$ the integral over y is uniformly bounded, so that $I_2(s)$ is regular for $\sigma > 0$. To deal with $I_1(s)$ we shall use Lemma 8.6, which says that

$$\tilde{L}(p) \ll 1,\ p = \frac{1}{T} + iu,\ T \ge T_0,\ 0 \le u \le (T^{1/2}\log T)^{-1},$$

where

$$\tilde{L}(p) := \int_0^\infty Z(t)H(\tfrac12 + it)\mathrm{e}^{-pt}\,\mathrm{d}t \qquad (\Re\mathrm{e}\, p > 0)$$

with $H(s)$ as in Section 8.2. The function $H(s)$ satisfies

$$H(\tfrac12+it)=1+O\left(\frac{1}{|t|+1}\right),\qquad H'(\tfrac12+it)=O\left(\frac{1}{(|t|+1)^2}\right).$$

If we set $k(t)=1-H(\frac12+it)$, then

$$I_1(s)=I_3(s)+B(s),$$

say, where $B(s)$ is regular for $\sigma>0$ and

$$\begin{aligned}I_3(s)&:=\int_1^\infty\left(\int_0^\infty Z(t)k(t)\,\mathrm{e}^{-t/x}\,\mathrm{d}t\right)x^{-1-s}\,\mathrm{d}x\\&=\int_1^\infty Z(t)k(t)t^{-s}\left(\int_0^\infty \mathrm{e}^{-u}u^{s-1}\,\mathrm{d}u\right)\mathrm{d}t=\Gamma(s)\int_1^\infty Z(t)k(t)t^{-s}\,\mathrm{d}t.\end{aligned}$$

Integration by parts and (7.24) show that the $\int_1^\infty Z(t)k(t)t^{-s}\,\mathrm{d}t$ represents a regular function even for $\sigma>-3/4$, implying that $I_3(s)$, and consequently $\mathcal{M}_1(s)$, admits analytic continuation to the region $\sigma>0$, where it is regular.

We continue with the analytic continuation of $\mathcal{M}_2(s)$. This is contained in the following theorem.

Theorem 9.7 *The function $\mathcal{M}_2(s)$ continues meromorphically to $\mathbb{C}$, having a double pole at $s=1$, and at most simple poles at $s=-1,-3,\ldots$. The principal part of its Laurent expansion at $s=1$ is given by*

$$\frac{1}{(s-1)^2}+\frac{2\gamma-\log(2\pi)}{s-1},\tag{9.36}$$

where $\gamma=-\Gamma'(1)=0.577\,215\ldots$ is Euler's constant.

Proof of Theorem 9.7 The method of proof is similar to the proof of Theorem 9.6. Let

$$\bar L_1(s):=\int_1^\infty|\zeta(\tfrac12+iy)|^2\mathrm{e}^{-ys}\,\mathrm{d}y=\int_1^\infty Z^2(y)\mathrm{e}^{-ys}\,\mathrm{d}y\quad(\mathbb{R}\mathrm{e}\,s>0).\tag{9.37}$$

Then we have by absolute convergence, taking $\sigma=\mathbb{R}\mathrm{e}\,s$ sufficiently large and making the change of variable $xy=t$,

$$\begin{aligned}\int_0^\infty \bar L_1(x)x^{s-1}\,\mathrm{d}x&=\int_0^\infty\left(\int_1^\infty|\zeta(\tfrac12+iy)|^2\mathrm{e}^{-yx}\,\mathrm{d}y\right)x^{s-1}\,\mathrm{d}x\\&=\int_1^\infty|\zeta(\tfrac12+iy)|^2\left(\int_0^\infty x^{s-1}\mathrm{e}^{-xy}\,\mathrm{d}x\right)\mathrm{d}y\\&=\int_1^\infty|\zeta(\tfrac12+iy)|^2y^{-s}\,\mathrm{d}y\int_0^\infty \mathrm{e}^{-t}t^{s-1}\,\mathrm{d}t=\mathcal{M}_2(s)\Gamma(s).\end{aligned}\tag{9.38}$$

Further we have

$$\int_0^\infty \bar{L}_1(x)x^{s-1}\,\mathrm{d}x = \int_0^1 \bar{L}_1(x)x^{s-1}\,\mathrm{d}x + \int_1^\infty \bar{L}_1(x)x^{s-1}\,\mathrm{d}x$$
$$= \int_1^\infty \bar{L}_1(1/x)x^{-1-s}\,\mathrm{d}x + A_1(s) \quad (\sigma > 1),$$

say, where $A_1(s)$ is an entire function. We write

$$\bar{L}_1(1/x) = L_1(1/x) - \int_0^1 |\zeta(\tfrac{1}{2}+iy)|^2 \mathrm{e}^{-y/x}\,\mathrm{d}y \qquad (x \geq 1),$$

where

$$L_1(s) := \int_0^\infty |\zeta(\tfrac{1}{2}+iy)|^2 \mathrm{e}^{-sy}\,\mathrm{d}y \qquad (\mathbb{R}\mathrm{e}\, s > 0)$$

is the Laplace transform of $|\zeta(\frac{1}{2}+iy)|^2$. It follows from (9.38) by analytic continuation that, for $\sigma > 1$,

$$\mathcal{M}_2(s)\Gamma(s) = \int_1^\infty L_1(1/x)x^{-1-s}\,\mathrm{d}x$$
$$- \int_1^\infty \Big(\int_0^1 |\zeta(\tfrac{1}{2}+iy)|^2 \mathrm{e}^{-y/x}\,\mathrm{d}y\Big)x^{-1-s}\,\mathrm{d}x + A_1(s)$$
$$= I_1(s) - I_2(s) + A_1(s), \tag{9.39}$$

say. Clearly, for any integer $M \geq 1$, we have

$$I_2(s) = \int_1^\infty \left\{ \int_0^1 |\zeta(\tfrac{1}{2}+iy)|^2 \left(\sum_{m=0}^M \frac{(-1)^m}{m!} \left(\frac{y}{x}\right)^m + O_M(x^{-M-1}) \right) \mathrm{d}y \right\} x^{-1-s}\,\mathrm{d}x$$
$$= \sum_{m=0}^M \frac{(-1)^m}{m!} h_m \cdot \frac{1}{m+s} + H_M(s), \tag{9.40}$$

say, where $H_M(s)$ is a regular function of s for $\sigma > -M-1$, and h_m is a constant. At this point we invoke the formula

$$L_1(2\sigma) = \frac{\gamma - \log(4\pi\sigma)}{2\sin\sigma} + \sum_{n=0}^N c_n \sigma^n + O(\sigma^{N+1}) \quad (\sigma \to 0+) \tag{9.41}$$

for any given integer $N \geq 1$, where the c_ns are effectively computable constants and γ is Euler's constant.

Note that, for $\sigma = 1/(2T)$ $(T \to \infty)$ and any integer $N \ge 0$, (9.41) gives

$$L_1\left(\frac{1}{T}\right) = \left(\log\left(\frac{T}{2\pi}\right) + \gamma\right)\sum_{n=0}^{N} a_n T^{1-2n} + \sum_{n=0}^{N} b_n T^{-2n} + O_N(T^{-1-2N}\log T)$$

with suitable a_n, b_n $(a_0 = 1)$. Inserting this formula in $I_1(s)$ in (9.39) we have

$$\begin{aligned} I_1(s) &= \int_1^\infty (\log\frac{x}{2\pi} + \gamma)\sum_{n=0}^{N} a_n x^{-2n-s}\,\mathrm{d}x + \int_1^\infty \sum_{n=0}^{N} b_n x^{-1-2n-s}\,\mathrm{d}x + K_N(s) \\ &= \sum_{n=0}^{N} a_n\left(\frac{1}{(2n+s-1)^2} + \frac{\gamma - \log 2\pi}{2n+s-1}\right) + K_N(s) \quad (\sigma > 1), \end{aligned} \tag{9.42}$$

say, where $K_N(s)$ is regular for $\sigma > -2N$. Taking $M = 2N$ it follows from (9.40)–(9.42) that

$$\begin{aligned} \mathcal{M}_2(s)\Gamma(s) &= \sum_{n=0}^{N} a_n\left(\frac{1}{(2n+s-1)^2} + \frac{\gamma - \log 2\pi}{2n+s-1}\right) \\ &\quad + \sum_{m=0}^{2N} \frac{(-1)^m}{m!} h_m \cdot \frac{1}{m+s} + R_N(s), \end{aligned} \tag{9.43}$$

say, where $R_N(s)$ is a regular function of s for $\sigma > -2N$. This holds initially for $\sigma > 1$, but by analytic continuation it holds for $\sigma > -2N$. Since N is arbitrary and $\Gamma(s)$ has no zeros, it follows that (9.43) provides meromorphic continuation of $\mathcal{M}_2(s)$ to $\mathbb{C}$. Taking into account that $\Gamma(s)$ has simple poles at $s = -m$ $(m = 0, 1, 2, \dots)$ we obtain then the analytic continuation of $\mathcal{M}_2(s)$ to $\mathbb{C}$, showing that besides $s = 1$ the only poles of $\mathcal{M}_2(s)$ can be simple poles at $s = 1 - 2n$ for $n \in \mathbb{N}$, as asserted by Theorem 9.7. With more care the residues at these poles could be explicitly evaluated. Finally using (9.43) and the series expansion

$$\frac{1}{\Gamma(s)} = 1 + \gamma(s-1) + \sum_{n=2}^{\infty} d_n (s-1)^n,$$

we obtain that the principal part of the Laurent expansion of $\mathcal{M}_2(s)$ at $s = 1$ is given by (9.36).

We pass to the case of $\mathcal{M}_3(s)$. This is presented in the following theorem.

Theorem 9.8 *We have*

$$\mathcal{M}_3(s) = \int_1^\infty Z^3(x)x^{-s}\mathrm{d}x = V_1(s) + V_2(s), \tag{9.44}$$

say, where $V_2(s)$ is regular for $\sigma > 3/4$ and for $\sigma > 1$ the function

$$V_1(s) = (2\pi)^{1-s}\sqrt{\frac{2}{3}}\sum_{n=1}^\infty d_3(n)n^{-\frac{1}{6}-\frac{2s}{3}}\cos\left(3\pi n^{\frac{2}{3}} + \tfrac{1}{8}\pi\right) \tag{9.45}$$

is regular, where $d_3(n) = \sum_{k\ell m=n} 1$.

Proof of Theorem 9.8 From (7.5) of Theorem 7.1 (with $k = 3$) we have

$$\begin{aligned} F_3(x) &= \int_0^x Z^3(y)\,\mathrm{d}y \\ &= 2\pi\sqrt{\frac{2}{3}}\sum_{n\le(\frac{x}{2\pi})^{3/2}} d_3(n)n^{-\frac{1}{6}}\cos\left(3\pi n^{\frac{2}{3}} + \tfrac{1}{8}\pi\right) + O_\varepsilon(x^{3/4+\varepsilon}), \end{aligned} \tag{9.46}$$

since the terms standing for $+\cdots+$ can be (see Remark 7.3) omitted. Inserting (9.46) in (7.3) we see that

$$\mathcal{M}_3(s) = V_1(s) + V_2(s),$$

where $V_2(s)$ (coming from the error term) is obviously regular for $\sigma > 3/4$ and satisfies $V_2(s) = O(|s| + 1)$. Therefore the main problem is the analytic continuation of

$$V_1(s) := 2\pi\sqrt{\frac{2}{3}}s\int_1^\infty x^{-s-1}\sum_{n\le(\frac{x}{2\pi})^{3/2}} d_3(n)n^{-\frac{1}{6}}\cos\left(3\pi n^{\frac{2}{3}} + \tfrac{1}{8}\pi\right)\mathrm{d}x. \tag{9.47}$$

If in (9.47) we invert the order of summation and integration we obtain

$$\begin{aligned} V_1(s) &= -2\pi\sqrt{\frac{2}{3}}\sum_{n=1}^\infty d_3(n)n^{-\frac{1}{6}}\cos\left(3\pi n^{\frac{2}{3}} + \tfrac{1}{8}\pi\right)\int_{2\pi n^{2/3}}^\infty \mathrm{d}(x^{-s}) \\ &= (2\pi)^{1-s}\sqrt{\frac{2}{3}}\sum_{n=1}^\infty d_3(n)n^{-\frac{1}{6}-\frac{2s}{3}}\cos\left(3\pi n^{\frac{2}{3}} + \tfrac{1}{8}\pi\right). \end{aligned} \tag{9.48}$$

The series in (9.48) converges absolutely for $\sigma > 5/4$, since $d_3(n) \ll_\varepsilon n^\varepsilon$. This is trivial, and we seek a better result. By considering the portion of the series in (9.48) over $[X, 2X]$ (for large X and $s = \sigma + it$ fixed) we want to show that it is $\ll_\varepsilon X^{-\varepsilon}$, which provides then the desired analytic continuation to the right of the σ–line. By using the Stieltjes integral representation and then integration by parts, we are led to two integrals, of which the relevant one is

$$J(s;X) := \int_X^{2X} \Delta_3(x)x^{-\frac{1}{2}-\frac{2s}{3}}\cos\big(3\pi x^{\frac{2}{3}} + \tfrac{1}{8}\pi\big)\,\mathrm{d}x. \tag{9.49}$$

On applying the truncated Perron inversion formula (see, e.g., the appendix of [Iv1]) we have

$$\Delta_3(x) = \frac{1}{2\pi i}\int_{\frac{1}{2}-iX}^{\frac{1}{2}+iX} \frac{\zeta^3(w)x^w}{w}\,\mathrm{d}w + O_\varepsilon(X^\varepsilon) \qquad (X \le x \le 2X), \tag{9.50}$$

where as usual

$$\Delta_3(x) := \sum_{n\le x} d_3(n) - xP_2(\log x) \tag{9.51}$$

is the error term in the asymptotic formula for the summatory function of $d_3(n)$. In (9.51) $P_2(y)$ is a suitable quadratic function in y.

The error term in (9.50) contributes to the integral in (9.49)

$$\ll_\varepsilon X^{\frac{1}{2}-\frac{2\sigma}{3}+\varepsilon} \ll_\varepsilon X^{-\varepsilon}$$

for $\sigma > 3/4$. The main term in (9.50) produces

$$\frac{1}{2\pi i}\int_{\frac{1}{2}-iX}^{\frac{1}{2}+iX} \frac{\zeta^3(w)}{w}\left(\int_X^{2X} x^{-2\sigma/3}\exp\big(iF_\pm(x)\big)\,\mathrm{d}x\right)\mathrm{d}w,$$

where

$$w = \tfrac{1}{2} + iv,\ s = \sigma + it,\ F_\pm(x) := \big(v - (2t)/3\big)\log x \pm 3\pi x^{2/3}.$$

Note that the saddle point of the integral over x is (the root of the equation $F'_\pm(x) = 0$)

$$x_0 = \left(\frac{|v-(2t)/3|}{2\pi}\right)^{3/2} \in [X, 2X] \qquad (\text{for } v \asymp X^{2/3}),$$

in which case $|F''_\pm(x_0)|^{-1/2} \asymp X^{2/3}$. Hence by the saddle-point method the total contribution to (3.18) is $\ll_\varepsilon X^{(2/3)(1-\sigma)+\varepsilon}$, and this provides the desired analytic continuation of $\mathcal{M}_3(s)$ only to $\sigma > 1$ as before. One can make the calculation of (9.49) simpler by making the change of variable $x^{2/3} = y$, after $\Delta_3(x)$ is replaced by (9.50). However, at present one does not see any better way to tackle the problem of the analytic continuation of $\mathcal{M}_3(s)$, although it seems certain that it can be done.

We shall now formulate the result concerning the analytic continuation of $\mathcal{M}_4(s) \equiv \mathcal{Z}_2(s)$ (see (7.2)). We recall that, for $k \in \mathbb{N}$, $\sigma \geq \sigma_0(k)$,

$$\mathcal{Z}_k(s) = \mathcal{M}_{2k}(s) = \int_1^\infty Z^{2k}(x)x^{-s}\,\mathrm{d}x = \int_1^\infty |\zeta(\tfrac{1}{2} + ix)|^{2k} x^{-s}\,\mathrm{d}x. \tag{9.52}$$

Theorem 9.9 *The function $\mathcal{Z}_2(s)$ has meromorphic continuation over $\mathbb{C}$. In the half-plane $\mathrm{Re}\, s > 0$ it has the following singularities: the pole $s = 1$ of order five, simple poles at $s = \frac{1}{2} \pm i\kappa_j$ ($\kappa_j = \sqrt{\lambda_j - \frac{1}{4}}$) and poles at $s = \frac{1}{2}\rho$, where ρ denotes complex zeros of $\zeta(s)$. The residue of $\mathcal{Z}_2(s)$ at $s = \frac{1}{2} + i\kappa_h$ equals*

$$R(\kappa_h) := \sqrt{\frac{\pi}{2}}\left(2^{-i\kappa_h}\frac{\Gamma(\frac{1}{4} - \frac{1}{2}i\kappa_h)}{\Gamma(\frac{1}{4} + \frac{1}{2}i\kappa_h)}\right)^3 \Gamma(2i\kappa_h)\cosh(\pi\kappa_h) \sum_{\kappa_j = \kappa_h} \alpha_j H_j^3(\tfrac{1}{2}),$$

and the residue at $s = \frac{1}{2} - i\kappa_h$ equals $\overline{R(\kappa_h)}$.

The proof of Theorem 9.9 is outside the scope of this text. Here, as usual, $H_j(s)$ is the Hecke series

$$H_j(s) = \sum_{n=1}^\infty t_j(n)n^{-s} \qquad (\sigma > 1),$$

associated with the Maass wave form $\psi_j(z)$, where $\rho_j(1)t_j(n) = \rho_j(n)$ and $\rho_j(n)$ is the nth Fourier coefficient of $\psi_j(z)$. The function $H_j(s)$ can be continued to an entire function. It satisfies the functional equation, similar to the functional equation for $\zeta(s)$, namely

$$\begin{aligned} H_j(s) = 2^{2s-1}\pi^{2s-2}\Gamma(1 - s + i\kappa_j)\Gamma(1 - s - i\kappa_j)(\varepsilon_j \cosh(\pi\kappa_j) \\ - \cos(\pi s))H_j(1 - s), \end{aligned}$$

where ε_j ($= \pm 1$) is the so-called parity sign of $\psi_j(z)$ ($z = x + iy$). This means that $\varepsilon_j = 1$ if $\psi_j(z)$ is an even function of x, and $\varepsilon_j = -1$ if $\psi_j(z)$ is an odd function of x. By

$$\left\{\lambda_j = \kappa_j^2 + \frac{1}{4}\right\} \cup \{0\}$$

we denote the eigenvalues (discrete spectrum) of the hyperbolic Laplacian

$$\Delta = -y^2\left(\left(\frac{\partial}{\partial x}\right)^2 + \left(\frac{\partial}{\partial y}\right)^2\right)$$

acting over the Hilbert space composed of all Γ-automorphic functions which are square integrable with respect to the hyperbolic measure ($\Gamma = \mathrm{PSL}(2, \mathbb{Z})$).

Further, $\alpha_j = |\rho_j(1)|^2(\cosh \pi\kappa_j)^{-1}$, where $\rho_j(1)$ is the first Fourier coefficient of the Maass wave form corresponding to the eigenvalue λ_j to which the Hecke L-function $H_j(s)$ is attached.

The principal part of $\mathcal{Z}_2(s)$ has the form (this may be compared with (9.36))

$$\sum_{j=1}^{5} \frac{A_j}{(s-1)^j},$$

where $A_5 = 12/\pi^2$, and the remaining A_js can be evaluated explicitly by following the analysis in Y. Motohashi's paper [Mot4].

Remark 9.10 It is curious that obviously the shapes of $\mathcal{M}_k(s)$ for $k = 1, 2, 3, 4$ (the cases when we know something relevant) are totally different! The fact that $Z(x)$ is an oscillating function, while $|\zeta(\frac{1}{2} + ix)|$ is non-negative is reflected in what we expect: $\mathcal{M}_{2\ell}(s) = \mathcal{Z}_\ell(s)$ should have a pole of order $\ell^2 + 1$ at $s = 1$, while $\mathcal{M}_{2\ell-1}(s)$ should be regular at $s = 1$, at least for $1 \le \ell \le 4$.

Notes

Mellin transforms (named after the Finnish mathematician Robert Hjalmar Mellin, June 19, 1854–April 5, 1933) of $\mathcal{M}_k(s)$ and $\mathcal{Z}_k(s)$ are a topic of ongoing research. The reader can consult the present author's papers [Iv9], [Iv11], [Iv12] [Iv13], [Iv15], [Iv17], [Iv18], [Iv22], the joint paper [IJM] of Jutila, Motohashi and the present author, M. Jutila's papers [Jut3], [Jut4], [Jut6], M. Lukkarinen's dissertation [Luk], and Y. Motohashi's works [Mot4], [Mot5].

For a comprehensive work on Fourier transforms, which contains all the material needed in this text, the reader is referred to E. C. Titchmarsh's classical monograph [Tit2]. For example, for (9.8) and (9.16) see [Tit2, theorem 73].

In mathematical analysis, Parseval's formula (or Parseval's identity) is a fundamental result on the summability of the Fourier series of a function. Marc-Antoine Parseval des Chênes (April 27, 1755–August 16, 1836) was a French mathematician. Geometrically, Parseval's formula is the Pythagorean theorem for inner-product spaces. Informally, the identity asserts that the sum of the squares of the Fourier coefficients of a function is equal to the integral of the square of the function, namely

$$\sum_{n=-\infty}^{\infty} |c_n|^2 = \frac{1}{2\pi}\int_{-\pi}^{\pi} |f(x)|^2 \,\mathrm{d}x,$$

where the Fourier coefficients c_n of f are given by

$$c_n = \frac{1}{2\pi}\int_{-\pi}^{\pi} f(x)\mathrm{e}^{-inx} \,\mathrm{d}x.$$

More formally, the result holds as stated provided $f(x)$ is square-integrable or, more generally, in $L^2[-\pi, \pi]$. A similar result is the *Plancherel theorem* (Michel Plancherel, January 16, 1885–March 4, 1967, a Swiss mathematician), which asserts that the integral of the square of the Fourier transform of a function is equal to the integral of the square of the function itself.

In [Jut8] M. Jutila proved that $\mathcal{M}_1(s)$ is actually entire, which improves on Theorem 9.6. He modified the method used in [Jut5], defining

$$f(s,w) = \zeta(w)\chi(w)^{-1/2}\left(-\left(w-\tfrac{1}{2}\right)^2\right)^{-s/2}, \tag{9.53}$$

and noting that, for $\sigma > 1$,

$$\mathcal{M}_1(s) = -i\int_{1/2+i}^{1/2+i\infty} f(s,w)\,\mathrm{d}w. \tag{9.54}$$

The line of integration in (9.54) is moved to the line $\mathbb{R}\mathrm{e}\, w = \frac{5}{4}$, and to ensure the convergence of the new integral one initially supposes that $\sigma > \frac{11}{8}$. It turns out that

$$\mathcal{M}_1(s) \sim \sum_{n=1}^{\infty} n^{-5/4} I_n(s),$$

where

$$I_n(s) := \int_1^{\infty} n^{-ix}\left(\chi(\tfrac{5}{4}+ix)\right)^{-1/2}\alpha(x,s)x^{-s}\,\mathrm{d}x$$

with

$$\alpha(x,s) := \left(1 - \frac{3}{2}ix^{-1} - \frac{9}{16}x^{-2}\right)^{-s/2}.$$

The assertion follows by careful estimation, after certain simplifications, of $I_n(s)$ by the saddle point method.

Theorem 9.7 was proved by the present author [Iv13]. A different proof is to be found in the dissertation of M. Lukkarinen [Luk]. Also in [Jut4] M. Jutila proved that $\mathcal{M}_2(s)$ continues meromorphically to $\mathbb{C}$, having as singularities a double pole at $s = 1$ and at most simple poles for $s = -1, -3, \ldots$. M. Lukkarinen showed that these points are actually simple poles and calculated, in term of the Bernoulli numbers, the residues. The residue at $s = -k$ ($k \in \mathbb{N}$) equals

$$\frac{i^{-k-1}(1-2^{-k})}{k+1}B_{k+1}.$$

E. C. Titchmarsh's well-known monograph [Tit3], chapter 7 gives a discussion of

$$L_k(s) \;:=\; \int_0^{\infty} |\zeta(\tfrac{1}{2}+ix)|^{2k}\mathrm{e}^{-sx}\,\mathrm{d}x \qquad (k \in \mathbb{N},\ \sigma = \mathbb{R}\mathrm{e}\, s > 0)$$

when $\sigma \to 0+$, especially detailed in the cases $k = 1$ and $k = 2$. The formula (9.41) is a classical result of H. Kober [Kob].

For complex values of s the function $L_1(s)$ was studied by F. V. Atkinson [Atk1], and more recently by M. Jutila [Jut3]. Atkinson [Atk2] obtained the asymptotic formula, as $\sigma \to 0+$,

$$L_2(\sigma) = \frac{1}{\sigma}\left(A\log^4\frac{1}{\sigma} + B\log^3\frac{1}{\sigma} + C\log^2\frac{1}{\sigma} + D\log\frac{1}{\sigma} + E\right) + \lambda_2(\sigma),$$

where

$$A = \frac{1}{2\pi^2},\quad B = \frac{1}{\pi^2}\left(2\log(2\pi) - 6\gamma + \frac{24\zeta'(2)}{\pi^2}\right),$$

and

$$\lambda_2(\sigma) \;\ll_{\varepsilon}\; \left(\frac{1}{\sigma}\right)^{\frac{13}{14}+\varepsilon}.$$

He also indicated how the exponent 13/14 can be replaced by 8/9. This is of historical interest, since it is one of the first applications of Kloosterman sums to zeta-function theory. The Kloosterman sums are

$$S(m,n;c) := \sum_{1\le d\le c} e\left(\frac{md+nd'}{c}\right) \qquad (e(z)=\exp(2\pi i z)),$$

where $dd' \equiv 1 \bmod c$, and they have played an important rôle in analytic number theory in the last thirty years.

Atkinson in fact showed that ($\sigma = \mathbb{R}\mathrm{e}\, s > 0$)

$$L_2(s) \;=\; 4\pi e^{-\frac{1}{2}s}\sum_{n=1}^{\infty} d_4(n)K_0(4\pi i\sqrt{n}e^{-\frac{1}{2}s}) + \phi(s), \tag{9.55}$$

where $d_4(n)$ is the divisor function generated by $\zeta^4(s)$, K_0 is the Bessel function, and the series in (9.55) as well as $\phi(s)$ are both analytic in the region $|s| < \pi$. When $s = \sigma \to 0+$ one can use the asymptotic formula

$$K_0(z) \;=\; \tfrac{1}{2}\sqrt{\pi}z^{-1/2}e^{-z}\left(1 - 8z^{-1} + O(|z|^{-2})\right) \quad (|\arg z| < \theta < \tfrac{3\pi}{2},\ |z| \ge 1)$$

to make a simplification of (9.55).

The present author [Iv5] gave explicit, albeit complicated expressions for the remaining coefficients C, D and E in (1.5). More importantly, he applied a result on the fourth moment of $|\zeta(\frac{1}{2}+it)|$, obtained jointly with Y. Motohashi [IvMo2], [IvMo3], to establish that

$$\lambda_2(\sigma) \;\ll\; \sigma^{-1/2} \qquad (\sigma \to 0+),$$

which is actually best possible. It was proved there that

$$\int_0^T E_2(t)\,\mathrm{d}t \;\ll\; T^{3/2}, \qquad \int_0^T E_2^2(t)\,\mathrm{d}t \;\ll\; T^2\log^C T, \tag{9.56}$$

where $E_2(T)$ denotes the error term in the fourth moment formula for $|\zeta(\frac{1}{2}+it)|$ (see (4.55) and (10.13)). In [Iv5] it was also proved that

$$\lambda_k\left(\frac{1}{T}\right) \;=\; O(T^{c_k-1}),$$

where in general one defines, for a fixed $k \in \mathbb{N}$,

$$\lambda_k\left(\frac{1}{T}\right) = \int_0^\infty |\zeta(\tfrac{1}{2}+it)|^{2k}\mathrm{e}^{-t/T}\,\mathrm{d}t - TQ_{k^2}(\log T)$$

for a suitable polynomial $Q_{k^2}(x)$ in x of degree k, provided that

$$\int_0^T E_k(t)\,\mathrm{d}t \;=\; O(T^{c_k}) \qquad (c_k > 0).$$

Here $E_k(T)$ is defined by (10.13)-(10.14). Since, by the first bound in (9.55), we have $c_2 = 3/2$ it follows that

$$\lambda_2\left(\frac{1}{T}\right) \;=\; O(T^{1/2}),$$

which is equivalent to $\lambda_2(\sigma) \ll \sigma^{-1/2}$ ($\sigma \to 0+$). Moreover the coefficients of $Q_{k^2}(y)$ can be expressed as linear combinations of the coefficients of $P_{k^2}(y)$.

Theorem 9.8 is proved by the author in [Iv15].

For an account of *the Dirichlet divisor problem* the reader is referred to chapter 13 of [Iv1] and to chapter 12 of [Tit3]. For the classical case (this corresponds to $k = 2$, and one usually denotes $\Delta_2(x)$ by $\Delta(x)$) see (4.86). In the general case this concerns the estimation of

$$\Delta_k(x) := \sum_{n\le x} d_k(n) - xP_{k-1}(\log x),$$

where $P_{k-1}(y)$ is a polynomial of degree $k-1$ in y, whose coefficients depend on k (≥ 2). One has

$$P_{k-1}(\log x) = \operatorname*{Res}_{s=1} x^{s-1}\zeta^k(s)s^{-1}. \tag{9.57}$$

Recall (see Chapter 1) that

$$\zeta(s) = \frac{1}{s-1} + \gamma + \sum_{k=1}^{\infty} \gamma_k(s-1)^k$$

is the Laurent expansion of $\zeta(s)$ at $s = 1$. In particular, using (9.57) we obtain

$$\begin{aligned} P_1(y) &= y + (2\gamma - 1),\\ P_2(y) &= \tfrac{1}{2}y^2 + (3\gamma - 1)y + (3\gamma^2 - 3\gamma + 3\gamma_1 + 1),\\ P_3(y) &= \tfrac{1}{6}y^3 + (2\gamma - 1/2)y^2 + (6\gamma^2 - 4\gamma + 4\gamma_1 + 1)y\\ &\quad + \left\{-1 + 4(\gamma - \gamma_1 + \gamma_2) - 6\gamma^2 + 4\gamma^3 + 12\gamma\gamma_1\right\}. \end{aligned}$$

General formulas for the coefficients of $P_{k-1}(y)$ may be found in the work of A. F. Lavrik [Lavr], and some integrals involving $\Delta_k(x)$ are evaluated by Lavrik *et al.* [LaIE].

The function $\mathcal{Z}_2(s)$ has quite a different analytic behavior from the function $\mathcal{Z}_1(s)$. It was introduced and studied by Y. Motohashi [Mot4], who proved Theorem 9.9. See also his monograph [Mot5], which contains an extensive account of spectral theory and its applications to $\zeta(s)$, and the monographs of H. Iwaniec [Iwa1], [Iwa2] for an account on spectral theory and automorphic forms. In [Mot3] and [Mot5], Motohashi obtained a spectral expansion of the important (weighted) integral

$$\int_{-\infty}^{\infty} |\zeta(\tfrac{1}{2} + iT + it)|^4 \mathrm{e}^{-(t/\Delta)^2}\,\mathrm{d}t \qquad (0 < \Delta \le T/\log T),$$

which is fundamental in the study of the (unweighted) moment $\int_0^T |\zeta(\frac{1}{2} + it)|^4\,\mathrm{d}t$ (see, beside Motohashi's fundamental work [Mot4], e.g., [Iv4], [Iv5], [Iv10], [IvMo1], [IvMo2], [IvMo3], [Mot4] and [Mot5]).

10

Further results on $\mathcal{M}_k(s)$ and $\mathcal{Z}_k(s)$

10.1 Some relations for $\mathcal{M}_k(s)$

We continue the investigations concerning the modified Mellin transforms

$$\mathcal{M}_k(s) := \int_1^\infty Z^k(x)x^{-s}\,\mathrm{d}x \qquad (k \in \mathbb{N}, \sigma \geq \sigma_0(k)), \tag{10.1}$$

and

$$\mathcal{Z}_k(s) := \mathcal{M}_{2k}(s) = \int_1^\infty |\zeta(\tfrac{1}{2} + ix)|^{2k} x^{-s}\,\mathrm{d}x \qquad (k \in \mathbb{N}, \sigma \geq \sigma_1(k)). \tag{10.2}$$

In this section we shall deal with some relations involving the function $\mathcal{M}_k(s)$ in the general case. We start with the following theorem.

Theorem 10.1 *For* $c \geq c_k > 0, k \geq 2$ *and* $\sigma = \mathbb{R}\mathrm{e}\, s \geq \sigma_1(k)\ (> 1)$ *we have*

$$\mathcal{M}_k(s) = \frac{1}{2\pi i}\int_{(c)} \mathcal{M}_{k-r}(w)\mathcal{M}_r(1 - w + s)\,dw \quad (r = 1, \ldots, k-1). \tag{10.3}$$

In particular, for $\sigma > 1 + \varepsilon$, *we have*

$$\mathcal{M}_3(s) = \frac{1}{2\pi i}\int_{(1+\varepsilon)} \mathcal{M}_1(w)\mathcal{M}_2(1 - w + s)\,dw. \tag{10.4}$$

Remark 10.2 Theorem 10.1 is a sort of recurrent relation between the functions $\mathcal{M}_k(s)$ with different indices. The relation (10.4) offers the possibility to obtain information on $\mathcal{M}_3(s)$ from the information on the (much better understood) functions $\mathcal{M}_1(s)$ and $\mathcal{M}_2(s)$.

Proof of Theorem 10.1 Consider

$$f(x) = Z^{k-r}\Big(\frac{1}{x}\Big)\frac{1}{x}, \quad g(x) = Z^{r}\Big(\frac{1}{x}\Big)\frac{1}{x} \qquad (0 < x \le 1),$$

and $f(x) = g(x) = 0$ if $x > 1$. With the change of variable $y = 1/x$ we have

$$\begin{aligned} F(s) &= \int_0^\infty f(x)x^{s-1}\,\mathrm{d}x = \int_0^1 Z^{k-r}\Big(\frac{1}{x}\Big)x^{s-2}\,\mathrm{d}x \\ &= \int_1^\infty Z^{k-r}(y)y^{-s}\,\mathrm{d}y = \mathcal{M}_{k-r}(s), \end{aligned}$$

and likewise $G(s) = \mathcal{M}_r(s)$. Hence (9.9) yields, with sufficiently large $c \ge c_k > 0$,

$$\begin{aligned} \frac{1}{2\pi i}\int_{(c)} \mathcal{M}_{k-r}(w)\mathcal{M}_r(s-w)\,\mathrm{d}w &= \int_0^1 Z^{k-r}\Big(\frac{1}{x}\Big)Z^{r}\Big(\frac{1}{x}\Big)x^{s-3}\,\mathrm{d}x \\ &= \mathcal{M}_k(s-1), \end{aligned}$$

again with the change of variable $y = 1/x$. Finally changing $s-1$ to s we obtain (10.3).

To establish (10.4), take $\sigma > c > 1$ and let I denote the integral on the right-hand side. Then by the Cauchy-Schwarz inequality and (9.18) of Lemma 9.4 we obtain

$$\begin{aligned} I^2 &\le \int_{-\infty}^{\infty} |\mathcal{M}_1(c+iv)|^2\,\mathrm{d}v \int_{-\infty}^{\infty} |\mathcal{M}_2(1-c+\sigma+i(v+t))|^2\,\mathrm{d}v \\ &\ll \int_1^{\infty} |\zeta(\tfrac{1}{2}+ix)|^2 x^{1-2c}\,\mathrm{d}x \int_1^{\infty} |\zeta(\tfrac{1}{2}+ix)|^4 x^{2c-2\sigma-1}\,\mathrm{d}x \\ &\ll 1, \end{aligned}$$

since $1-2c < -1$, $2c-2\sigma-1 < -1$. Here we used the well-known bounds

$$\int_0^T |\zeta(\tfrac{1}{2}+it)|^{2k}\,\mathrm{d}t \ll T(\log T)^{k^2} \qquad (k = 1, 2). \tag{10.5}$$

Therefore I converges absolutely and (10.4) holds, providing incidentally the analytic continuation of $\mathcal{M}_3(s)$ to $\sigma > 1$ (this also follows directly from the defining relation (10.1) when $k = 3$).

Theorem 10.3 *If $k = 1, 2, 3, 4$ and $c > 1$ is fixed, then for $U \gg x$ and $\varepsilon > 0$ sufficiently small we have*

$$Z^k(x) = \frac{1}{2\pi i}\int_{c-iU}^{c+iU} x^{s-1}\mathcal{M}_k(s)\,ds + O_{\varepsilon,k}(x^{c-1}U^{-\varepsilon/2}). \tag{10.6}$$

Proof of Theorem 10.3 First note that, in view of (10.5), $\mathcal{M}_k(s)$ ($k \le 4$) converges absolutely for $\sigma > 1$. From (10.1) we obtain by the Mellin inversion formula (Lemma 9.1)

$$Z^k(x) = \frac{1}{2\pi i}\int_{(c)} \mathcal{M}_k(s)x^{s-1}\,\mathrm{d}s \tag{10.7}$$

for suitable c (> 0). However, (10.7) is not always easy to use in practice, and (10.6) is a truncated form, which is also flexible since the parameter U ($\gg x$) can be freely chosen.

Applying the residue theorem to (10.7) we obtain, for $U \gg x$ and a suitable c satisfying $0 < c < 1$,

$$\begin{aligned} Z^k(x) &= \frac{1}{2\pi}\int_{(c)} x^{s-1}\mathcal{M}_k(s)\,\mathrm{d}s \\ &= \frac{1}{2\pi}\left(\int_{c-iU}^{c+iU} + \int_{c-i\infty}^{c-iU} + \int_{c+iU}^{c+i\infty}\right) + O_{\varepsilon,k}(x^{\varepsilon}) \\ &= \frac{1}{2\pi}(I_1 + I_2 + I_3) + O_{\varepsilon,k}(x^{\varepsilon}), \end{aligned}$$

say. The O-term comes from the residue at $s = 1$ (for $k = 1$ the function $\mathcal{M}_1(s)$ is regular for $s = 1$, while for $k = 3$ very likely $\mathcal{M}_3(s)$ is also regular at $s = 1$, but this has not been proved yet). Therefore to prove (10.6) it suffices to show that

$$I_3 \ll_{\varepsilon,k} x^{c-1}U^{-\varepsilon/2}, \tag{10.8}$$

since the estimation of I_2 is analogous to the estimation of I_3. For $\sigma > 1$, $T_1 \le t \le 2T_1$ (with the aim of taking later $T_1 = U$, $T_1 = 2U$, etc.) we have

$$\mathcal{M}_k(s) = \int_1^{T_1^{1-\varepsilon}} Z^k(u)\varphi(u)u^{-s}\,\mathrm{d}u + \int_{\frac{1}{2}T_1^{1-\varepsilon}}^{\infty} Z^k(u)(1-\varphi(u))u^{-s}\,\mathrm{d}u = I_4 + I_5,$$

say. Here $\varphi(u)$ (≥ 0) is a smooth function supported in $[1, T_1^{1-\varepsilon}]$ such that $\varphi(u) = 1$ for $1 \le u \le \frac{1}{2}T_1^{1-\varepsilon}$ and

$$\varphi^{(r)}(u) \ll_r T_1^{-r(1-\varepsilon)} \qquad (r = 0, 1, 2, \dots). \tag{10.9}$$

Repeated integration by parts shows that, for $N \geq N_0(\varepsilon, k)$,

$$I_4 = \frac{c_{1,k}}{s-1} + \frac{1}{s-1}\int_1^{T_1^{1-\varepsilon}} u^{1-s}\left(\varphi(u)Z^k(u)\right)' \,du = \cdots$$
$$= \frac{c_{1,k}}{s-1} + \cdots + \frac{c_{N,k}}{(s-1)^N} + O_{N,k}(T_1^{-\frac{1}{2}\varepsilon N}) \tag{10.10}$$

since, for $\ell_j, m_j \geq 0$ and $m_1 + \cdots + m_N = k$ (for a formula for $Z^{(m)}(x)$ see (5.7) of Theorem 5.2),

$$\int_1^X \left(Z^{m_1}(x)\right)^{(\ell_1)} \ldots \left(Z^{m_N}(x)\right)^{(\ell_N)} \,dx \ll_{\varepsilon,k,N} X^{1+\varepsilon}. \tag{10.11}$$

One obtains (10.11) similarly as (10.5), using Hölder's inequality, the defining relation for $Z(t)$ and the asymptotic formula (2.17) for the χ-function. The reason that we do not have (yet) Theorem 10.3 for $k > 4$ is essentially the fact that we do not have yet the bound

$$\int_0^T |\zeta(\tfrac{1}{2} + it)|^m \,dt \ll_\varepsilon T^{1+\varepsilon}$$

for any fixed $m > 4$.

Continuing with the proof, note that by the first derivative test

$$\int_{c+iT_1}^{c+i2T_1} x^{s-1} I_4 \,ds = \frac{c_{1,k}}{2\pi i}\int_{c+iT_1}^{c+i2T_1} \frac{x^{s-1}}{s-1}\,ds + O\left(\frac{x^{c-1}}{T_1}\right) = O\left(\frac{x^{c-1}}{T_1}\right).$$

On the other hand

$$\int_{c+iT_1}^{c+i2T_1} x^{s-1} I_5 \,ds = i\int_{T_1}^{2T_1} x^{c+it-1}\left(\int_{\frac{1}{2}T_1^{1-\varepsilon}}^{\infty} Z^k(u)(1-\varphi(u))u^{-c-it}\,du\right)dt$$
$$= ix^{c-1}\int_{\frac{1}{2}T^{1-\varepsilon}}^{\infty} Z^k(u)(1-\varphi(u))u^{-c}\left(\int_{T_1}^{2T_1} e^{it\log(x/u)}\,dt\right)du.$$

For $T_1 \gg x$ it follows, by direct integration, that the last integral over t is bounded. Thus the last expression, for some constant $c_k \geq 0$, is

$$\ll x^{c-1}(\log T_1)^{c_k} T_1^{(1-\varepsilon)(1-c)}.$$

Therefore we have

$$I_3 \ll \frac{x^{c-1}}{U} + x^{c-1}(\log U)^{c_k} U^{(1-\varepsilon)(1-c)} \ll x^{c-1}U^{-\varepsilon/2}$$

if $\varepsilon > 0$ is sufficiently small, and (10.6) follows. Theorem 10.3 is proved.

The next result expresses $\mathcal{M}_k^2(s)$ as a sort of convolution of $Z^k(t)$ with itself. This is

Theorem 10.4 *In the region of absolute convergence we have*

$$\mathcal{M}_k^2(s) = 2\int_1^\infty x^{-s}\left(\int_{\sqrt{x}}^x Z^k(u)Z^k\left(\frac{x}{u}\right)\frac{\mathrm{d}u}{u}\right)\mathrm{d}x. \tag{10.12}$$

Proof of Theorem 10.4 Set $f(x) = Z^k(x)$ and make the change of variables $xy = X$, $x/y = Y$, so that the absolute value of the Jacobian of the transformation is equal to $1/(2Y)$. Therefore

$$\begin{aligned}\mathcal{M}_k^2(s) &= \int_1^\infty\int_1^\infty (xy)^{-s} f(x)f(y)\,\mathrm{d}x\,\mathrm{d}y\\ &= \frac{1}{2}\int_1^\infty X^{-s}\int_{1/X}^X \frac{1}{Y} f(\sqrt{XY})f(\sqrt{X/Y})\,\mathrm{d}Y\,\mathrm{d}X.\end{aligned}$$

But as we have $(y = 1/u)$

$$\int_{1/x}^x f(\sqrt{xy})f(\sqrt{x/y})\frac{\mathrm{d}y}{y} = \int_{1/x}^1 + \int_1^x = 2\int_1^x f(\sqrt{x/u})f(\sqrt{xu})\frac{\mathrm{d}u}{u},$$

we obtain that, in the region of absolute convergence, the identity

$$\mathcal{M}_k^2(s) = \int_1^\infty x^{-s}\left(\int_1^x f(\sqrt{xy})f(\sqrt{x/y})\frac{\mathrm{d}y}{y}\right)\mathrm{d}x$$

is valid. The inner integral here becomes, after the change of variable $\sqrt{xy} = u$,

$$2\int_{\sqrt{x}}^x f(u)f\left(\frac{x}{u}\right)\frac{\mathrm{d}u}{u},$$

and (10.12) follows. The argument also shows that, for $0 < a < b$ and any integrable function f on $[a, b]$,

$$\left(\int_a^b f(x)x^{-s}\,\mathrm{d}x\right)^2 = 2\int_{a^2}^{b^2} x^{-s}\left\{\int_{\sqrt{x}}^{\min(x/a,b)} f(u)f\left(\frac{x}{u}\right)\frac{\mathrm{d}u}{u}\right\}\mathrm{d}x.$$

10.2 Mean square identities for $\mathcal{M}_k(s)$

In this section we shall deal with some mean square identities for $\mathcal{M}_k(s)$ that shed light on the behavior of these functions. To formulate our results, we need the definition of the function $E_k(T)$, the error term in the asymptotic formula for the $2k$th moment of $|\zeta(\frac{1}{2} + it)|$. Namely, for any fixed $k \in \mathbb{N}$, we expect (since the lower bound of integration in (10.1) is unity, it is convenient to have it also in the integral in (10.13))

$$\int_1^T |\zeta(\tfrac{1}{2}+it)|^{2k}\,\mathrm{d}t = T P_{k^2}(\log T) + E_k(T) \tag{10.13}$$

to hold, where it is generally assumed that

$$P_{k^2}(y) = \sum_{j=0}^{k^2} a_{j,k} y^j \tag{10.14}$$

is a polynomial in y of degree k^2. The function $E_k(T)$ is to be considered as the error term in (10.13), namely one supposes that

$$E_k(T) \;=\; o(T) \qquad (T \to \infty). \tag{10.15}$$

So far the formulas (10.13)-(10.15) are known to hold only for $k = 1$ and $k = 2$ (see [Iv1] and [Iv4] for a detailed account). For higher moments one has the bound

$$\int_1^T |\zeta(\tfrac{1}{2}+it)|^{2k}\,\mathrm{d}t \;\ll\; T^{\frac{1}{4}(k+2)} \log^{C_k} T \qquad (2 \le k \le 6), \tag{10.16}$$

and more complicated bounds for $k > 6$. Therefore in view of the existing knowledge on the higher moments of $|\zeta(\frac{1}{2}+it)|$, embodied essentially in (10.16), at present the really important cases of (10.13) are $k = 1$ and $k = 2$.

For $k = 1$ the relation (10.13) becomes (see (8.5)) the well-known mean square formula

$$\int_1^T |\zeta(\tfrac{1}{2}+it)|^{2}\,\mathrm{d}t = T\Big(\log\frac{T}{2\pi} + 2\gamma - 1\Big) + E(T). \tag{10.17}$$

We have (cf. (4.54)) $E(T) \ll_\varepsilon T^{131/416+\varepsilon}$, $E(T) = \Omega_\pm(T^{1/4})$, and

$$\int_1^T E(t)\,\mathrm{d}t = \pi T + G(T), \quad G(T) = O(T^{3/4}), \quad G(T) = \Omega_\pm(T^{3/4}). \tag{10.18}$$

For $k = 2$ we have

$$E_2(T) \;=\; \int_1^T |\zeta(\tfrac{1}{2}+it)|^{4}\,\mathrm{d}t - T P_4(\log T).$$

Recall that $P_4(x)$ is a polynomial of degree four in x with leading coefficient $1/(2\pi^2)$. We have $E_2(T) \ll T^{2/3}\log^9 T$ and $E_2(T) = \Omega_\pm(T^{1/2})$. It is conjectured that $E_1(T) \ll_\varepsilon T^{1/4+\varepsilon}$ and that $E_2(T) \ll_\varepsilon T^{1/2+\varepsilon}$, which is in tune with the known omega-results.

Theorem 10.5 *For $\sigma > 1$ we have (γ is Euler's constant)*

$$\frac{1}{\pi}\int_0^\infty |\mathcal{M}_1(\sigma+it)|^2\,dt = \frac{2\gamma-\log(2\pi)}{2\sigma-2} + \frac{1}{(2\sigma-2)^2} + 2\gamma - 1 - \log(2\pi)$$
$$+2\pi\sigma + 2\sigma(2\sigma-1)\int_1^\infty G(x)x^{-1-2\sigma}\,dx, \tag{10.19}$$

where the integral on the right-hand side of (10.19) *converges absolutely for* $\sigma > 3/8$.

Corollary 10.6 *We have*

$$\lim_{\sigma\to 1+0}\left\{\frac{1}{\pi}\int_0^\infty \left|\mathcal{M}_1(\sigma+it)\right|^2 \mathrm{d}t - \frac{2\gamma-\log(2\pi)}{2\sigma-2} - \frac{1}{(2\sigma-2)^2}\right\}$$
$$= 2\pi + 2\gamma - 1 - \log(2\pi) + 2\int_1^\infty G(x)x^{-3}\mathrm{d}x.$$

Proof of Theorem 10.5 We begin with (9.17) of Lemma 9.3, which in case when

$$f(x) = g(x) = Z^k(x)$$

reduces to

$$\int_1^\infty |\zeta(\tfrac{1}{2}+ix)|^{2k}x^{1-2\sigma}\,\mathrm{d}x = \frac{1}{2\pi}\int_{-\infty}^\infty |\mathcal{M}_k(\sigma+it)|^2\,\mathrm{d}t$$
$$= \frac{1}{\pi}\int_0^\infty |\mathcal{M}_k(\sigma+it)|^2\,\mathrm{d}t, \tag{10.20}$$

since $\overline{\mathcal{M}_k(s)} = \mathcal{M}_k(\bar{s})$. To evaluate the left-hand side of (10.20) note that, using (10.13), we have

$$\int_1^\infty |\zeta(\tfrac{1}{2}+ix)|^{2k}x^{1-2\sigma}\,\mathrm{d}x = \int_1^\infty x^{1-2\sigma}\,\mathrm{d}\Big\{xP_{k^2}(\log x) + E_k(x)\Big\}$$
$$= \int_1^\infty \Big(P_{k^2}(\log x) + P'_{k^2}(\log x)\Big)x^{1-2\sigma}\,\mathrm{d}x - E_k(1)$$
$$+ (2\sigma-1)\int_1^\infty E_k(x)x^{-2\sigma}\,\mathrm{d}x. \tag{10.21}$$

But for $\sigma > 1$, change of the variable $\log x = t$ gives

$$\int_1^\infty \Big(P_{k^2}(\log x) + P'_{k^2}(\log x)\Big)x^{1-2\sigma}\,\mathrm{d}x$$
$$= \int_1^\infty \left\{\sum_{j=0}^{k^2} a_{j,k}\log^j x + \sum_{j=0}^{k^2-1}(j+1)a_{j+1,k}\log^j x\right\} x^{1-2\sigma}\,\mathrm{d}x$$
$$= \int_0^\infty \left\{\sum_{j=0}^{k^2} a_{j,k}t^j + \sum_{j=0}^{k^2-1}(j+1)a_{j+1,k}t^j\right\} \mathrm{e}^{-(2\sigma-2)t}\,\mathrm{d}t$$
$$= \frac{a_{k^2,k}(k^2)!}{(2\sigma-2)^{k^2+1}} + \sum_{j=0}^{k^2-1}(a_{j,k}j! + a_{j+1,k}(j+1)!)(2\sigma-2)^{-j-1}. \tag{10.22}$$

When $k = 1$ we have, by (10.17),

$$P_1(y) = y + 2\gamma - 1 - \log(2\pi),$$

hence for $\sigma > 1$

$$\int_1^\infty \big(P_1(\log x) + P_1'(\log x)\big)\,x^{1-2\sigma}\,\mathrm{d}x$$
$$= \int_1^\infty (\log x + 2\gamma - \log(2\pi))x^{1-2\sigma}\,\mathrm{d}x$$
$$= \frac{2\gamma - \log(2\pi)}{2\sigma-2} + \frac{1}{(2\sigma-2)^2}. \tag{10.23}$$

Next note that

$$E_1(1) \equiv E(1) = -P_1(0) = \log(2\pi) + 1 - 2\gamma. \tag{10.24}$$

Finally integration by parts yields, on using (10.18),

$$\int_1^\infty E_1(x)x^{-2\sigma}\,\mathrm{d}x$$
$$= \int_1^x E(y)\,\mathrm{d}y\cdot x^{-2\sigma}\Big|_1^\infty + 2\sigma\int_1^\infty\int_1^x E(y)\,\mathrm{d}y\cdot x^{-1-2\sigma}\,\mathrm{d}x$$
$$= 2\sigma\int_1^\infty (\pi x + G(x))\,x^{-1-2\sigma}\,\mathrm{d}x$$
$$= \frac{2\pi\sigma}{2\sigma-1} + 2\sigma\int_1^\infty G(x)x^{-1-2\sigma}\,\mathrm{d}x, \tag{10.25}$$

and the last integral converges absolutely for $\sigma > 3/8$ in view of the O-bound in (10.18). The assertion of Theorem 10.1 then follows from (10.20)-(10.25).

Remark 10.7 Theorem 10.5 could have been formulated, analogously to (10.26) below, as

$$\frac{1}{\pi}\int_0^\infty |\mathcal{M}_1(\sigma+it)|^2\,\mathrm{d}t = C_0 + \frac{C_1}{\sigma-1} + \frac{C_2}{(\sigma-1)^2} + (2\sigma-1)\int_1^\infty E(x)x^{-2\sigma}\,\mathrm{d}x,$$

where the integral on the right-hand side converges absolutely for $\sigma > 5/8$. This follows by the Cauchy-Schwarz inequality for integrals from (8.14), but (10.19) is more precise.

Theorem 10.8 *For $\sigma > 1$ we have*

$$\frac{1}{\pi}\int_0^\infty |\mathcal{M}_2(\sigma+it)|^2\mathrm{d}t = \sum_{j=0}^{5}\frac{c_j}{(\sigma-1)^j} + (2\sigma-1)\int_1^\infty E_2(x)x^{-2\sigma}\mathrm{d}x, \tag{10.26}$$

where the constants c_j can be evaluated explicitly, and the integral on the right-hand side of (10.26) *converges absolutely for $\sigma > 3/4$.*

Corollary 10.9

$$\lim_{\sigma\to 1+0}\left\{\frac{1}{\pi}\int_0^\infty \left|\mathcal{M}_2(\sigma+it)\right|^2\mathrm{d}t - \sum_{j=0}^{5}\frac{c_j}{(\sigma-1)^j}\right\} = \int_1^\infty E_2(x)x^{-2}\mathrm{d}x.$$

Proof of Theorem 10.8 For $k = 2$ write (10.14) as

$$P_4(y) = \sum_{j=0}^{4} a_{j,4}y^j = \sum_{j=0}^{4} A_j y^j,\ A_4 = 1/(2\pi^2), \tag{10.27}$$

so that $E_2(1) = -P_4(0) = -A_0$. From (10.21) and (10.27) we infer that, for $\sigma > 1$,

$$\int_1^\infty \Big(P_4(\log x) + P_4'(\log x)\Big)x^{1-2\sigma}\,\mathrm{d}x = \sum_{j=1}^{5}\frac{B_j}{(\sigma-1)^j}$$

with

$$B_5 = \frac{3}{8\pi^2},\ B_j = A_{j-1}(j-1)!2^{-j} + A_j j!2^{-j}\quad (j = 1, 2, 3, 4).$$

This clearly gives (10.26) of Theorem 10.8 with

$$c_0 = A_0,\ c_j = A_{j-1}(j-1)!2^{-j} + A_j j!2^{-j}\ (j = 1, 2, 3, 4),\ c_5 = \frac{3}{8\pi^2}.$$

The integral on the right-hand side of (10.26) converges absolutely for $\sigma > 3/4$, since by the Cauchy-Schwarz inequality for integrals and the mean square bound

$$\int_0^T E_2^2(t)\,\mathrm{d}t \ll T^2 \log^C T$$

we have

$$\int_X^{2X} E_2(x)x^{-2\sigma}\,\mathrm{d}x \ll \left\{\int_X^{2X} E_2^2(x)\,\mathrm{d}x \int_X^{2X} x^{-4\sigma}\,\mathrm{d}x\right\}^{1/2} \ll_\varepsilon X^{3/2-2\sigma+\varepsilon} \le X^{-\varepsilon}$$

for $\sigma > 3/4 + \varepsilon$.

We turn now to another related problem on mean values of $\mathcal{M}_k(s)$. For a fixed integer k such that $k \ge 3$, let us define

$$\theta_k := \inf\left\{ a_k : \int_{-\infty}^{\infty} |\mathcal{M}_k(\sigma+it)|^2\,\mathrm{d}t < \infty \text{ for } \sigma > a_k\right\}.$$

It is clear that θ_k always exists (e.g., since $\zeta(\frac{1}{2}+it) \ll |t|^{1/6}$) and that it has an intrinsic connection with power moments of $|\zeta(\frac{1}{2}+it)|$, as can be seen from e.g. (10.20). Of course, the above definition of θ_k makes sense for $k = 1, 2$ as well, but in these cases we have much more precise information in view of Theorem 10.5 and Theorem 10.8.

Theorem 10.10 *We have*

$$\theta_k \ge 1 \quad (\forall k), \qquad \theta_k \le \frac{3}{4} + \frac{k}{8} \quad (3 \le k \le 6), \tag{10.27}$$

and

$$\int_1^T |\zeta(\tfrac{1}{2}+it)|^{2k}\mathrm{d}t \ll_\varepsilon T^{2\theta_k - 1+\varepsilon} \qquad (k \ge 3). \tag{10.28}$$

Proof of Theorem 10.10 To prove Theorem 10.10 recall that we have the lower bound

$$\int_1^T |\zeta(\tfrac{1}{2}+it)|^{2k}\,\mathrm{d}t \gg_k T(\log T)^{k^2}. \tag{10.29}$$

This implies that $\mathcal{M}_k(s)$ diverges for $s = 1$, hence $\theta_k \ge 1$ must hold for $k \in \mathbb{N}$. On the other hand we use (10.16) to deduce that, for $3 \le k \le 6$,

$$\int_X^{2X} |\zeta(\tfrac{1}{2}+ix)|^{2k}x^{1-2\sigma}\,\mathrm{d}x \ll X^{1-2\sigma}X^{(k+2)/4}(\log X)^{C_k}$$

and $1 - 2\sigma + (k+2)/4 < 0$ for $\sigma > 3/4 + k/8$. This means that, in this range for σ, the integral

$$\int_1^\infty |\zeta(\tfrac{1}{2} + ix)|^{2k} x^{1-2\sigma}\,\mathrm{d}x$$

converges. But then (10.20) holds and hence $\theta_k \le 3/4 + k/8$ for $3 \le k \le 6$, proving Theorem 10.10. The conjecture $\theta_k = 1\ (\forall k)$ is clearly equivalent to the Lindelöf hypothesis that $\zeta(\frac{1}{2} + it) \ll_\varepsilon |t|^\varepsilon$ (see (7.2) of [Tit3] and theorem 13.2 of [Iv1]).

10.3 Estimates for $\mathcal{M}_k(s)$

We shall now give estimates for $\mathcal{M}_1(s)$, $\mathcal{M}_2(s)\ (= \mathcal{Z}_1(s))$ and $\mathcal{M}_4(s)\ (= \mathcal{Z}_2(s))$, both pointwise and in the mean square sense. The case of $\mathcal{M}_3(s)$ is omitted, since we have no way yet of obtaining the analytic continuation of $\mathcal{M}_3(s)$ outside the region of absolute convergence, namely $\mathbb{R}\mathrm{e}\, s > 1$, of the defining integral (9.10) when $k = 3$.

Theorem 10.11 *For fixed σ such that $\frac{1}{4} < \sigma \le \frac{5}{4}$ we have*

$$\mathcal{M}_1(\sigma + it) \ll_\varepsilon t^{\frac{3}{4}-\sigma+\varepsilon}(1 + t^{\frac{3}{4}-\sigma}) \qquad (t \ge t_0 > 0). \tag{10.30}$$

We also have, for fixed σ such that $\frac{1}{2} < \sigma \le 1$,

$$\int_1^T |\mathcal{M}_1(\sigma + it)|^2 \mathrm{d}t \ll_\varepsilon T^{2-2\sigma+\varepsilon}, \tag{10.31}$$

$$\int_1^T |\mathcal{M}_1(\sigma + it)|^2 \mathrm{d}t \gg_\varepsilon T^{2-2\sigma-\varepsilon}. \tag{10.32}$$

Remark 10.12 The bounds in (10.31) and (10.32) determine, up to "ε" in the exponent of T, the true order of the mean square integral of $|\mathcal{M}_1(\sigma + it)|$. A true asymptotic formula for the integral in question seems difficult to obtain.

Proof of Theorem 10.11 To obtain the pointwise bound (10.30) we suppose $T \le t \le 2T$ and use

$$\mathcal{M}_1(s) = O\Big(\frac{1}{t}\Big) + \int_{T^{1-\varepsilon}}^X Z(x)x^{-s}\,\mathrm{d}x + \int_X^\infty Z(x)x^{-s}\,\mathrm{d}x, \tag{10.33}$$

which is valid initially for $\sigma > 1$ and where $X(\gg T)$ is a parameter to be chosen a little later. One obtains (10.33) by successive integrations by parts of

$$\int_1^{T^{1-\varepsilon}} Z(x)x^{-s}\,\mathrm{d}x,$$

similarly as in (10.10). Integration by parts and (8.18) show that

$$\int_X^\infty Z(x)x^{-s}\,\mathrm{d}x \ll_\varepsilon t^{1+\varepsilon}X^{1/4-\sigma} \qquad (\sigma > 1/4,\ X \ll t^C). \tag{10.34}$$

The remaining integral in (10.33) is split into $O(\log t)$ integrals of the form

$$\begin{aligned}\int_Y^{Y'} Z(x)x^{-s}\,\mathrm{d}x &= 2\int_Y^{Y'} \sum_{n\le\sqrt{\frac{x}{2\pi}}} n^{-1/2}\cos\Big(x\log\frac{\sqrt{x/(2\pi)}}{n} - \tfrac{1}{2}x - \tfrac{\pi}{8}\Big)x^{-s}\,\mathrm{d}x \\ &\quad + O\Big(\int_Y^{Y'} x^{-1/4-\sigma}\,\mathrm{d}x\Big),\end{aligned}$$

where $T^{1-\varepsilon} \le Y < Y' \le 2Y \le X$, and we used the Riemann-Siegel formula in the form of (4.1). Interchanging summation and integration it is seen that the expression on the right-hand side above is

$$2\sum_{n\le\sqrt{\frac{Y'}{2\pi}}} n^{-1/2}\int_{\max(Y,2\pi n^2)}^{Y'} x^{-\sigma}\mathrm{e}^{iG_\pm(x)}\,\mathrm{d}x + O(Y^{3/4-\sigma}), \tag{10.35}$$

with

$$\begin{aligned}G_\pm(x) &:= x\log\frac{\sqrt{x/(2\pi)}}{n} - \frac{1}{2}x - \frac{\pi}{8} \pm t\log x,\\ G'_\pm(x) &= \log\frac{\sqrt{x/(2\pi)}}{n} \pm \frac{t}{x},\quad G''_\pm(x) = \frac{1}{2x} \mp \frac{t}{x^2}.\end{aligned}$$

Consider the contribution of $G_+(x)$, when $G'_+(x) > 0$. If $Y > 4t$ then $1/(2x) > 2t/(x^2)$, hence by the second derivative test (Lemma 2.3) the sum in (10.35) is $\ll Y^{3/4-\sigma}$. If $Y < t/2$ then $t/(x^2) > 1/x$, hence again by the second derivative test we obtain a contribution which is

$$\ll Yt^{-1/2}\cdot Y^{1/4-\sigma} \ll Y^{3/4-\sigma}.$$

If $t/2 \le Y \le 4t$, then $G'_+(x) \gg 1$, hence by the first derivative test we obtain again a contribution which is $\ll Y^{3/4-\sigma}$. A similar analysis holds for the contribution of $G_-(x)$, when $G''_-(x) \gg 1/x$. Therefore we have

$$\int_Y^{Y'} Z(x)x^{-s}\,\mathrm{d}x \ll_\varepsilon t^{(3/4-\sigma)(1-\varepsilon)} + X^{3/4-\sigma}. \tag{10.36}$$

Choosing $X = t^2$ and noting that $t^{3/2-2\sigma} \gg t^{-1}$ for $\sigma \le 5/4$ we obtain (10.30) from (10.34) and (10.36).

To prove the mean square bound (10.31), we start from (10.33). Then we note that integration by parts gives

$$\int_X^\infty Z(x)x^{-s}\,\mathrm{d}x = O_\varepsilon(X^{1/4-\sigma+\varepsilon}) + s\int_X^\infty F(x)x^{-s-1}\,\mathrm{d}x.$$

Here the integral with $F(x)$ (see (8.1)) converges absolutely for $\sigma > 1/4$, and the O-term makes a negligible contribution to (10.31). We also have

$$\int_T^{2T}\left|s\int_X^\infty F(x)x^{-s-1}\,\mathrm{d}x\right|^2\mathrm{d}t \ll T^2\int_X^\infty F^2(x)x^{-2\sigma-1}\,\mathrm{d}x \ll T^2X^{1/2-2\sigma},$$

where we used (8.4) and Lemma 9.4. If $X = T^2$, then the above contribution is $\ll T^{2-2\sigma}$, as desired. In a similar vein it is seen that we have yet to consider $O(\log T)$ contributions to the mean square integral from expressions of the form

$$s\int_Y^{Y'} F(x)x^{-s-1}\,\mathrm{d}x = s\int_Y^{Y'}\Big(S_1(x) + S_2(x) + O(\log^{5/4}x)\Big)x^{-s-1}\,\mathrm{d}x, \tag{10.37}$$

where $T^{1-\varepsilon} \le Y < Y' \le 2Y \le T^2$ and we used Theorem 8.2. Since $S_2(x)$ is essentially a zeta-sum, by the *mean value theorem for Dirichlet polynomials* the portion of the right-hand side of (10.37) containing $S_2(x)$ and the error term will be $\ll T^{2-2\sigma}$. The other portion with $S_1(x)$ is a multiple of the sum of two expressions of the form

$$s\int_Y^{Y'} x^{-\sigma-3/4}\sum_{0\le n\le\sqrt{cY}}(-1)^{n(n+1)/2}(n+\tfrac12)^{-1}\exp\Big(iL_\pm(x,n)\Big)\,\mathrm{d}x, \tag{10.38}$$

where

$$L_\pm(x,n) := \sqrt{2\pi}(n+\tfrac12)x^{1/2} + c_3(n+\tfrac12)^3x^{-1/2} + \ldots \pm t\log x.$$

It is the contribution of $L_-(x,n)$ which is more difficult, so we shall consider it in detail. We have

$$L'(x,n) = \frac{\partial L(x,n)}{\partial x} \gg nY^{-1/2}$$

if $n \gg TY^{-1/2}$. Thus by the first derivative test the expression in (10.38) is

$$\ll TY^{-3/4-\sigma} \sum_{n \gg TY^{-1/2}} n^{-2} Y^{1/2} \ll Y^{1/4-\sigma} \ll_\varepsilon T^{1/4-\sigma+\varepsilon}$$

for $Y \gg T, \sigma > 1/4$. This makes a contribution of $O_\varepsilon(T^{3/2-2\sigma+\varepsilon})$ to the mean square integral.

Likewise, if $n \ll T/Y$, then $L'_-(x,n) \gg T/Y$, and the contribution is again small. Finally, when $T/Y \ll n \ll T/\sqrt{Y}$, then by Lemma 9.4 we have

$$\begin{aligned}
&\int_T^{2T} \Bigg| s \int_Y^{Y'} x^{-3/4-\sigma} \sum_{T/Y \ll n \ll T/\sqrt{Y}} (-1)^{n(n+1)/2} (n+\tfrac{1}{2})^{-1} \\
&\quad \times \exp\Big(iL_\pm(x,n)\Big)\,\mathrm{d}x \Bigg|^2 \mathrm{d}t \\
&\ll T^2 \int_Y^{Y'} Y^{-1/2-2\sigma} \Bigg| \sum_{T/Y \ll n \ll T/\sqrt{Y}} (-1)^{n(n+1)/2} (n+\tfrac{1}{2})^{-1} \\
&\quad \times \exp\Big(\sqrt{2\pi}(n+\tfrac{1}{2})x^{1/2} + c_3(n+\tfrac{1}{2})^3 x^{-1/2} + \dots\Big)\Bigg|^2 \mathrm{d}x \\
&\ll T^2 Y^{-1/2-2\sigma}\Big(Y \sum_{n \ll T/Y} n^{-2} + \sum_{m,n \ll T/Y; m \neq n} (mn|m-n|)^{-1} Y^{1/2}\Big) \\
&\ll TY^{1/2-2\sigma} \ll_\varepsilon T^{3/2-2\sigma+\varepsilon}.
\end{aligned}$$

Thus we obtain

$$\int_T^{2T} \Big|\mathcal{M}_1(\sigma+it)\Big|^2 \mathrm{d}t \ll_\varepsilon T^{2-2\sigma+\varepsilon} \qquad (\tfrac{1}{2} \le \sigma \le 1),$$

and then (10.31) follows on replacing T by $T2^{-j}$ and summing the above bounds for $j = 1, 2, \dots$.

It remains for us to establish (10.32). From Theorem 10.3 (with $c \le \frac{5}{4}$, $U = X, x \asymp X$) we have

$$Z(x) = \frac{1}{2\pi i} \int_{\frac{5}{4}-iX}^{\frac{5}{4}+iX} x^{s-1} \mathcal{M}_1(s)\,\mathrm{d}s + O(X^{-1/4}).$$

Therefore we obtain

$$\int_X^{2X} Z^2(x)\,\mathrm{d}x \ll \int_X^{2X} \left|\int_{c-iX}^{c+iX} x^{s-1}\mathcal{M}_1(s)\,\mathrm{d}s\right|^2 \mathrm{d}x + X^{1-2\varepsilon}.$$

Since

$$Z^2(x) = |\zeta(\tfrac{1}{2}+ix)|^2, \qquad \int\limits_X^{2X} |\zeta(\tfrac{1}{2}+ix)|^2\,\mathrm{d}x \gg X\log X,$$

it follows that

$$X\log X \ll \int_{X/2}^{5X/2} \varphi(x)\left|\int_1^X x^{c+it-1}\mathcal{M}_1(c+it)\,\mathrm{d}t\right|^2 \mathrm{d}x, \tag{10.39}$$

as

$$\int_0^1 x^{s-1}\mathcal{M}_1(s)\,\mathrm{d}s \ll 1 \qquad (x \asymp X,\ \tfrac{1}{2} < c \le 1).$$

Here $\varphi(x)\,(\ge 0)$ is a smooth function supported in $[X/2, 5X/2]$ and equal to unity in $[X, 2X]$. When we develop the square on right-hand side of (10.39) and integrate sufficiently many times by parts we obtain that

$$\begin{aligned} X\log X &\ll \int_{X/2}^{5X/2} x^{2c-2}\int_1^X\int_{1,|u-t|\le X^\varepsilon}^X |\mathcal{M}_1(c+it)\mathcal{M}_1(c+iu)|\,\mathrm{d}u\,\mathrm{d}t\,\mathrm{d}x \\ &\ll X^{2c-1}\int_1^X\int_{1,|u-t|\le X^\varepsilon}^X \left(|\mathcal{M}_1(c+it)|^2 + |\mathcal{M}_1(c+iu)|^2\right)\mathrm{d}u\,\mathrm{d}t \\ &\ll_\varepsilon X^{2c-1+\varepsilon}\int_1^X |\mathcal{M}_1(c+it)|^2\,\mathrm{d}t, \end{aligned}$$

since the contribution of $|u-t| \ge X^\varepsilon$ will be negligibly small. This implies the assertion (10.32) with $\sigma = c \ge \frac{1}{2} + 2\varepsilon$.

For $\mathcal{Z}_1(s) \equiv \mathcal{M}_2(s)$ we have the following result

Theorem 10.13 *For* $0 \le \sigma \le 1,\ t \ge t_0 > 0$*, and* $T \ge 1$*, we have*

$$\mathcal{Z}_1(\sigma+it) \ll_\varepsilon t^{1-\sigma+\varepsilon}, \tag{10.40}$$

and

$$\int_1^T |\mathcal{Z}_1(\sigma+it)|^2\mathrm{d}t \ll_\varepsilon \begin{cases} T^{3-4\sigma+\varepsilon} & \text{if } 0 \le \sigma \le \frac{1}{2}, \\ T^{2-2\sigma+\varepsilon} & \text{if } \frac{1}{2} \le \sigma \le 1. \end{cases} \tag{10.41}$$

Proof of Theorem 10.13 From the defining relation

$$\mathcal{Z}_1(s) = \int_1^\infty |\zeta(\tfrac{1}{2}+ix)|^2 x^{-s}\,\mathrm{d}x$$

and the defining relation (8.5) of $E(T)$ we have, initially for $\sigma > 1$,

$$\begin{aligned}\mathcal{Z}_1(s) &= \int_1^\infty (\log x + 2\gamma - \log(2\pi))x^{-s}\,\mathrm{d}x + \int_1^\infty E'(x)x^{-s}\,\mathrm{d}x\\ &= \frac{1}{(s-1)^2} + \frac{2\gamma - \log(2\pi)}{s-1} - E(1) + s\int_1^\infty E(x)x^{-s-1}\,\mathrm{d}x.\end{aligned}$$

However, if we apply the Cauchy-Schwarz inequality, we see that the last integral converges absolutely for $\sigma > \frac{1}{4}$ in view of (8.14). Thus we have

$$\mathcal{Z}_1(s) = \frac{1}{(s-1)^2} + \frac{2\gamma - \log(2\pi)}{s-1} - E(1) + s\int_1^\infty E(x)x^{-s-1}\,\mathrm{d}x \quad (\sigma > \tfrac{1}{4}). \tag{10.42}$$

We shall show that for the function

$$Y(s) := s\int_1^\infty E(x)x^{-s-1}\,\mathrm{d}x$$

in (10.42) we have

$$Y(\sigma + it) \ll_\varepsilon T^{1-\sigma+\varepsilon} \qquad (t \asymp T) \tag{10.43}$$

and

$$\int_T^{2T} |Y(\sigma + it)|^2\,\mathrm{d}t \ll_\varepsilon T^{3-4\sigma+\varepsilon} \qquad (0 \le \sigma \le \tfrac{1}{2}), \tag{10.44}$$

from which (10.40) and the first bound in (10.41) follow. Although (10.42) is valid for $\sigma > \frac{1}{4}$, the arguments that follow will show that the range $\sigma > 0$ may be treated.

We split up the integral in the definition of $Y(s)$ in subintegrals over $[X, 2X]$, so that $Y(s)$ will be the sum of the entire functions of the form

$$Y_X(s) := s\int_X^{2X} E(x)x^{-s-1}\,\mathrm{d}x. \tag{10.45}$$

Consider separately three cases: $X < T$, $T \le X \le T^2$, and $X > T^2$. In the first case, the assertions (10.43) and (10.44) for $Y_X(s)$ in place of $Y(s)$ are verified as follows. We apply integration by parts in (10.44), use the bound $E(x) \ll x^{1/3}$, and apply Lemma 9.4 to obtain

$$\begin{aligned}\int_T^{2T} \left|\int_X^{2X} |\zeta(\tfrac{1}{2} + ix)|^2 x^{-s}\,\mathrm{d}x\right|^2 \mathrm{d}t &\ll \int_X^{2X} |\zeta(\tfrac{1}{2} + ix)|^4 x^{1-2\sigma}\,\mathrm{d}x\\ &\ll X^{2-2\sigma}\log^4 X \ll T^{2-2\sigma}\log^4 T.\end{aligned} \tag{10.46}$$

In the case $X > T$ we invoke Atkinson's formula for $E(T)$ (Lemma 8.1), which reads as follows:

$$E(x) = \Sigma_1(x) + \Sigma_2(x) + O(\log^2 X) \qquad (X \le x \le 2X),$$

and we choose $N = cX$ with $0 < c < 1$ a small constant. Here the term $\Sigma_2(x)$ is of logarithmic order in a mean square sense, so that its contribution together with that of the error term can be treated easily as above.

The contribution of $\Sigma_1(x)$ to $Y_X(s)$ is a constant multiple of

$$s \sum_{n \le cX} (-1)^n d(n) n^{-3/4} \int_X^{2X} x^{-3/4-\sigma} e(x, n) \mathrm{e}^{iF_\pm(x)}\, \mathrm{d}x,$$

where

$$e(x, n) := \left(1 + \frac{\pi n}{2x}\right)^{-1/4} \left(\sqrt{\frac{2x}{\pi n}} \operatorname{arsinh} \sqrt{\frac{\pi n}{2x}}\right)^{-1} = 1 + O\left(\frac{n}{x}\right),$$

$$F_\pm(x) = f(x, n) \pm t \log x,$$

and $f(x, n)$ is as in (8.9). We may consider only $F := F_-$, since F_+ will have no saddle points, and its contribution is easily handled by the first derivative test. Note that

$$F'(x) = 2\log\left(\sqrt{\frac{\pi n}{2x}} + \sqrt{1 + \frac{\pi n}{2x}}\right) \pm \frac{t}{x} \gg \sqrt{\frac{n}{X}}$$

for $X \gg t^2$ and sufficiently small c. Hence by the first derivative test the contribution of such X is

$$\ll T \sum_{n \le cX} d(n) n^{-3/4} X^{-3/4-\sigma} n^{-1/2} X^{1/2} \ll T X^{-1/4-\sigma}.$$

Thus the series over the "dyadic" values $X > T^2$ converges, giving a contribution $\ll T^{1/2-2\sigma}$ to $Y(s)$, and the contribution to the mean square is $\ll T^{2-4\sigma}$.

Finally let $T \le X \le T^2$. If $n \ge c_1 T^2/X$ with c_1 sufficiently large, then again $F'(x) \gg \sqrt{n/X}$, and the preceding estimations apply to show that the corresponding sum over n is $\ll T^{1/2} X^{-\sigma} \ll T^{1/2-\sigma}$, which suffices for our purposes. The first derivative test applies even if $n \le c_2 T^2/X$ with c_2 a sufficiently small positive constant; in this case we have $|F'(x)| \gg T/X$.

The critical range for n is $n \asymp T^2/X$, when a saddle point for the integral in $Y_X(s)$ may occur, and this saddle point will be of the order t^2/n. And if there is a point $x_0 \in [X, 2X]$ such that $F'(x_0)$ vanishes or at least attains a value which is small in comparison with T/X, then $|F''(x)| \asymp T/X^2$ in a certain vicinity of x_0, and elsewhere in the interval $[X, 2X]$ we have $|F'(x)| \asymp T/X$. To see this, note that n/X is small, so the logarithmic term in the above formula for $F'(x)$ behaves in the first approximation like the function $(2\pi n/x)^{1/2}$. Thus, if the first derivative test applies, the argument is as above, and otherwise for $n \asymp T^2/X$ we use the second derivative test to obtain a contribution to $Y_X(s)$ which is

$$\ll T \sum_{n \asymp T^2/X} (T/X^2)^{-1/2} d(n) n^{-3/4} X^{-\frac{3}{4}-\sigma} \ll_\varepsilon TX^{-\sigma+\varepsilon} \ll_\varepsilon T^{1-\sigma+\varepsilon},$$

and this completes the proof of (10.43).

For a proof of (10.44), we still have to estimate the integral

$$T^2 \int_T^{2T} \left| \int_X^{2X} \tilde{\Sigma}_1(x) x^{-\sigma-it-1} \, dx \right|^2 dt, \tag{10.47}$$

where $\tilde{\Sigma}_1(x)$ stands for the critical part of $\Sigma_1(x)$, where $n \asymp T^2/X$, which implies $T \ll X \ll T^2$. Applying Lemma 9.4, we have

$$\int_X^{2X} |\tilde{\Sigma}_1(x)|^2 \, dx \ll_\varepsilon X^\varepsilon (X + X^{3/2}(T^2/X)^{-1/2}) \ll_\varepsilon X^2 T^{-1+\varepsilon}.$$

Hence we see that the expression (10.46) is

$$\ll_\varepsilon T^{1+\varepsilon} X^{1-2\sigma} \ll_\varepsilon T^{3-4\sigma+\varepsilon},$$

as required for finishing the proof of (10.44).

To prove the second mean square bound in (10.44) we start from the representation

$$\begin{aligned} \mathcal{Z}_1(s) = &\int_1^X |\zeta(\tfrac{1}{2} + ix)|^2 x^{-s} \, dx + \frac{X^{1-s}}{s-1} \left(\frac{1}{s-1} + \log X + 2\gamma - \log 2\pi \right) \\ &- E(X) X^{-s} + s \int_X^\infty E(x) x^{-s-1} \, dx. \end{aligned} \tag{10.48}$$

This is valid for $X > 1$, $\sigma > \frac{1}{4}$, and follows analogously to the proof of (10.42). Thus for $\frac{1}{2} \le \sigma \le 1$ we obtain from (10.42), using (10.34) and Lemma 9.4,

$$\begin{aligned} \int_T^{2T} |\mathcal{Z}_1(s)|^2 \, dt \ll &\int_T^{2T} \left| \int_1^X |\zeta(\tfrac{1}{2} + ix)|^2 x^{-s} \, dx \right|^2 dt + T^{-1} X^{2-2\sigma} \log^2 X \\ &+ TX^{\frac{2}{3}-2\sigma} + T^2 \int_T^{2T} \left| \int_X^\infty E(x) x^{-s-1} \, dx \right|^2 dt \\ \ll &\int_1^X |\zeta(\tfrac{1}{2} + ix)|^4 x^{1-2\sigma} \, dx + T^{-1} X^{2-2\sigma} \log^2 X \\ &+ TX^{\frac{2}{3}-2\sigma} + T^2 \int_X^\infty E^2(x) x^{-1-2\sigma} \, dx \end{aligned}$$

$$\ll X^{2-2\sigma}\log^5 X + T^{-1}X^{2-2\sigma}\log^2 X + TX^{\frac{2}{3}-2\sigma} + T^2X^{\frac{1}{2}-2\sigma}$$
$$\ll T^{8(1-\sigma)/3}\log^5 T$$

with the choice $X = T^{4/3}$. The second mean square bound in (10.41) follows now from the first bound in (10.41) at $\sigma = \frac{1}{2}$, and the above bound at $\sigma = 1$ by the convexity of mean values for Dirichlet series.

For a mean square bound involving $\mathcal{M}_4(s) \equiv \mathcal{Z}_2(s)$ we recall that $E_2(T)$ denotes the error term in the formula

$$\int_0^T |\zeta(\tfrac{1}{2}+it)|^4\,\mathrm{d}t \;=\; TP_4(\log T) \;+\; E_2(T).$$

Let c be a constant such that

$$E_2(T) \;\ll\; T^c. \tag{10.49}$$

Then the results on $\mathcal{Z}_2(s)$ may be formulated as

Theorem 10.14 *For $s = \sigma + it$, $\frac{1}{2} < \sigma < 1$, $t \ge t_0$, we have*

$$\mathcal{Z}_2(s) \;\ll\; t^{2-2\sigma}(\log t)^{18-14\sigma}. \tag{10.50}$$

If (10.49) *holds then we have*

$$\int_0^T |\mathcal{Z}_2(\sigma+it)|^2\mathrm{d}t \;\ll_\varepsilon\; T^\varepsilon\left(T + T^{\frac{2-2\sigma}{1-c}}\right) \qquad (\tfrac{1}{2} < \sigma < 1), \tag{10.51}$$

while we also have unconditionally

$$\int_0^T |\mathcal{Z}_2(\sigma+it)|^2\mathrm{d}t \ll T^{(10-8\sigma)/3}\log^C T \qquad (\tfrac{1}{2} < \sigma \le 1, C > 0). \tag{10.52}$$

Proof of Theorem 10.14 For $\sigma = \mathbb{R}\mathrm{e}\, s > 1$ we have

$$\mathcal{Z}_2(s) = \int_1^X |\zeta(\tfrac{1}{2}+ix)|^4 x^{-s}\,\mathrm{d}x + \int_X^\infty |\zeta(\tfrac{1}{2}+ix)|^4 x^{-s}\,\mathrm{d}x = I' + I'',$$

say, where $X > 1$ is a parameter to be suitably determined later. We have

$$I' := \int_1^X |\zeta(\tfrac{1}{2}+ix)|^4 x^{-s}\,\mathrm{d}x \;\ll\; X^{1-\sigma}\log^4 X \quad (\tfrac{1}{2} < \sigma < 1), \tag{10.53}$$

which follows by integration by parts and the classical bound for the fourth moment of $|\zeta(\frac{1}{2}+ix)|$. Next, for $s = \sigma + it$, $\sigma > 1$, $t \ge t_0 > 0$,

$Q_4 = P_4 + P_4'$,

$$I'' := \int_X^\infty |\zeta(\tfrac{1}{2}+ix)|^4 x^{-s}\,dx = \int_X^\infty (Q_4(\log x) + E_2'(x))x^{-s}\,dx$$
$$= X^{1-s}\sum_{j=1}^{5}\frac{Q_4^{(j-1)}(\log X)}{(s-1)^j} - E_2(X)X^{-s} + s\int_X^\infty E_2(x)x^{-s-1}\,dx, \tag{10.54}$$

where we used the defining property (10.13) with $k = 2$. Suppose now that (10.49) holds. Currently we know, as mentioned after (10.18), that (10.49) holds with $c = \frac{2}{3}$ and that $c < \frac{1}{2}$ cannot hold, and it is conjectured that $c = \frac{1}{2}$ holds. Since we have

$$\int_0^T E_2^2(t)\,dt \ll T^2\log^C T \qquad (C \le 22), \tag{10.55}$$

it follows that

$$\int_T^{2T} |E_2(x)|x^{-\sigma-1}\,dx \le \Big(\int_T^{2T} E_2^2(x)\,dx \int_T^{2T} x^{-2\sigma-2}\,dx\Big)^{1/2}$$
$$\ll T^{\frac{1}{2}-\sigma}\log^{11} T. \tag{10.56}$$

Thus the last integral in (10.54) converges absolutely for $\sigma > \frac{1}{2}$, thereby providing analytic continuation of $\mathcal{Z}_2(s)$ in the half-plane $\sigma > \frac{1}{2}$. From (10.53)-(10.56) we then obtain

$$\mathcal{Z}_2(s) \ll X^{1-\sigma}\log^4 X + tX^{\frac{1}{2}-\sigma}\log^{11} X,$$

and the choice $X = t^2\log^{14} t$ (σ is assumed throughout to be fixed) gives

$$\mathcal{Z}_2(s) \ll t^{2-2\sigma}(\log t)^{18-14\sigma} \qquad (\tfrac{1}{2} < \sigma < 1). \tag{10.57}$$

We pass now to mean square estimates, starting from

$$I(s) := \int_T^{2T} |\mathcal{Z}_2(\sigma+it)|^2\,dt \le \int_{T/2}^{5T/2} \varphi(t)|\mathcal{Z}_2(\sigma+it)|^2\,dt,$$

where $\frac{1}{2} < \sigma < 1$ as before, and $\varphi(t)$ (≥ 0) is a smooth function supported in $[T/2,\ 5T/2]$ such that $\varphi(t) = 0$ for $t \in [T, 2T]$. We have

$$I(s) \ll I_1(s) + I_2(s),$$

where using (10.53) and (10.54) we have

$$I_1(s) := \int_{T/2}^{5T/2} \varphi(t) \left| \int_1^X |\zeta(\tfrac{1}{2}+ix)|^4 x^{-s}\,\mathrm{d}x \right|^2 \mathrm{d}t,$$

$$I_2(s) := \int_{T/2}^{5T/2} \varphi(t) \Bigg| X^{1-s} \sum_{j=1}^{5} \frac{Q_4^{(j-1)}(\log X)}{(s-1)^j} - E_2(X)X^{-s} + s\int_X^\infty E_2(x)x^{-s-1}\,\mathrm{d}x \Bigg|^2 \mathrm{d}t.$$

Note now that by r integrations by parts we have, since $\varphi^{(r)}(t) \ll_r T^{-r}$,

$$\int_{T/2}^{5T/2} \varphi(t)\left(\frac{y}{x}\right)^{it} \mathrm{d}t = (-1)^r \int_{T/2}^{5T/2} \varphi^{(r)}(t) \frac{\left(\frac{y}{x}\right)^{it}}{\left(i\log\frac{y}{x}\right)^r}\,\mathrm{d}t$$
$$\ll_r T^{1-r}\left|\log\frac{y}{x}\right|^{-r} \ll T^{-A} \tag{10.58}$$

for any fixed $A > 0$ and any given $\varepsilon > 0$, provided that $|y-x| \ge xT^{\varepsilon-1}$ and $r = r(A,\varepsilon)$ is large enough. By using (10.55) and (10.58) it follows that

$$\begin{aligned}
I_1(s) &= \int_1^X\int_1^X |\zeta(\tfrac{1}{2}+ix)|\zeta(\tfrac{1}{2}+iy)|^4 (xy)^{-\sigma} \int_{T/2}^{5T/2} \varphi(t)\left(\frac{y}{x}\right)^{it} \mathrm{d}t\,\mathrm{d}x\,\mathrm{d}y \\
&\ll_\varepsilon 1 + \int_{T/2}^{5T/2}\int_1^X |\zeta(\tfrac{1}{2}+ix)|^4 x^{-2\sigma} \int_{x-xT^{\varepsilon-1}}^{x+xT^{\varepsilon-1}} |\zeta(\tfrac{1}{2}+iy)|^4\,\mathrm{d}y\,\mathrm{d}x\,\mathrm{d}t \\
&\ll_\varepsilon 1 + T\int_1^X |\zeta(\tfrac{1}{2}+ix)|^4 x^{-2\sigma}(xT^{\varepsilon-1}\log^4 x + E_2(x+xT^{\varepsilon-1}) \\
&\quad - E_2(x-xT^{\varepsilon-1}))\,\mathrm{d}x \\
&\ll_\varepsilon 1 + T\int_1^X |\zeta(\tfrac{1}{2}+ix)|^4 x^{-2\sigma}(xT^{\varepsilon-1}\log^4 x + x^{c+\varepsilon})\,\mathrm{d}x \\
&\ll_\varepsilon T^\varepsilon(X^{2-2\sigma} + T + TX^{1+c-2\sigma})
\end{aligned}$$

if $X \ll T^B$ $(B > 0)$, since $\int_1^X |\zeta(\frac{1}{2}+ix)|^4 x^{-a}\,\mathrm{d}x$ is bounded for $a > 1$. To bound $I_2(s)$ we shall use again (10.58) to obtain

$$\begin{aligned}
I_2(s) &\ll_\varepsilon X^{2-2\sigma} + TX^{2c-2\sigma+\varepsilon} \\
&\quad + T^2 \int_{T/2}^{5T/2} \varphi(t) \int_X^\infty |E_2(x)|x^{-\sigma-1} \int_{x-xT^{\varepsilon-1}}^{x+xT^{\varepsilon-1}} |E_2(y)|y^{-\sigma-1}\,\mathrm{d}y\,\mathrm{d}x\,\mathrm{d}t.
\end{aligned}$$

We use the elementary inequality $|ab| \le \frac{1}{2}(|a|^2+|b|^2)$, and the mean square bound (10.55) for $E_2(T)$ to infer that the inner integrals in the above bound are

$$\begin{aligned}
&\ll_\varepsilon \int_X^\infty x^{-2\sigma-2} \int_{x-xT^{\varepsilon-1}}^{x+xT^{\varepsilon-1}} \left(E_2^2(x)+E_2^2(y)\right) \mathrm{d}y\,\mathrm{d}x \\
&\ll_\varepsilon X^{1+\varepsilon-2\sigma}T^{-1} + \int_{X/2}^\infty y^{-2\sigma-2}E_2^2(y) \int_{y/(1+T^{\varepsilon-1})}^{y/(1-T^{\varepsilon-1})} \mathrm{d}x\,\mathrm{d}y \\
&\ll_\varepsilon X^{1+\varepsilon-2\sigma}T^{-1}.
\end{aligned}$$

Consequently

$$I_2(s) \ll_\varepsilon X^{2-2\sigma} + TX^{2c-2\sigma+\varepsilon} + T^{2+\varepsilon}X^{1-2\sigma}$$

and

$$I(s) \ll_\varepsilon T^\varepsilon(X^{2-2\sigma} + T + TX^{1+c-2\sigma} + T^2X^{1-2\sigma}).$$

Hence taking $X = T^{1/(1-c)}$ it follows that

$$I(s) \ll_\varepsilon T^\varepsilon(T + T^{\frac{2-2\sigma}{1-c}}) \qquad (\tfrac{1}{2} < \sigma < 1), \tag{10.59}$$

since $\frac{1}{2} < c < 1$ yields

$$T^2X^{1-2\sigma} = T^2T^{\frac{1}{1-c}}X^{-2\sigma} \le T\cdot T^{\frac{c}{1-c}}X^{-2\sigma} = TX^{1+c-2\sigma}.$$

The above discussion shows that

$$\int_1^T |\mathcal{Z}_2(\sigma+it)|^2\,\mathrm{d}t \ll_\varepsilon T^\varepsilon(T + T^{\frac{2-2\sigma}{1-c}}) \qquad (\tfrac{1}{2} < \sigma < 1). \tag{10.60}$$

We shall obtain now another mean square bound for $\mathcal{Z}_2(s)$. We start from (10.53) and (10.54) to obtain

$$\int_T^{2T} |\mathcal{Z}_2(s)|^2\,\mathrm{d}t \ll J_1(s) + T^2J_2(s) + X^{2-2\sigma}\log^8 X + TE_2^2(X)X^{-2\sigma},$$

where

$$\begin{aligned}
J_1(s) &:= \int_T^{2T} \left|\int_1^X |\zeta(\tfrac{1}{2}+ix)|^4x^{-s}\,\mathrm{d}x\right|^2 \mathrm{d}t, \\
J_2(s) &:= \int_T^{2T} \left|\int_X^\infty E_2(x)x^{-s-1}\,\mathrm{d}x\right|^2 \mathrm{d}t.
\end{aligned}$$

To bound $J_1(s)$ and $J_2(s)$ we use Lemma 9.4, (10.55) and the mean square bound

$$\int_0^T |\zeta(\tfrac{1}{2}+it)|^8\,\mathrm{d}t \ll T^{3/2}\log^C T. \tag{10.61}$$

It follows that

$$\int_T^{2T} |\mathcal{Z}_2(s)|^2 \,\mathrm{d}t \ll \Big(X^{\frac{5}{2}-2\sigma} + T^2X^{1-2\sigma} + X^{2-2\sigma}\log^8 X + TE_2^2(X)X^{-2\sigma}\Big)\log^C X.$$

Since $E_2(T) \ll T^{2/3}\log^C T$, the choice $X = T^{4/3}$ gives then

$$\int_1^T |\mathcal{Z}_2(\sigma+it)|^2 \,\mathrm{d}t \ll T^{(10-8\sigma)/3}\log^C T \qquad (\tfrac{1}{2} \le \sigma \le 1),$$

If in (10.61) for the eighth moment one had the conjectural exponent $1+\varepsilon$, then we would improve this bound to

$$\int_1^T |\mathcal{Z}_2(\sigma+it)|^2 \,\mathrm{d}t \ll_\varepsilon T^{4-4\sigma+\varepsilon} \qquad (\tfrac{1}{2} \le \sigma \le 1).$$

10.4 Natural boundaries

If a Dirichlet series $F(s)$ has a (meromorphic) continuation to the half-plane $\mathbb{R}\mathrm{e}\, s > \sigma_0$, then the line $\mathbb{R}\mathrm{e}\, s = \sigma_0$ is said to be the *natural boundary* (or *natural barrier*) of $F(s)$ if $F(s)$ cannot be continued analytically to the region $\mathbb{R}\mathrm{e}\, s \le \sigma_0$. The history of natural boundaries for Dirichlet series goes at least back to T. Estermann [Est1]. For example, one has

$$\sum_{n=1}^{\infty} d_k^2(n)n^{-s} = \zeta^{k^2}(s)\prod_p P_k(p^{-s}) \qquad (\mathbb{R}\mathrm{e}\, s > 1),$$

$$P_k(u) := (1-u)^{2k-1}\sum_{n=0}^{k}\binom{k-1}{n}^2 u^k,$$

and Estermann showed that the above Euler product has meromorphic continuation to $\mathbb{R}\mathrm{e}\, s > 0$, but has the line $\mathbb{R}\mathrm{e}\, s = 0$ as the natural boundary when $k > 2$.

If $A(s) = \sum_{n=1}^{\infty} a_n n^{-s}$ in its region of absolute convergence, say $\mathbb{R}\mathrm{e}\, s > \sigma_a$, then by Perron's inversion formula

$$\sum_{n\le x} a_n = \frac{1}{2\pi i}\int_{c-i\infty}^{c+i\infty} A(s)\frac{x^s}{s}\,\mathrm{d}s \qquad (x \notin \mathbb{N},\ c > \sigma_a). \tag{10.62}$$

In practice one wants to shift the line of integration in (10.62) to the left, to reduce the contribution of the term x^s. This is possible only if $A(s)$ is holomorphic on the new path. If $\sigma = \sigma_0\ (< \sigma_a)$ is the natural boundary of

$F(s)$, then we cannot have $c \le \sigma_0$, hence the usefulness of (10.62) is limited if σ_0 exists. This is one of the reasons which makes the study of natural boundaries of Dirichlet series important.

Note that $\mathcal{Z}_k(s)$ does not have an Euler product, which makes the problem more difficult. It is conjectured that the analytic continuation of $\mathcal{Z}_3(s)$ ($\equiv \mathcal{M}_6(s)$) has $\mathbb{R}\mathrm{e}\, s = \frac{1}{2}$ as the natural boundary, and that, more generally, $\mathcal{Z}_k(s)$ for $k \ge 3$ has also the line $\mathbb{R}\mathrm{e}\, s = \frac{1}{2}$ as the natural boundary. A full proof of this important claim concerning $\mathcal{Z}_k(s)$ would be most welcome. The basic idea that leads to it is simple, and is open to generalizations. Namely the analytic continuation of (s_1, s_2 and w are complex variables, $m, n \in \mathbb{N}$)

$$\int_1^\infty \left(\frac{m}{n}\right)^{ix} L(s_1 + ix)L(s_2 - ix)\, x^{-w}\, \mathrm{d}x \tag{10.63}$$

produces the analytic continuation of

$$\int_1^\infty |F(\sigma + ix)L(\sigma + ix)|^2\, x^{-w}\, \mathrm{d}x, \qquad F(s) = \sum_{n=1}^\infty f(n)n^{-s} \tag{10.64}$$

under some reasonable conditions, simply by squaring out $|F|^2$ and summing over the relevant m, n.

If in (10.64) we take $F = \zeta, L = \zeta^2, \sigma = \frac{1}{2}$, then we have to recall (see Theorem 9.9) that $\mathcal{Z}_2(s)$ has (see Section 10.1) infinitely many poles at $s = \frac{1}{2} \pm i\kappa_j, \kappa_j = \sqrt{\lambda_j - \frac{1}{4}}$. Heuristically, when we sum over various m, n in (10.63) to get the analytic continuation of $\mathcal{Z}_3(w)$, each of the poles $\frac{1}{2} \pm i\kappa_j$ will be somewhat perturbed. Their totality will be dense on the $\frac{1}{2}$-line, hence $\mathcal{Z}_3(w)$ cannot be regular on $\mathbb{R}\mathrm{e}\, s = \frac{1}{2}$, and the $\frac{1}{2}$-line will be the natural boundary for $\mathcal{Z}_3(w)$. Inasmuch as this seems plausible, a rigorous proof is in order.

Suppose that one has found the analytic continuation of $\mathcal{Z}_3(w)$ to the region $\mathbb{R}\mathrm{e}\, w > \frac{1}{2}$. Then it is plausible that $\mathcal{Z}_3(w)$ (being more complex that $\mathcal{Z}_2(w)$) will have infinitely many poles as well. Where are these poles located? One does not expect them be too near the $\frac{1}{2}$-line, so the $\frac{3}{4}$-line is a very good candidate to contain infinitely many poles of $\mathcal{Z}_3(w)$. But by the principle inherent in (10.63)-(10.64), then the $\frac{3}{4}$-line would be a natural boundary for $\mathcal{Z}_4(w)$, and so on – each $\mathcal{Z}_k(w)$ would, with increasing k, have poles nearing the 1-line.

The recent work of Y. Motohashi [Mot6] on the explicit spectral decomposition of the mean value

$$\int_{-\infty}^\infty |\zeta(\tfrac{1}{2} + it)|^4 |A(\tfrac{1}{2} + it)|^2 g(t)\, \mathrm{d}t$$

supports the claim that $\mathcal{Z}_3(s)$ has $\sigma = 1/2$ as the natural boundary. Here $A(s) = \sum_n \alpha_n n^{-s}$ is a finite Dirichlet series (i.e. a Dirichlet polynomial) and $g(t)$ is assumed to be even, regular, real-valued on $\mathbb{R}$, and of fast decay in a sufficiently wide horizontal strip. The author says: "Our theorem suggests that the Mellin transform $\int_1^\infty |\zeta(\frac{1}{2}+ix)|^6 x^{-s}\,dx$ should have the line $\mathbb{R}e\, s = 1/2$ as a natural boundary... The same was speculated also by a few people other than us, but it appears that our theorem is so far the sole explicit evidence supporting the observation."

If the line $\mathbb{R}e\, s = 1/2$ is a natural boundary of $\mathcal{Z}_3(s)$, then this certainly indicates a complicated structure of the error term $E_3(T)$ for the sixth moment of $|\zeta(\frac{1}{2}+it)|$. If $E_3(T) \ll_\varepsilon T^{\theta+\varepsilon}$ with θ as small as possible, then $\mathcal{Z}_3(s)$ would have singularities on $\mathbb{R}e\, s = \theta$, if $1/2 < \theta < 1$. Inasmuch as it seems plausible to the author that $\theta = 3/4$, this is a major unsolved problem.

Remark 10.15 The present author believes that

$$\int_0^T |\zeta(\tfrac{1}{2}+it)|^6\,dt = TP_9(\log T) + E_3(T),$$
$$E_3(T) = O_\varepsilon(T^{3/4+\varepsilon}), \qquad E_3(T) = \Omega(T^{3/4})$$

holds, where the main term $TP_9(\log T)$ ($P_9(y)$ is an explicit polynomial in y of degree nine, in fact this is (10.14) with $k = 3$) is the one predicted by Conrey *et al.* [CFKRS1], [CFKRS2], who gave explicit coefficients for the general polynomial $P_{k^2}(y)$. Conjectures for the higher moments of the zeta-function were also given by J.B. Conrey and S.M. Gonek [CoGo]. However, the error term is indicated by them to be (in all cases) $O_\varepsilon(T^{1/2+\varepsilon})$, which does not seem to the present author, in general, to be true.

In what concerns the true order of higher moments of $|\zeta(\frac{1}{2}+it)|$, the situation is even more unclear. Already for the eighth moment it is hard to ascertain what goes on, much less for the higher moments. The main term for the general $2k$th moment should involve a main term of the type suggested by Conrey et al. (*op. cit.*), but it could turn out that the error term (see (10.13)-(10.15))

$$E_k(T) = \int_0^T |\zeta(\tfrac{1}{2}+it)|^{2k}\,dt - TP_{k^2}(\log T) \qquad (k \in \mathbb{N})$$

in the general case (when $k \ge 4$) contains expressions which make it *larger* than the term $TP_{k^2}(\log T)$. For this see the discussion in [Iv12] (also [Mot5], pp. 218-219). Essentially the argument is as follows. In general, from the knowledge about the order of $E_k(T)$ one can deduce a bound for $\zeta(\frac{1}{2}+iT)$ via the elementary estimate

$$\zeta(\tfrac{1}{2}+iT) \ll_k (\log T)^{(k^2+1)/(2k)} + \Big(\log T \max_{t\in[T-1,T+1]} |E_k(t)|\Big)^{1/(2k)}, \tag{10.65}$$

which is lemma 4.2 of [Iv4]. The conjectured bounds

$$E_k(T) \ll_{\varepsilon,k} T^{k/4+\varepsilon} \qquad (k \le 4) \tag{10.66}$$

by (10.65) all imply $\zeta(\frac{1}{2}+it) \ll_\varepsilon |t|^{1/8+\varepsilon}$, which is out of reach at present, but is still much weaker than the Lindelöf hypothesis that $\zeta(\frac{1}{2}+it) \ll_\varepsilon |t|^\varepsilon$. On the other hand, we know that the omega-result

$$E_k(T) = \Omega(T^{k/4}) \tag{10.67}$$

holds for $k = 1, 2$, and as already explained, there are reasons to believe that (10.67) holds for $k = 3$. Perhaps it holds for $k = 4$ also, but the truth of (10.67) for any $k > 4$ would imply that the Lindelöf hypothesis is false, and *ipse facto* the falsity of the Riemann hypothesis. Namely it is well-known (see, e.g., [Iv1] or [Tit3]) that the RH implies even $\log|\zeta(\frac{1}{2}+it)| \ll \log|t|/\log\log|t|$, which is stronger than the Lindelöf hypothesis, namely that $\log|\zeta(\frac{1}{2}+it)| \ll_\varepsilon \varepsilon\log|t|$. The reason why, in general, (10.66) makes sense is that a bound $E_k(T) \ll T^{c_k}$ for some fixed k (> 4) with $c_k < k/4$ would imply (by (10.65)) the bound $\zeta(\frac{1}{2}+it) \ll_\varepsilon |t|^{c_k/(2k)+\varepsilon}$ with $c_k/(2k) < 1/8$. But the most one can get (by using (10.65)) from the error term in the mean square and the fourth moment of $|\zeta(\frac{1}{2}+it)|$ is the bound

$$\zeta(\tfrac{1}{2}+it) \ll_\varepsilon |t|^{1/8+\varepsilon}.$$

It does not appear likely to the author that, say from the twelfth moment ($k = 6$), one will get a better pointwise estimate from (10.65) for $\zeta(\frac{1}{2}+it)$ than what one can get from the mean square formula ($k = 1$). Nothing, of course, precludes yet that this does not happen, just that it appears to me not to be likely. As in all such dilemmas, only rigorous proofs will reveal in due time the real truth.

Notes

Plausible heuristic arguments for the values of the coefficients $a_{j,k}$ in the general case of (10.13)-(10.14) were given by Conrey *et al.* [CFKRS1], [CFKRS2], by using methods from random matrix theory. In particular they showed that the value of the leading coefficient of P_{k^2} should be

$$a_k \prod_{j=0}^{k-1} \frac{j!}{(j+k)!}, \qquad a_k = \prod_p (1-p^{-1})^{k^2} {}_2F_1(k;k;1;1/p), \tag{10.68}$$

where ${}_2F_1(a;b;c;z)$ is the *Gauss hypergeometric function* in standard notation,

$${}_2F_1(a;b;c;z) = \frac{\Gamma(c)}{\Gamma(a)\Gamma(b)} \sum_{n=0}^{\infty} \frac{\Gamma(a+n)\Gamma(b+n)}{\Gamma(c+n)} \frac{z^n}{n!} \qquad (|z| < 1),$$

which agrees with the earlier value proposed by J. Keating and N. Snaith [KeSn].

We can get (at least in principle) the information about the sixth moment of $\zeta(\frac{1}{2}+it)$ from $\mathcal{M}_3(s)$. Namely from (10.6) of Theorem 10.3 with $k=3$, or from the method of proof of Lemma 9.4, we get that

$$\int_T^{2T} |\zeta(\tfrac{1}{2}+it)|^6 \,\mathrm{d}t \ll_\varepsilon T^{2\sigma-1}\int_1^{T^{1+\varepsilon}} |\mathcal{M}_3(\sigma+it)|^2 \,\mathrm{d}t + T^{1+\varepsilon} \quad (\tfrac{1}{2}<\sigma\le 1), \tag{10.69}$$

provided that $\mathcal{M}_3(s)$ can be continued to $\mathbb{R}\mathrm{e}\, s \ge \sigma$ (and that is the catch!). Heuristically, we should be able to have $\sigma = 3/4+\varepsilon$, and then the integral on the right-hand side of (10.69) should be $\ll_\varepsilon T^{1/2+\varepsilon}$, giving a weak form of the sixth moment. Note that (see [Iv12], eq. (4.7)) for the eighth moment we have

$$\int_T^{2T} |\zeta(\tfrac{1}{2}+it)|^8 \,\mathrm{d}t \ll_\varepsilon T^{2\sigma-1}\int_1^{T^{1+\varepsilon}} |\mathcal{Z}_2(\sigma+it)|^2 \,\mathrm{d}t + T^{1+\varepsilon} \quad (\tfrac{1}{2}<\sigma\le 1), \tag{10.70}$$

and the analogs of (10.69) and (10.70) hold also for the mean square and fourth power of $|\zeta(\frac{1}{2}+it)|$. In these cases, however, the results are not of particular interest, since we have precise information which has been obtained by other methods (see (10.17) and the discussion thereafter). The bounds for the sixth moment of $|\zeta(\frac{1}{2}+it)|$ are intricately connected with the problem of the analytic continuation of $\mathcal{M}_3(s)$ to the region $\sigma \le 1$. It should be noted that the bounds

$$\int_0^T |\zeta(\tfrac{1}{2}+it)|^8 \,\mathrm{d}t \ll_\varepsilon T^{1+\varepsilon}$$

and

$$\int_T^{2T} |\mathcal{Z}_2(\sigma+it)|^2 \,\mathrm{d}t \ll_\varepsilon T^{4-4\sigma+\varepsilon} \qquad (\tfrac{1}{2}<\sigma\le 1),$$
$$\int_T^{2T} |\mathcal{Z}_2(\sigma+it)|^2 \,\mathrm{d}t \ll_\varepsilon T^{2-2\sigma+\varepsilon} + T^{-1} \quad (\sigma \ge 1)$$

are equivalent (see [Iv12], eqs. (4.3) and (4.8)).

For a proof of (10.29) see the monograph of K. Ramachandra [Ram]. We could have used (2.7) or Lemma 2.10 as well. Ramachandra proved (10.29) for $2k \in \mathbb{N}$. This result was extended by D. R. Heath-Brown [Hea4] to the case when $k>0$ is rational. Under the RH, both Ramachandra and Heath-Brown (*op. cit.*) showed that (10.29) holds. Recently M. Radziwill and K. Soundararajan [RaSo] showed that (unconditionally)

$$\int_0^T |\zeta(\tfrac{1}{2}+it)|^{2k} \,\mathrm{d}t \ge \mathrm{e}^{-30k^4}T(\log T)^{k^2}$$

holds for any real $k>1$ and $T \ge T_0$.

The bound (10.30) of Theorem 10.11 was sharpened by M. Jutila [Jut7]. He proved the following. Let σ_1 and σ_2 be fixed numbers such that $\frac{1}{4}<\sigma_1<\sigma_2$. Then for $\sigma_1 \le \sigma \le \sigma_2$ and any fixed $\varepsilon>0$ we have

$$\mathcal{M}(\sigma+it) \ll_\varepsilon \tau^{5/6-(4/3)\sigma+\varepsilon} + \tau^{2/3-\sigma} + \tau^{-1},$$

where $\tau = \max(|t|,1)$. This is sharper than (10.30). His proof is based on a complex integration technique, similar to the one he used in [Jut3] to obtain the analytic continuation of $\mathcal{M}_2(s)$ to $\mathbb{C}$. This consists of writing (see (9.53) and (9.54))

$$f(s,w) := \zeta(w)\chi(w)^{-1/2}\left(-\left(w-\tfrac{1}{2}\right)^2\right)^{-s/2},$$

and noting that for $s = \sigma + it$ with $\sigma > 1$ we have

$$\mathcal{M}_1(s) = -i \int_{1/2+i}^{1/2+i\infty} f(s, w)\,\mathrm{d}w.$$

The contour in the above integral is then suitably modified, and the ensuing integrals are carefully estimated.

For $\sigma > 1/2$ note that the bound in (10.30) is better than the bound

$$\mathcal{Z}_1(\sigma + it) \ll_\varepsilon t^{1-\sigma+\varepsilon} \qquad (0 \le \sigma \le 1,\ t \ge t_0 > 0),$$

proved in [IJM], and for $\sigma > 2/3$ the bound with the exponent $5/6 - \sigma + \varepsilon$ proved by M. Jutila [Jut4].

It was proved (see [Iv11] and [IJM]) that

$$\int_1^T |\mathcal{Z}_k(\sigma + it)|^2\,\mathrm{d}t \gg_\varepsilon T^{2-2\sigma-\varepsilon} \qquad (k = 1, 2; \tfrac{1}{2} < \sigma \le 1),$$

and the lower bound in (10.32) is the analog of this result for $\mathcal{M}_1$.

The fact that $\Sigma_2(x)$ in Atkinson's formula (Lemma 8.1) is of logarithmic order in the mean square sense is equation (15.61) of [Iv1].

The convexity of mean values for Dirichlet series is standard, see e.g., [Iv1], Lemma 8.6. This says the following: Let $F(s)$ be regular in the region

$$\Big\{ \mathcal{D} : \alpha \le \sigma \le \beta, s = \sigma + it, t > 1 \Big\},$$

and let $F(s) \ll \mathrm{e}^{Ct^2}$ for $s \in \mathcal{D}$ and some $C > 0$. Then, for any fixed $q > 0$ and γ satisfying $\alpha < \gamma < \beta$, we have

$$\int_2^T |F(\gamma + it)|^q\,\mathrm{d}t$$
$$\le \left(\int_1^{2T} |F(\alpha + it)|^q\,\mathrm{d}t + 1\right)^{(\beta-\gamma)/(\beta-\alpha)} \left(\int_1^{2T} |F(\beta + it)|^q\,\mathrm{d}t + 1\right)^{(\gamma-\alpha)/(\beta-\alpha)}.$$

The mean value theorem for Dirichlet polynomials states that, for arbitrary complex numbers $a_1, \dots, a_N$,

$$\int_0^T \Big|\sum_{n\le N} a_n n^{it}\Big|^2\,\mathrm{d}t = T\sum_{n\le N} |a_n|^2 + O\Big(\sum_{n\le N} n|a_n|^2\Big),$$

and the above formula remains valid for $N = \infty$, provided that the series on the right-hand side converge. For a proof, see e.g., theorem 5.2 of [Iv1].

A. Laurinčikas [Lau6] discusses, by using methods from probabilistic number theory, limit theorems for the Mellin transforms of the Riemann zeta-function.

T. Estermann's result (Theodor Estermann, February 5, 1902-November 29, 1991) mentioned in the text holds for a class of Dirichlet series of which the above product is a special case. Estermann's results were generalized by G. Dahlquist [Dah], and recent investigations include the works of G. Bhowmik and J.-C. Schlage-Puchta [BhS1], [BhS2].

For the works on natural boundaries for $\mathcal{Z}_k(s)$ and in general for *multiple Dirichlet series*, see the papers of A. Diaconu *et al.* [DGG], [DGH] and Y. Motohashi [Mot6]. In [DGH] Diaconu *et al.* consider the series

$$Z(s_1, \cdots, s_{2m}, w) = \int_1^\infty \zeta(s_1 + it)\cdots\zeta(s_m + it)\zeta(s_{m+1} - it)\cdots\zeta(s_{2m} - it)t^{-w}\,\mathrm{d}t \tag{10.71}$$

connected with moments of the Riemann zeta-function. Analytic properties of this function, closely connected to our function $\mathcal{Z}_k(s)$, may be put to advantage to deal with the important problem of the analytic continuation of the function $\mathcal{Z}_k(s)$ itself. It is shown in [DGH] that (10.71) has meromorphic continuation (as a function of $2m+1$ complex variables $s_1, \ldots, s_{2m}, w$) slightly beyond the region of absolute convergence, with a polar divisor at $w = 1$. It is also shown that (10.71) satisfies certain quasi-functional equations, which are used to obtain meromorphic continuation to an even larger region. Under the assumption that

$$Z(\tfrac{1}{2}, \cdots, \tfrac{1}{2}, w) \equiv \mathcal{Z}_m(w)$$

has holomorphic continuation to the region $\Re e\, w \geq 1$ (except for the pole at $w = 1$ of order m^2+1), the authors derive the conjecture on the moments of the zeta-function on the critical line in the form

$$\int_0^T |\zeta(\tfrac{1}{2}+it)|^{2k}\,\mathrm{d}t = (c_k + o(1))T\log^{k^2} T \qquad (T \to \infty), \tag{10.72}$$

where $k \geq 2$ is a fixed integer and

$$c_k = \frac{a_k g_k}{\Gamma(1+k^2)}, \quad a_k = \prod_p (1-1/p)^{k^2}\sum_{j=0}^{\infty} d_k^2(p^j), \quad g_k = (k^2)!\prod_{j=0}^{k-1}\frac{j!}{(j+k)!}. \tag{10.73}$$

The formulas (10.72)-(10.73) coincide with the conjectures of [KeSn] and [CFKRS1], and a_k is the same as in (10.68).

In [DGH] the conjectural formula (10.73) for the coefficients c_k, in the final step, is derived from the following *Tauberian theorem*, named after Alfred Tauber (born November 5, 1866 in Bratislava, died July 26, 1942 in the Theresienstadt concentration camp). Let $F(x)$ $(x \geq 1)$ be a non-decreasing continuous function, and

$$f(s) := \int_1^{\infty} F(x)x^{-s-1}\,\mathrm{d}x,$$

so that $f(s)$ is the modified Mellin transform of $F(x)x^{-1}$. Let

$$P(s) = \gamma_M + \gamma_{M-1}(s-1) + \cdots + \gamma_0(s-1)^M \qquad (\gamma_M \neq 0),$$

and suppose that $f(s) - P(s)(s-1)^{-M-1}$ is holomorphic for $\mathbb{R}\mathrm{e}\, s > 1$ and continuous for $\mathbb{R}\mathrm{e}\, s = 1$. Then

$$F(x) \sim \frac{\gamma_M}{M!}x(\log x)^M \qquad (x \to \infty).$$

When $M = 0$ this is the classical *Wiener-Ikehara Tauberian theorem* (Norbert Wiener, November 26, 1894-March 18, 1964; Shikao Ikehara, April 11, 1904-October 10, 1984). For a proof of the above result see, for example, J. Korevaar [Kore], theorem III.4.1. Note that (*op. cit.*, eq. (4.1)) $\mathrm{d}s(v) = \mathrm{d}(s(v) - B)$ for any constant B, hence we may in fact assume that $F(x)$ above is actually non-negative. J. Korevaar kindly pointed out to me that M. A. Subhankulov [Sub] derives the general case under some more stringent conditions than the ones given above (but quite sufficient for applications involving the moments of $|\zeta(\frac{1}{2}+it)|$). A generalized version of the Wiener-Ikehara Tauberian theorem was obtained long ago by H. Delange [Del] (Hubert Delange, 1913-2003, French mathematician), but it is not obvious whether his arguments can be modified to yield the above result. Professor Korevaar also pointed out to me how the general case, stated above, can be reduced to follow from his proof of the Wiener-Ikehara Tauberian theorem, given in [Kore]. It is based on the following proposition: Let $\sigma(t)$ vanish for $t < 0$, be non-negative for

$t \geq 0$ and such that the Laplace transform

$$F(z) = \mathcal{L}[\sigma(z)] = \int_0^\infty \sigma(t) e^{-zt}\, dt$$

exists for $\mathbb{R}e\, z = x > 0$. Suppose that for $x \to 0+$ the function

$$G(z) = F(z) - A/z, \quad z = x + iy,$$

converges to a boundary function $G(iy)$ in $L^1(-\lambda, \lambda)$. Then the integral

$$\int_{-\infty}^{\infty} K_\lambda(u - t)\sigma(t)\, dt = \int_{-\infty}^{\lambda u} \sigma(u - v/\lambda) K(v)\, dv$$

exists, and as $u \to \infty$ tends to $A \int_{-\infty}^{\infty} K(v)\, dv = A$, where K is the well-known *Fejér kernel* (named after the Hungarian mathematician Lipót Fejér (or Leopold Fejér), February 9, 1880-October 15, 1959)

$$K_\lambda(t) = \lambda K(\lambda t) = \frac{\lambda}{2\pi} \left(\frac{\sin \lambda t/2}{\lambda t/2} \right)^2.$$

11

On some problems involving Hardy's function and zeta-moments

11.1 The distribution of values of Hardy's function

This last chapter is devoted to some open problems involving Hardy's function and zeta-moments. It is hoped that this will be of interest to all who want to do research in areas connected to these topics. The aim of this section is to study the distribution of positive and negative values of $Z(t)$. This is an important problem, since $Z(t)$ is a highly oscillating function, with $\gg T \log T$ zeros in $[0,\ T]$ (see (1.49)). Also recall that (see (1.53))

$$|Z(t)| = \Omega\left\{\exp\left[(1+o(1))\sqrt{\frac{\log t}{\log\log t}}\right]\right\} \qquad (t \to \infty),$$

so that $|Z(t)|$ takes some fairly large values.

To study the distribution of values of $Z(t)$ we introduce the integrals

$$I_+(T) := \int_{T,Z(t)>0}^{2T} Z(t)\,\mathrm{d}t, \quad I_-(T) := \int_{T,Z(t)<0}^{2T} Z(t)\,\mathrm{d}t$$

and set

$$\begin{aligned}\mathcal{J}_+(T) &:= \mu\Big\{T < t \le 2T\ :\ Z(t) > 0\Big\},\\ \mathcal{J}_-(T) &:= \mu\Big\{T < t \le 2T\ :\ Z(t) < 0\Big\},\end{aligned} \tag{11.1}$$

where $\mu(\cdot)$ denotes measure. Note that by the (7.24) we have

$$I_+(T) + I_-(T) = \int_T^{2T} Z(t)\,\mathrm{d}t = O_\varepsilon(T^{1/4+\varepsilon}). \tag{11.2}$$

On the other hand

$$I_+(T) - I_-(T) = \int_{T,Z(t)>0}^{2T} Z(t)\,\mathrm{d}t - \int_{T,Z(t)<0}^{2T} Z(t)\,\mathrm{d}t$$
$$= \int_T^{2T} |Z(t)|\,\mathrm{d}t = \int_T^{2T} |\zeta(\tfrac{1}{2}+it)|\,\mathrm{d}t, \tag{11.3}$$

since $|Z(t)| = |\zeta(\frac{1}{2}+it)|$. But we know that

$$T(\log T)^{1/4} \ll \int_T^{2T} |\zeta(\tfrac{1}{2}+it)|\,\mathrm{d}t \ll T(\log T)^{1/4}, \tag{11.4}$$

and from (11.2)-(11.3) it follows that

$$I_+(T) = \tfrac{1}{2}\int_T^{2T} |\zeta(\tfrac{1}{2}+it)|\,\mathrm{d}t + O_\varepsilon(T^{1/4+\varepsilon}),$$
$$-I_-(T) = \tfrac{1}{2}\int_T^{2T} |\zeta(\tfrac{1}{2}+it)|\,\mathrm{d}t + O_\varepsilon(T^{1/4+\varepsilon}).$$

In fact, as shown by (8.18), we can dispense with "ε" in the above two formulas. In view of (11.4) we then obtain the following.

Theorem 11.1 *We have*

$$T(\log T)^{1/4} \ll I_+(T) \ll T(\log T)^{1/4},$$
$$T(\log T)^{1/4} \ll -I_-(T) \ll T(\log T)^{1/4}. \tag{11.5}$$

If one could sharpen (11.4) to an asymptotic formula, then we could sharpen (11.5) and solve the following

Problem 11.2 Is it true that there is a constant $A > 0$ such that, for $T \to \infty$,

$$I_+(T) = (A + o(1))T(\log T)^{1/4},$$
$$-I_-(T) = (A + o(1))T(\log T)^{1/4}?$$

If such A exists, is it true that $A = 1/2$?

11.2 The order of the primitive of Hardy's function

As in Chapters 7 and 8, let

$$F(T) := \int_1^T Z(t)\,\mathrm{d}t \qquad (T \ge 1) \tag{11.6}$$

denote a primitive of Hardy's function $Z(t)$. We recall Theorem 8.3, which gives an explicit expression for $F(T)$, and note that the main contribution to

$F(T)$ comes from the expression

$$\bar{F}(T) := \left(\frac{T}{2\pi}\right)^{1/4}(-1)^L K(\theta) = 2\pi\left(\frac{T}{2\pi}\right)^{1/4}(-1)^{[\sqrt{T/(2\pi)}\,]}.$$

We shall show now that, on the average, the function $\bar{F}(T)$ is small. This is provided by the following

Theorem 11.3 *We have*

$$\int_1^T \bar{F}(t)\mathrm{d}t = O(T^{3/4}),$$
$$\int_1^T \bar{F}(t)\mathrm{d}t = \Omega_\pm(T^{3/4}). \tag{11.7}$$

Proof of Theorem 11.3. Setting $u(x) = (-1)^{[x]}K(\{x\})$ ($[x]$ is the integer part of x and $\{x\}$ is its fractional part) we see that $u(x)$ is an odd function satisfying $u(x+2) = u(x)$. It follows that we have the Fourier series (for $x - \frac{1}{4} \notin \mathbb{Z}$, $x - \frac{3}{4} \notin \mathbb{Z}$)

$$u(x) = 4\sum_{n=1}^{\infty} \frac{\cos(\frac{\pi n}{4}) - \cos(\frac{3\pi n}{4})}{n} \sin(\pi n x).$$

Setting

$$a(n) := \cos(\frac{\pi n}{4}) - \cos(\frac{3\pi n}{4})$$

we see that $a(2k) = 0$ and that $a(n) = \sqrt{2}$ for $n \equiv 1, 7 \pmod 8$, and $a(n) = -\sqrt{2}$ for $n \equiv 3, 5 \pmod 8$. This gives

$$u(x) = 4\sum_{k=1}^{\infty} \frac{a(2k-1)}{2k-1} \sin\big(\pi(2k-1)x\big) \quad (x - \tfrac{1}{4} \notin \mathbb{Z},\ x - \tfrac{3}{4} \notin \mathbb{Z}).$$

Hence with the change of variable $t = 2\pi x^2$ and integration by parts we obtain, since the above series is boundedly convergent and can be integrated termwise,

$$\begin{aligned}\int_1^T \bar{F}(t)\,\mathrm{d}t &= \int_1^T \left(\frac{t}{2\pi}\right)^{1/4} u\left(\left(\frac{t}{2\pi}\right)^{1/2}\right)\mathrm{d}t\\ &= 16\pi \int_1^{\sqrt{T/2\pi}} x^{3/2} \sum_{k=1}^{\infty} \frac{a(2k-1)}{2k-1} \sin\big(\pi(2k-1)x\big)\,\mathrm{d}x\\ &= 16\pi \sum_{k=1}^{\infty} \frac{a(2k-1)}{2k-1} \int_1^{\sqrt{T/2\pi}} x^{3/2} \sin\big(\pi(2k-1)x\big)\,\mathrm{d}x\\ &= -16\left(\frac{T}{2\pi}\right)^{3/4} \sum_{k=1}^{\infty} \frac{a(2k-1)}{(2k-1)^2} \cos\left(\pi(2k-1)\sqrt{\frac{T}{2\pi}}\right) + O(T^{1/4}).\end{aligned}$$

Taking into account that the last series is absolutely convergent, we obtain the O-estimate of (11.7). The omega-results follow if we take $T = 2\pi(\frac{3}{4} + 2m)^2$ and $T = 2\pi(\frac{1}{4} + 2m)^2$ with $m \in \mathbb{N}$ and $m \to \infty$. In fact, the last O-term in the formula above stands for a function that is $O(T^{1/4})$ and also $\Omega_\pm(T^{1/4})$. This proves (11.7) and settles the question of the true order of $\bar{F}(T)$. However, the true order of the primitive of $F(T)$ remains elusive, since it is not obvious how small will be, when integrated, the expression standing for the O-term in Jutila's expression for $F(T)$ in Theorem 8.2 and Theorem 8.3. Observe that, on integrating by parts, we have

$$\int_1^T F(t)\,\mathrm{d}t = \int_1^T (T - t)Z(t)\,\mathrm{d}t,$$

which shows the connection between the primitive of $F(t)$ and $Z(t)$.

Problem 11.4 What is the true order of

$$\int_1^T F(t)\,\mathrm{d}t?$$

Is it perhaps true that

$$\int_1^T F(t)\,\mathrm{d}t = O(T^{3/4}), \quad \int_1^T F(t)\,\mathrm{d}t = \Omega_\pm(T^{3/4})?$$

11.3 The cubic moment of Hardy's function

The analogous problem when $Z(t)$ in $I_\pm(T)$ is replaced by $Z^3(t)$ is much harder. The cubic moment of $Z(t)$ was discussed in Chapter 7. As for its order, we state it here as follows.

Problem 11.5 Does there exist a constant $0 < c < 1$ such that

$$\int_1^T Z^3(t)\,\mathrm{d}t = O(T^c)? \tag{11.8}$$

One may naturally ask for bounds for higher moments of $Z(t)$. However, only odd moments are interesting in this context, because of their oscillating nature, when one expects considerable cancelation to occur in the integrand. Unfortunately when $k > 4$ not much, in general, is known on the moments of $|\zeta(\frac{1}{2} + it)|^k$ (see, e.g., [Iv1]). No one has been able to prove (11.8) yet with some $c < 1$, although one feels that there must be a lot of cancelation between the positive and negative values of $Z(t)$ and that it should be true. Recall that what can be proved is (this is (7.5) with $k = 3$ and, as mentioned in Section 7.2, the terms standing for $+\cdots+$ may be omitted))

$$\int_T^{2T} Z^3(t)\,\mathrm{d}t = 2\pi\sqrt{\frac{2}{3}} \sum_{(\frac{T}{2\pi})^{3/2}\le n\le(\frac{T}{\pi})^{3/2}} d_3(n)n^{-\frac{1}{6}}\cos\big(3\pi n^{\frac{2}{3}}+\tfrac{1}{8}\pi\big) + O_\varepsilon(T^{3/4+\varepsilon}), \tag{11.9}$$

where as usual $d_3(n)$ is the divisor function

$$d_3(n) = \sum_{k\ell m=n} 1,$$

generated by $\zeta^3(s)$. The difficulty in obtaining (11.8) with some $3/4 < c < 1$ is in the estimation of the exponential sum on the right-hand side of (11.9). It can be shown that

$$\sum_{N\le n\le 2N} d_3(n)n^{-1/6}\cos\big(3\pi n^{\frac{2}{3}}+\tfrac{1}{8}\pi\big) \ll_\varepsilon N^{2/3+\varepsilon}. \tag{11.10}$$

However, this just gives the bound $O_\varepsilon(T^{1+\varepsilon})$ for the integral on the left-hand side of (11.8). This is unfortunately weak, since by the Cauchy-Schwarz inequality for integrals we easily obtain a better result, namely

$$\Big|\int_T^{2T} Z^3(t)\,\mathrm{d}t\Big| \le \left(\int_T^{2T} |\zeta(\tfrac{1}{2}+it)|^2\,\mathrm{d}t \int_T^{2T} |\zeta(\tfrac{1}{2}+it)|^4\,\mathrm{d}t\right)^{1/2} \ll T(\log T)^{5/2} \tag{11.11}$$

on using the well-known elementary bounds

$$\int_T^{2T} |\zeta(\tfrac{1}{2}+it)|^2\,\mathrm{d}t \ll T\log T,$$
$$\int_T^{2T} |\zeta(\tfrac{1}{2}+it)|^4\,\mathrm{d}t \ll T\log^4 T.$$

For the cubic moment of $Z(t)$ there seems to exist no better bound than (11.11), and any improvement thereof would be interesting. A nice feature of the exponential sum in (11.10) is that it is "pure" in the sense that the argument of the cosine depends only on n, and not on any other quantity. A reasonable conjecture is that

$$\int_1^T Z^3(t)\,\mathrm{d}t = O_\varepsilon(T^{3/4+\varepsilon}), \qquad \int_1^T Z^3(t)\,\mathrm{d}t = \Omega_\pm(T^{3/4}).$$

We note that actually (11.10) holds if $d_3(n)$ is replaced by the general divisor function $d_k(n)$ (the number of ways n can be written as a product of k factors, so that $d_1(n) \equiv 1$), provided that we have the estimate

$$\int_0^T |\zeta(\tfrac{1}{2} + it)|^k \,\mathrm{d}t \ll_{\varepsilon,k} T^{1+\varepsilon} \qquad (k \in \mathbb{N}). \tag{11.12}$$

In (11.12) it is assumed that k is fixed. If (11.12) holds for any k ($\ge k_0$), then this is equivalent to the Lindelöf hypothesis that $|\zeta(\frac{1}{2} + it)| \ll_\varepsilon |t|^\varepsilon$.

The assertion related to (11.10) is contained in the next theorem.

Theorem 11.6 *If* (11.12) *holds for some fixed* $k \in \mathbb{N}$, *then*

$$\sum_{N \le n \le 2N} d_k(n) n^{-1/6} \cos\big(3\pi n^{\frac{2}{3}} + \tfrac{1}{8}\pi\big) \ll_\varepsilon N^{2/3+\varepsilon}. \tag{11.13}$$

The bound (11.12) is at present known to hold only when $k \le 4$, in which case we have a non-trivial result. It is unclear whether (11.13), in the general case, has an arithmetic meaning as in the case $k = 3$ (cf. (11.9)).

Proof of Theorem 11.6. From the Perron inversion formula we have

$$\sum_{n \le x} d_k(n) = \frac{1}{2\pi i} \int_{1+\varepsilon-iT}^{1+\varepsilon+iT} \frac{\zeta^k(w) x^w}{w} \,\mathrm{d}w + O_\varepsilon(x^\varepsilon) \qquad (T \asymp x, x \asymp N). \tag{11.14}$$

We replace the segment of integration by $[\frac{1}{2} - iT, \frac{1}{2} + iT]$, passing over the pole $w = 1$ of order k of $\zeta^k(w)$. The residue is $xP_{k-1}(\log x)$, with $P_{k-1}(\log x)$ a polynomial of degree $k - 1$ in $\log x$, whose coefficients depend on k. It follows by the residue theorem that

$$\begin{aligned}\Delta_k(x) &:= \sum_{n \le x} d_k(n) - xP_{k-1}(\log x)\\ &= \frac{1}{2\pi i} \int_{\frac{1}{2}-iT}^{\frac{1}{2}+iT} \frac{\zeta^k(w) x^w}{w} \,\mathrm{d}w + R_k(x, T),\end{aligned} \tag{11.15}$$

say. Here $\Delta_k(x)$ (cf. (9.57)) is the error term in the asymptotic formula for the summatory function of $d_k(n)$, and $R_k(x, T)$ stands for the error term in (11.14) plus the contribution over the segments $[\frac{1}{2} \pm iT,\ 1 + \varepsilon \pm iT]$, so that

$$R_k(x, T) \ll_\varepsilon \int_{\frac{1}{2}}^{1+\varepsilon} |\zeta(\sigma + iT)|^k x^\sigma T^{-1} \,\mathrm{d}\sigma + N^\varepsilon. \tag{11.16}$$

If we integrate (11.16) over T from T_1 to $2T_1$, we obtain

$$\begin{aligned}\int_{T_1}^{2T_1} R_k(x, T) \,\mathrm{d}T &\ll_\varepsilon \int_{\frac{1}{2}}^{1+\varepsilon} \Big(\int_{T_1}^{2T_1} |\zeta(\sigma + iT)|^k x^\sigma T^{-1} \,\mathrm{d}T\Big) \mathrm{d}\sigma + N^{1+\varepsilon}\\ &\ll_\varepsilon N^{1+\varepsilon}.\end{aligned}$$

Here we used the bound

$$\int_0^T |\zeta(\sigma+it)|^k \,\mathrm{d}t \ll_{\varepsilon,k} T^{1+\varepsilon} \qquad (\tfrac12 \le \sigma \le 1),$$

which follows by (11.12) and the convexity of mean values (see Notes of Chapter 10). This means that one can find a value of T ($\asymp N$) such that $R_k(x,T) \ll_{\varepsilon,k} N^\varepsilon$ uniformly in x when $N \le x \le 2N$. It is actually this value of T that is taken in (11.14).

We write now the sum in (11.13) as

$$\int_N^{2N} x^{-1/6}\cos\bigl(3\pi x^{\frac23} + \tfrac18\pi\bigr)\,\mathrm{d}\Bigl(\sum_{n\le x} d_k(n)\Bigr)$$

and use (11.14). The contribution of $xP_{k-1}(\log x)$, by the first derivative test, will be

$$\int_N^{2N} x^{-1/6}\Bigl(P_{k-1}(\log x) + P_{k-1}'(\log x)\Bigr)\cos\bigl(3\pi x^{\frac23} + \tfrac18\pi\bigr)\,\mathrm{d}x$$
$$\ll N^{-1/6}\log^{k-1} N \cdot N^{1/3} = N^{1/6}\log^{k-1} N.$$

In the portion pertaining to $\Delta_k(x)$ we integrate by parts the term $R_k(x,T)$. Since $R_k(x,T) \ll_{\varepsilon,k} N^\varepsilon$, trivial estimation will yield a contribution which is $\ll_\varepsilon N^{1/2+\varepsilon}$, and this is probably optimal. After differentiating the remaining expression over x and writing $w = \frac12 + iv$ and the cosine as a sum of exponentials, we are left with a multiple of

$$\int_{-T}^{T} \zeta^k(\tfrac12 + iv)\left\{\int_N^{2N} x^{-2/3}\exp\left(iv\log x \pm 3\pi i x^{2/3}\right)\,\mathrm{d}x\right\}\mathrm{d}v.$$

The function in the exponential is $iF_\pm(x)$ with

$$F_\pm(x) = F_\pm(x,v) := v\log x \pm 3\pi x^{2/3},$$

so that

$$F_\pm'(x) = \frac{v}{x} \pm 2\pi x^{-1/3}, \quad F_\pm''(x) = -\frac{v}{x^2} \mp \frac23\pi x^{-4/3}.$$

We suppose that $v > 0$, since the other case is analogous. The saddle point x_0 (root of $F_-'(x) = 0$ in this case) is $x_0 = (v/(2\pi))^{3/2}$, and $x_0 \in [N, 2N]$ for $v \asymp N^{2/3}$. When this is not satisfied, the contribution is, by the first derivative test, $\ll_\varepsilon N^{1/3+\varepsilon}$. For $v \asymp N^{2/3}$ we have

$$|F_-''(x_0)|^{-1/2} \asymp v \asymp N^{2/3}.$$

Hence by the second derivative test and (11.12) we have that this contribution is, with suitable constants $0 < C_1 < C_2$,

$$\ll \int_{C_1 N^{2/3}}^{C_2 N^{2/3}} |\zeta(\tfrac{1}{2} + iv)|^k \, \mathrm{d}v \cdot N^{-2/3} N^{2/3} \ll_\varepsilon N^{2/3+\varepsilon}.$$

Therefore (11.13) follows, and Theorem 11.6 is proved. If one wants to have explicitly the main term produced by this method, one can use the saddle-point method (see, e.g., [Iv1, chapter 2]).

11.4 Further problems on the distribution of values

If (11.8) holds with some $0 < c < 1$ then, similarly as in the case of $I_\pm(T)$, we obtain

$$\begin{aligned} \int_{T, Z(t)>0}^{2T} Z^3(t) \, \mathrm{d}t &= \tfrac{1}{2} \int_T^{2T} |\zeta(\tfrac{1}{2} + it)|^3 \, \mathrm{d}t + O(T^c), \\ \int_{T, Z(t)<0}^{2T} Z^3(t) \, \mathrm{d}t &= \tfrac{1}{2} \int_T^{2T} |\zeta(\tfrac{1}{2} + it)|^3 \, \mathrm{d}t + O(T^c). \end{aligned} \tag{11.17}$$

Note that no asymptotic formula exists yet for the integral on the right-hand side of (11.17). In general it is conjectured that (known to be true only when $k = 2, 4$)

$$\int_0^T |\zeta(\tfrac{1}{2} + it)|^k \, \mathrm{d}t = (c_k + o(1)) T (\log T)^{k^2/4} \qquad (k \in \mathbb{N}, \ T \to \infty), \tag{11.18}$$

where the value of c_k for even k is given by (10.72)-(10.73). In the case $k = 3$ all that is currently known is

$$T(\log T)^{9/4} \ll \int_0^T |\zeta(\tfrac{1}{2} + it)|^3 \, \mathrm{d}t \ll T(\log T)^{5/2}.$$

The upper bound follows as in (11.11), and the lower bound is a special case of the general lower bound (10.29) of K. Ramachandra [Ram], namely

$$\int_0^T |\zeta(\tfrac{1}{2} + it)|^k \, \mathrm{d}t \gg_k T(\log T)^{k^2/4} \qquad (k \in \mathbb{N}).$$

Note that this lower bound is of the order given by the conjectural formula (11.18).

The problems of the asymptotic evaluation of $I_{\pm}(T)$ are clearly connected to the distribution of the values of $\zeta(\frac{1}{2}+it)$ and $|\zeta(\frac{1}{2}+it)|$. It is known that

$$\mu(A_c(T)) = \frac{T}{2} + O\left(T\frac{(\log\log\log T)^2}{\sqrt{\log\log T}}\right), \tag{11.19}$$

where $c > 0$ is any constant and

$$A_c(T) := \left\{0 < t \le T : |\zeta(\tfrac{1}{2}+it)| \le c\right\}.$$

Unfortunately one does not see how to put (11.19) to use in connection with the distribution of positive values of $Z(t)$. It is hard to determine asymptotically the order of $\mathcal{J}_+(T)$, $\mathcal{J}_-(T)$, defined by (11.1). We formulate the following.

Problem 11.7 Is it true that there exist constants $A_+ > 0$, $A_- > 0$ such that

$$\mathcal{J}_+(T) := \mu\Big\{T < t \le 2T \;:\; Z(t) > 0\Big\} = \big(A_+ + o(1)\big)T \quad (T \to \infty), \tag{11.20}$$

$$\mathcal{J}_-(T) := \mu\Big\{T < t \le 2T \;:\; Z(t) < 0\Big\} = \big(A_- + o(1)\big)T \quad (T \to \infty)? \tag{11.21}$$

Obviously $A_+ + A_- = 1$ (if A_+, A_- exist). The asymptotic formula (11.18) gives rise to the thought that in that case maybe $A_+ = A_- = 1/2$. On the other hand, things may not be that simple. If one assumes the RH and the simplicity of zeta-zeros (these very strong conjectures seem to be independent in the sense that it is not known whether either of them implies the other one) then (since $Z(0) = \zeta(1/2) < 0$)

$$\mu\Big\{T < t \le 2T : Z(t) > 0\Big\} = \sum_{T<\gamma_{2n}\le 2T} (\gamma_{2n} - \gamma_{2n-1}) + O(1), \tag{11.22}$$

where $0 < \gamma_1 < \gamma_2 < \cdots$ are the ordinates of complex zeros of $\zeta(s)$. The sum in (11.22) is connected to the sum ($\alpha \ge 0$ is fixed)

$$\sum\nolimits_\alpha(T) := \sum_{\gamma_n \le T} (\gamma_n - \gamma_{n-1})^\alpha. \tag{11.23}$$

The sum $\sum_\alpha(T)$ in turn can be connected to the Gaussian Unitary Ensemble hypothesis (see A. M. Odlyzko [Odl1], [Odl3]) and the pair correlation conjecture of H. L. Montgomery [Mon1]. Both of these conjectures assume the RH and, for example, the former states that, for

$$0 \le \alpha < \beta < \infty, \quad \delta_n = \frac{1}{2\pi}(\gamma_{n+1} - \gamma_n)\log\left(\frac{\gamma_n}{2\pi}\right),$$

we have

$$\sum_{\gamma_n \le T, \delta_n \in [\alpha,\beta]} 1 = \left(\int_\alpha^\beta p(0,u)\,\mathrm{d}u + o(1) \right) \frac{T}{2\pi} \log\left(\frac{T}{2\pi}\right) \qquad (T \to \infty),$$

where $p(0,u)$ is a certain probabilistic density, given by complicated functions defined in terms of prolate spheroidal functions. We have

$$\begin{aligned} 1 - \left(\frac{\sin \pi u}{\pi u}\right)^2 &= \sum_{k=0}^{\infty} p(k,u), \\ p(0,u) &= \frac{1}{3}\pi^2 u^2 - \frac{2}{15}\pi^4 u^4 + \frac{1}{315}\pi^6 u^6 + \cdots \qquad (u \to 0+), \\ \log p(0,u) &= -\frac{\pi^2}{8} + o(1) \qquad (u \to \infty). \end{aligned}$$

In fact, in [Iv6] it was proved that, if the RH and the Gaussian unitary ensemble hypothesis hold, then for $\alpha \ge 0$ fixed and $T \to \infty$,

$$\sum\nolimits_\alpha (T) = \left(\int_0^\infty p(0,u) u^\alpha \,\mathrm{d}u + o(1) \right) \left(\frac{2\pi}{\log(\frac{T}{2\pi}) - 1} \right)^{\alpha - 1} T.$$

Also note that, since

$$\mathbb{R}\mathrm{e} \log \zeta(\tfrac{1}{2} + it) = \log |Z(t)|,$$

a classical result of A. Selberg (see [Sel]) states that, for any real $\alpha < \beta$,

$$\lim_{T\to\infty} \frac{1}{T} \mu \left\{ t : t \in [T, 2T],\ \alpha < \frac{\log |Z(t)|}{\sqrt{\frac{1}{2}\log\log T}} < \beta \right\} = \frac{1}{\sqrt{2\pi}} \int_\alpha^\beta \mathrm{e}^{-\frac{1}{2}x^2}\,\mathrm{d}x, \tag{11.24}$$

but here we are interested in the distribution of values of $Z(t)$ and not $|Z(t)|$.

Recently J. Kalpokas and J. Steuding [KaS] proved, among other things, that for $\phi \in [0, \pi)$,

$$\sum_{0<t\le T, \zeta(\frac{1}{2}+it) \in \mathrm{e}^{i\phi}\mathbb{R}} \zeta(\tfrac{1}{2} + it) = \left(2\mathrm{e}^{i\phi} \cos\phi\right) \frac{T}{2\pi} \log \frac{T}{2\pi \mathrm{e}} + O_\varepsilon(T^{1/2+\varepsilon}). \tag{11.25}$$

An analogous result holds for the sums of $|\zeta(\frac{1}{2}+it)|^2$, namely

$$\sum_{0<t\le T,\zeta(\frac{1}{2}+it)\in e^{i\phi}\mathbb{R}} |\zeta(\tfrac{1}{2}+it)|^2$$
$$= \frac{T}{2\pi}\left(\log\frac{T}{2\pi e}\right)^2 + \left(2\gamma + 2\cos(2\phi)\right)\frac{T}{2\pi}\log\frac{T}{2\pi e} + \frac{T}{2\pi} + O_\varepsilon(T^{1/2+\varepsilon}),$$

where γ is Euler's constant. It is unclear whether (11.25) and the other approaches mentioned above can be put to use in connection with our problems.

Although (11.20) and (11.21) seem difficult to prove, one can at least show that

$$\mathcal{J}_+(T) \gg T(\log T)^{-1/2}, \tag{11.26}$$

and a similar lower bound holds for $\mathcal{J}_-(T)$. Namely from (11.5) we have, by the Cauchy-Schwarz inequality,

$$T(\log T)^{1/4} \ll I_+(T) \le \left(\int_{T,Z(t)>0}^{2T} 1\,\mathrm{d}t \int_T^{2T} |\zeta(\tfrac{1}{2}+it)|^2\,\mathrm{d}t\right)^{1/2},$$

which easily gives (11.26), since

$$\int_{T,Z(t)>0}^{2T} 1\,\mathrm{d}t = \mu\Big\{T < t \le 2T : Z(t) > 0\Big\} = \mathcal{J}_+(T).$$

We shall consider now another problem related to the distribution of values of $Z(t)$. Henceforth let $0 < \gamma_1 \le \gamma_2 \le \cdots$ denote positive ordinates of complex zeros of $\zeta(s)$; as already mentioned in Chapter 2 it is known that $\gamma_1 = 14.13\ldots$, and all known ($> 10^9$ in number) zeros are simple and lie on the critical line $\mathbb{R}\mathrm{e}\, s = \frac{1}{2}$. We define $\gamma_-(t) = \gamma_n$ if $\gamma_n \le t < \gamma_{n+1}$, $\gamma_+(t) = \gamma_{n+1}$ if $\gamma_n < t \le \gamma_{n+1}$, $\gamma_-(t) = \gamma_+(t) = \gamma_n$ if $t = \gamma_n$,

$$\mathcal{A}(T) = \Big\{0 < t \le T : |Z(t)| \le \gamma_+(t) - \gamma_-(t)\Big\},$$

$$\mathcal{B}(T) = [0,\, T] \setminus \mathcal{A}(T) = \Big\{0 < t \le T : |Z(t)| > \gamma_+(t) - \gamma_-(t)\Big\}.$$

Natural problems are to evaluate asymptotically $\mu(\mathcal{A}(T))$ and $\mu(\mathcal{B}(T))$. We shall prove the following

Theorem 11.8 *We have*

$$\mu\big(\mathcal{B}(T)\big) = T + O\left(T\frac{(\log\log\log T)^2}{\sqrt{\log\log T}}\right). \tag{11.27}$$

Proof of Theorem 11.8 We shall employ a method based on the value distribution result (11.38). This leads to (11.27), but it is possible to replace $(\log\log\log T)^2$ by $\log\log\log T$ in (11.27). Let

$$\mathcal{C}_1(T) := \left\{0 < t \le T \ :\ \gamma_+(t) - \gamma_-(t) < \frac{(\log\log T)^6}{\log T}\right\},$$
$$\mathcal{C}_2(T) := \left\{0 < t \le T \ :\ |\zeta(\tfrac{1}{2} + it)| > \exp(-(\log\log T)^{3/4})\right\},$$

and let $\bar{S}$ denote the complement of S in $[0,\ T]$. From (11.38) with

$$a = -\infty, \quad b = -(\log\log T)^{3/4}$$

we obtain

$$\begin{aligned}\mu(\bar{\mathcal{C}}_2(T)) &\ll T \int_{\frac{1}{2}(\log\log T)^{1/4}}^{\infty} \mathrm{e}^{-\pi v^2}\,\mathrm{d}v + T\frac{(\log\log\log T)^2}{\sqrt{\log\log T}}\\ &\ll T\frac{(\log\log\log T)^2}{\sqrt{\log\log T}},\end{aligned}$$

hence

$$\mu\big(\mathcal{C}_2(T)\big) \ = \ T + O\left(T\frac{(\log\log\log T)^2}{\sqrt{\log\log T}}\right). \tag{11.28}$$

On the other hand

$$\mu(\mathcal{C}_2(T)) \ = \ \mu(\bar{\mathcal{C}}_1(T) \cap \mathcal{C}_2(T)) + \mu(\mathcal{C}_1(T) \cap \mathcal{C}_2(T)). \tag{11.29}$$

However, we have

$$\mu(\bar{\mathcal{C}}_1(T) \cap \mathcal{C}_2(T)) \ \le \ \mu(\bar{\mathcal{C}}_1(T)) \ll T\frac{(\log\log\log T)^2}{\sqrt{\log\log T}}. \tag{11.30}$$

The second bound in (11.30) is a consequence of a bound that follows from the following lemma.

Lemma 11.9 *Let* $0 < \gamma_1 \le \gamma_2 \le \cdots$ *denote imaginary parts of complex zeros of* $\zeta(s)$*, and let* $l \ge 2$*. Then there exists a constant* $C > 0$ *such that uniformly*

$$\sum_{T<\gamma_n\le T+H,\,\gamma_{n+1}-\gamma_n\ge l/\log T} 1 \ll \Big(N(T+H) - N(T)\Big)\exp(-Cl) + 1, \tag{11.31}$$

where $N(T)$ *is the number of zeros of* $\zeta(s)$ *with imaginary parts in* $(0,\ T]$*, and* $T^a < H \le T,\ a > \frac{1}{2}$.

Proof of Lemma 11.9 The basic result is the asymptotic formula [Tsa1, theorem 4] of K.-M. Tsang, whose proof is beyond the scope of this text. This says that, for $T^a < H \le T$, $a > \frac{1}{2}$, $0 < h < 1$ and any $k \in \mathbb{N}$, we have uniformly

$$\int_T^{T+H} \Big(S(t+h) - S(t)\Big)^{2k} \mathrm{d}t = \frac{H(2k)!}{(2\pi^2)^k k!} \log^k(2 + h\log T) + O\left\{H(ck)^k \left(k^k + \log^{k-\frac{1}{2}}(2 + h\log T)\right)\right\}, \tag{11.32}$$

where $c > 0$ is a constant, and as usual $S(T) = \frac{1}{\pi} \arg\zeta(\frac{1}{2} + iT)$. Recall the Riemann-von Mangoldt formula (1.31), namely

$$N(T) = \frac{T}{2\pi}\log\left(\frac{T}{2\pi}\right) - \frac{T}{2\pi} + S(T) + \frac{7}{8} + O\left(\frac{1}{T}\right).$$

This gives $\gamma_{n+1} - \gamma_n \ll 1$, similarly as in the proof of (2.13) (where the RH was assumed). We have also

$$S(t+h) - S(t) = N(t+h) - N(t) - \frac{h}{2\pi}\log t + O\left(\frac{h^2+1}{t}\right). \tag{11.33}$$

If

$$\gamma_n < t < \tfrac{1}{2}(\gamma_n + \gamma_{n+1}),\ \gamma_{n+1} - \gamma_n \ge \frac{l}{\log T},\ T \le t \le T + H,\ h = \frac{l}{2\log T}, \tag{11.34}$$

then $N(t+h) - N(t) = 0$, and $h \ll 1$ will hold in view of $\gamma_{n+1} - \gamma_n \ll 1$. For t satisfying (11.34) we have, in view of (11.33),

$$|S(t+h) - S(t)| \ge \frac{h}{4\pi}\log t \ge \frac{l}{8\pi},$$

and (11.32) will in fact hold for $0 < h \ll 1$. We obtain from (11.32)

$$\sum_{T<\gamma_n<\gamma_{n+1}\le T+H,\gamma_{n+1}-\gamma_n\ge l/\log T} \left(\frac{l}{8\pi}\right)^{2k} (\gamma_{n+1} - \gamma_n) \ll H(Ak(k + \log l))^k$$

with suitable $A > 0$, which implies that $(B = (8\pi)^2 A)$

$$\sum_{T<\gamma_n\le T+H,\gamma_{n+1}-\gamma_n\ge l/\log T} 1 \ll (N(T+H) - N(T))\left(Bk\frac{(k + \log l)}{l^2}\right)^k + 1. \tag{11.35}$$

We take

$$k = \left[\frac{l}{2\sqrt{B}}\right],$$

and (11.31) follows from (11.35) for $l \ge l_0 (\ge 2)$, while for $l < l_0$ the bound in (11.31) is trivial.

We continue with the proof of Theorem 11.8. To obtain (11.30) write

$$\begin{aligned}\bar{\mathcal{C}}_1(T) &= \bigcup_{k=1}^{\infty} D_k(T),\ D_k(T) := \{0 < t \le T : V_k(T) \\ &\le \gamma_+(t) - \gamma_-(t) < 2V_k(T)\}, \\ V_k(T) &:= \frac{2^{k-1}(\log\log T)^6}{\log T}.\end{aligned}$$

Hence with $\lambda = \lambda(k, T) = 2^{k-1}(\log\log T)^6$ we have, on using (11.31),

$$\begin{aligned}\mu(D_k(T)) &\le 2V_k(T) \sum_{\gamma_n \le T, \gamma_{n+1} - \gamma_n \ge \lambda/\log T} 1 \\ &\ll T\exp(-2^k(\log\log T)^2),\end{aligned}$$

which gives

$$\mu(\bar{\mathcal{C}}_1(T)) \ll T\sum_{k=1}^{\infty}\exp\Big(-2^k(\log\log T)^2\Big) \ll T\frac{(\log\log\log T)^2}{\sqrt{\log\log T}},$$

as asserted.

We therefore have, from (11.28)-(11.30),

$$\mu(\mathcal{C}_1(T) \cap \mathcal{C}_2(T)) = T + O\left(T\frac{(\log\log\log T)^2}{\sqrt{\log\log T}}\right),$$

and (11.27) with the error term $O\left(T\frac{(\log\log\log T)^2}{\sqrt{\log\log T}}\right)$ follows from

$$T \ge \mu(\mathcal{B}(T)) \ge \mu(\mathcal{C}_1(T) \cap \mathcal{C}_2(T)),$$

since for $t \in \mathcal{C}_1(T) \cap \mathcal{C}_2(T)$ we have

$$|Z(t)| > e^{-(\log\log T)^{3/4}} > \frac{(\log\log T)^6}{\log T} > \gamma_+(t) - \gamma_-(t).$$

Notes

The bounds in (11.4) are to be found in the monograph of K. Ramachandra [Ram], where many more results on mean values of powers of $\zeta(s)$ and its derivatives may be found. However, the asymptotic formula for

$$\int_0^T |\zeta(\tfrac{1}{2} + it)|\,\mathrm{d}t = \int_0^T |Z(t)|\,\mathrm{d}t$$

has not been established yet, and represents an interesting open problem. A major difficulty in the evaluation of the above integral is that the obvious identity $|z|^2 = z \cdot \bar{z}$, used in the evaluation of the even moments of $|\zeta(\frac{1}{2}+it)|$, seems to be of no use in this case.

A more elementary approach to Theorem 11.3, kindly suggested by M. Jutila, consists briefly of the following. The integral of $\bar{F}(t)$ over the interval $[2\pi m^2, 2\pi(m+1)^2]$ $(m \in \mathbb{N})$ is $\asymp \sqrt{m}$ in absolute value, and the sign depends on the parity of m. Thus summation over integers $m \ll \sqrt{T}$ of the same parity gives $\asymp T^{3/4}$, either positive or negative.

For a proof that the truth of (11.12) for every k implies the LH, see, for example, E. C. Titchmarsh [Tit3]. His book contains several other statements equivalent to the LH. For example, the LH is equivalent to $\alpha_k \le 1/2$ $(k = 2, 3, \ldots)$ or $\beta_k \le 1/2$ $(k = 2, 3, \ldots)$, where α_k, β_k denote the least numbers such that

$$\Delta_k(x) \ll_\varepsilon x^{\alpha_k+\varepsilon}, \qquad \int_1^x \Delta_k^2(y)\,\mathrm{d}y \ll_\varepsilon x^{1+2\beta_k+\varepsilon}$$

for any given $\varepsilon > 0$, and the error term in the general divisor problem $\Delta_k(x)$ is defined by (11.15). It is conjectured that $\forall k \ge 2$ one has $\alpha_k = \beta_k = (k-1)/(2k)$, but this is a very deep open problem. What is known is that $\alpha_k \ge (k-1)/(2k)$, $\beta_k \ge (k-1)/(2k)$ and various upper bounds exist for α_k and β_k (see [Iv1, chapter 13]). The conjecture $\alpha_k = \beta_k = (k-1)/(2k)$, even for a fixed k, does not seem to follow from other well-known conjectures such as the LH or the RH.

The Gaussian unitary ensemble is the most studied and basic ensemble in random matrix theory (see Notes to Chapter 2). It is a Gaussian measure on the set of Hermitian matrices (named after Charles Hermite, December 24, 1822-January 14, 1901, a French mathematician). The joint probability density for the eigenvalues of $n \times n$ Hermitian matrices is given by

$$\frac{1}{Z_n}\prod_{k=1}^{n} \mathrm{e}^{-\frac{1}{2}nx_k^2}\prod_{i<j}(x_j - x_i)^2,$$

where Z_n is a normalization constant. Random matrix theory has found applications in random tilings, multivariate statistics, quantum chaos and mesoscopic physics, among others. Recently, this theory also found has important applications concerning moments of L-functions, and conjectural formulas such as (10.72)-(10.73) (see, e.g., [KeSn], [CFKRS1], [CFKRS2]) are obtained by the use of this theory. These works also contain a discussion on the values of the coefficients c_k in (11.18) (see also the discussion in Notes of Chapter 10). With the exception of the Euler product, all the properties of the functions from the Selberg class $\mathcal{S}$ have a natural analogue in the characteristic polynomials of unitary matrices. This fact is expounded in detail in [CFKRS1], and it serves as a basis for modeling various conjectures involving L-functions in [CFKRS1] and [CFKRS2] by J. B. Conrey *et al.* For prolate spheroidal wave functions (a set of functions derived by timelimiting and lowpassing in mathematical physics), see, for example, C. Flammer [Fla].

For a proof of (11.19) see the present author's paper [Iv14]. It is based on an unpublished result of A. Selberg (see D. A. Hejhal [Hej2]) on the distribution of values of $\zeta(s)$. If we are given a measurable set $\mathcal{E}$ $(\subseteq \mathbb{C})$ with a positive *Jordan content* (Marie Ennemond Camille Jordan, January 5, 1838-January 22, 1922, a French mathematician), then Selberg showed that

$$\lim_{T\to\infty}\frac{1}{T}\mu\left(0 \le t \le T\ :\ \frac{\log\zeta(\frac{1}{2}+it)}{\sqrt{\log\log t}} \in \mathcal{E}\right) = \frac{1}{\pi}\iint\limits_{\mathcal{E}} \mathrm{e}^{-x^2-y^2}\,\mathrm{d}x\,\mathrm{d}y, \tag{11.36}$$

where $\mu(\cdot)$ denotes measure. Roughly speaking, this result says that $\log\zeta(\frac{1}{2}+it)/\sqrt{\log\log t}$ is approximately normally distributed. From (11.36) one can deduce the asymptotic formula

$$\int_0^T \log\left|\zeta(\tfrac{1}{2}+it) - a\right|\mathrm{d}t = (2\pi)^{1/2}T(\log\log T)^{1/2} + O_a(T),$$

where $a \neq 0$ is fixed. The book of D. Joyner [Joy] contains the extension to a rather wide class of L-functions of several well-known results by Selberg and Montgomery concerning the distribution of values of the Riemann zeta-function. In particular, it contains results about the distribution of values of $\zeta(s)$ and the pair correlation of the zeta-zeros. Some unpublished results of A. Selberg, like (11.36), are included with proof.

Independently of Selberg's unpublished work, A. Laurinčikas obtained results on the value distribution of $|Z(t)|$ and $|L(\frac{1}{2}+it,\chi)|$ in a series of papers (see [Lau1]-[Lau4]). As a special case of (11.36), he proved

$$\lim_{T\to\infty}\frac{1}{T}\mu\left(\left\{0\le t\le T\ :\ \frac{\log|Z(t)|}{\sqrt{\frac{1}{2}\log\log t}}\le y\right\}\right)=\frac{1}{\sqrt{2\pi}}\int_{-\infty}^{y}\mathrm{e}^{-u^2/2}\,\mathrm{d}u, \tag{11.37}$$

which is (11.24). It is interesting that Laurinčikas obtains (11.37) by applying techniques from probabilistic number theory to the asymptotic formula

$$\int_0^T|\zeta(\tfrac{1}{2}+it)|^{2\kappa}\,\mathrm{d}t=T(\log T)^{\kappa^2}\left\{1+O((\log\log T)^{-1/4})\right\},$$

which holds uniformly for $k_T\le k\le k_0$, where $k_0>0$ is arbitrary,

$$k_T:=\exp(-\sqrt{\log\log T}),\qquad \kappa:=([k^{-1}\sqrt{2\log\log T}]\pm 5)^{-1}.$$

D. A. Hejhal [Hej2], following Selberg's ideas, obtained a result similar to (11.28), but with an error term (see his equation (4.21)). He proved a general result, which in the case of $|Z(t)|=|\zeta(\frac{1}{2}+it)|$ gives

$$\mu(\{\,T\le t\le 2T\ :\ C_1\le|Z(t)|\le C_2\,\})=O\left(T\frac{(\log\log\log T)^2}{\sqrt{\log\log T}}\right)$$

for $0<C_1<C_2$, and for any fixed $c>0$

$$\mu(\{\,T\le t\le 2T\ :\ |Z(t)|\le c\,\})=\frac{1}{2}T+O\left(T\frac{(\log\log\log T)^2}{\sqrt{\log\log T}}\right).$$

This is in fact (11.19); note that the main term on the right-hand side does not depend on c. In particular, his equation (4.21), specialized to $\zeta(s)$, says that

$$\mu\left(\{\,T\le t\le 2T\ :\ \mathrm{e}^a\le|Z(t)|\le\mathrm{e}^b\,\}\right)=T\int_{a/\sqrt{\pi\psi}}^{b/\sqrt{\pi\psi}}\mathrm{e}^{-\pi v^2}\,\mathrm{d}v+O\left(\frac{T\log^2\psi}{\sqrt{\psi}}\right) \tag{11.38}$$

uniformly in $a,b\in\mathbb{R}$, where

$$\psi\ =\ \log\log T+O(\log\log\log T).$$

A discrete analogue of (11.37) holds for $Z'(t)$. First note that from (1.20) we have

$$Z'(t)=i\theta'(t)\mathrm{e}^{i\theta(t)}\zeta(t)+i\mathrm{e}^{i\theta(t)}\zeta'(\tfrac{1}{2}+it). \tag{11.39}$$

If $\gamma_n>0$ denotes ordinates of non-trivial zeros of $\zeta(s)$, and one assumes the RH, then (11.39) yields

$$|Z'(\gamma_n)|=|\zeta'(\tfrac{1}{2}+i\gamma_n)|.$$

D. A. Hejhal [Hej1] assumed the RH and a weak version of H. L. Montgomery's pair correlation conjecture [Mon1], namely that

$$\limsup_{N\to\infty}\left|\left\{n\ :\ N\le n\le 2N,(\gamma_{n+1}-\gamma_n)\log\gamma_n<c\right\}\right|\le Bc^{\tau}$$

for some $\tau > 0$, $B > 0$ and all $c \in (0,1)$. Then he proved, for $\alpha < \beta$,

$$\lim_{N\to\infty} \frac{1}{N}\left|\left\{ n \,:\, N \le n \le 2N, \frac{\log|Z'(\gamma_n)| + \log\Big(2\pi/(\log(\gamma_n/2\pi))\Big)}{\sqrt{\frac{1}{2}\log\log N}} \in (\alpha,\beta] \right\}\right|$$

$$= \frac{1}{\sqrt{2\pi}}\int_\alpha^\beta \mathrm{e}^{-u^2/2}\,\mathrm{d}u.$$

A numerical study of the derivative of the Riemann zeta-function at its zeros is to be found in the recent work of G. A. Hiary and A. M. Odlyzko [HiOd2]. It is connected to the study of the sum

$$J_\lambda(T) := \frac{1}{N(T)}\sum_{0<\gamma_n\le T} |\zeta'(\tfrac{1}{2}+i\gamma_n)|^{2\lambda} \qquad (\lambda > 0).$$

If the RH is assumed, then

$$J_\lambda(T) = \frac{1}{N(T)}\sum_{0<\gamma_n\le T} |Z'(\gamma_n)|^{2\lambda},$$

and if, additionally, one assumes the simplicity of zeta-zeros, then the sum is defined for all $\lambda \in \mathbb{R}$. Under the RH, S. M. Gonek [Gon] proved that $J_1(T) \sim \frac{1}{12}\log^3 T$ $(T \to \infty)$. Further, on the RH, it is conjectured by the use of random matrix methods (see the paper [HKN] by C. P. Hughes *et al.*) that, for $\lambda > -3/2$,

$$J_\lambda(T) \sim a(\lambda)\frac{G^2(\lambda+2)}{G(2\lambda+3)}\left(\log\frac{T}{2\pi}\right)^{\lambda(\lambda+2)} \qquad (T\to\infty).$$

Here $a(k)$ is the "arithmetic factor"

$$a(k) := \prod_p (1-1/p)^{k^2}\sum_{m=0}^\infty\left(\frac{\Gamma(m+k)}{m!\Gamma(k)}\right)^2 p^{-m}.$$

The function $G(z)$ is the *Barnes function* (named after the English mathematician Ernest William Barnes, 1874-1953). It satisfies the difference equation $G(z+1) = \Gamma(z)G(z)$, $G(1) = 1$ and may be defined as

$$G(z+1) = (2\pi)^{z/2}\exp\Big(-\tfrac{1}{2}(z^2+z+\gamma z^2)\Big) \times \prod_{n=1}^\infty\Big\{(1+z/n)^n\exp(-z+z^2/(2n))\Big\}.$$

In [Iv6] it was proved that, for $\alpha \ge 0$ fixed and $T \to \infty$, under the truth of the RH and the Gaussian unitary ensemble (GUE) hypothesis we have

$$\sum\nolimits_\alpha(T) = \Big(C_1(\alpha)+o(1)\Big)\left(\frac{2\pi}{\log\left(\frac{T}{2\pi}\right)-1}\right)^{\alpha-1} T,$$

$$c_1(\alpha) = \int_0^\infty p(0,u)u^\alpha\,\mathrm{d}u,$$

where $\sum_\alpha(T)$ is defined by (11.23), and $p(0,u)$ is the function appearing in the GUE hypothesis. It is easy to see, by using Hölder's inequality and

$$\sum_{\gamma_n\le T} 1 \ll T\log T, \qquad \sum_{\gamma_n\le T}(\gamma_{n+1}-\gamma_n) \sim T \qquad (T\to\infty)$$

that, for any fixed $\alpha > 1$, one has unconditionally

$$\sum\nolimits_{\alpha}(T) \gg T(\log T)^{\alpha-1}T.$$

In the other direction A. Fujii [Fuj2] in 1975 obtained the upper bound

$$\sum\nolimits_{\alpha}(T) \ll T(\log T)^{\alpha-1}T,$$

but getting an asymptotic formula for $\sum_{\alpha}(T)$ is difficult. The best unconditional upper bound for $\sum_{\gamma_n \le T}(\gamma_{n+1} - \gamma_n)^3$, where γ_n denotes zeros of $Z(t)$ (i.e. ordinates of zeros of $\zeta(s)$ on the critical line) is given by the bound in (2.54). The problems involving the zeros of $Z(t)$ are more difficult to deal with than the problems involving ordinates of all zeros of $\zeta(s)$, since for the former we do not have an analog of the Riemann-von Mangoldt formula ((1.30)-(1.31) of Theorem 1.6). The notation γ_n is commonly used in both cases.

Theorem 11.8 is from the present author's work [Iv14]. The improvement of the error term mentioned in the text is explained there. It follows from the work [IvPe] of A. Perelli and the present author.

We note that weaker results than Lemma 11.9 are given in A. Fujii [Fuj1], [Fuj2] and (without proof) in E. C. Titchmarsh [Tit3], p. 246.

Let

$$M_T(V) := \{\, t \in (0, T] \;:\; |Z(t)| \ge V \,\}.$$

Then, for $T \ge 2$, $1 \le V \le \log T$, M. Jutila [Jut1] proved that

$$\mu(M_T(V)) \ll T \exp\left\{-\frac{\log^2 V}{\log\log T}\left(1 + O\left(\frac{\log V}{\log\log T}\right)\right)\right\}$$

and also, for some constant $c > 0$,

$$\mu(M_T(V)) \ll T \exp\left(-c\frac{\log^2 V}{\log\log T}\right).$$

It follows that there exist positive constants a_1, a_2 and a_3 such that, for $T \ge 10$, one has

$$\exp\Big(a_1(\log\log T)^{1/2}\Big) \le |Z(t)| \le \exp\Big(a_2(\log\log T)^{1/2}\Big)$$

in a subset of measure at least $a_3 T$ of the interval $[0, T]$. Detailed proofs of the above results may be also found in chapter 6 of [Iv4]. Naturally, stronger bounds may be obtained if the RH is assumed. Thus (assuming the RH) K. Soundararajan [Sou4] proves, if $T, V \ge T_0\ (> 0)$, $\log_3 T = \log\log\log T$,

$$S(T, V) := \{\, t \in [T, 2T] \;:\; |Z(t)| \ge V \,\},$$

the following result (note that $S(T, V) = M_{2T}(V) \setminus M_T(V)$ in Jutila's notation). If

$$10\sqrt{\log\log T} \le V \le \log\log T,$$

then

$$\mu(S(T, V)) \ll T\frac{V}{\sqrt{\log\log T}} \exp\left(-\frac{V^2}{\log\log T}\left(1 - \frac{4}{\log_3 T}\right)\right).$$

If

$$\log\log T < V \le \tfrac{1}{2}(\log\log T)\log_3 T,$$

then

$$\mu(S(T,V)) \ll T\frac{V}{\sqrt{\log\log T}}\exp\left(-\frac{V^2}{\log\log T}\left(1-\frac{7V}{4(\log\log T)\log_3 T}\right)^2\right),$$

while if

$$V > \tfrac{1}{2}(\log\log T)\log_3 T,$$

then we have

$$\mu(S(T,V)) \ll T\exp\left(-\frac{1}{33}V\log V\right).$$

These bounds enabled Soundararajan (*op. cit.*) to obtain a result compatible with K. Ramachandra's lower bound (10.29), which is unconditional. Namely, under the RH, Soundararajan proved that, for every positive real k,

$$\int_0^T |\zeta(\tfrac{1}{2}+it)|^{2k}\,\mathrm{d}t \ll_{k,\varepsilon} T(\log T)^{k^2+\varepsilon}. \tag{11.40}$$

The present author [Iv21] generalized Soundararajan's result to short intervals, and sharpened (11.40). He proved that under the RH, for fixed $k>0$ and $H=T^\theta$ with fixed $0<\theta\le 1$,

$$\int_T^{T+H} |\zeta(\tfrac{1}{2}+it)|^{2k}\,\mathrm{d}t \ll H(\log T)^{k^2(1+O(1/\log_3 T))}.$$

Recently M. Radziwill [Rad] showed that, under the RH,

$$\int_0^T |\zeta(\tfrac{1}{2}+it)|^{2k}\,\mathrm{d}t \ll_k T(\log T)^{k^2} \tag{11.41}$$

holds for $0\le k\le 2+2/11$. In this range this result sharpens (11.40), showing that the upper bound is of the conjectured order of magnitude. Since (11.41) is known to be true for $k=2$, the interest in this result comes from extending the range to $k<2+2/11$.

D. W. Farmer *el al.* [FGH] conjecture that, for $d>0$ fixed and $T\to\infty$, one has

$$\frac{1}{T}\mu\left\{0<t<T \;:\; |Z(t)|>\exp(d\sqrt{\log T\log\log T})\right\} = \exp\Big(-2(1+o(1))d^2\log T\Big).$$

This conjecture is related to their conjecture concerning the maximal order of $|Z(t)|$, which was discussed in the Notes to Chapter 1.

Some problems on Hardy's function are discussed in the present author's work [Iv21].

References

[And] R. J. Anderson, On the function $Z(t)$ associated with the Riemann zeta-function, *J. Math. Anal. Appl.* **118** (1986), 323-340.

[AGZ] G. W. Anderson, A. Guionnet and O. Zeitouni, *An Introduction to Random Matrices*, Cambridge University Press, Cambridge, 2010.

[Ape] R. Apéry, Interpolation de fractions continues et irrationalité de certaines constantes, *Math. CTHS Bull. Sec. Sci. II (Bibl. Nat. Paris)*, 1981, 37-53.

[Atk1] F. V. Atkinson, The mean value of the zeta-function on the critical line, *Quart. J. Math. Oxford* **10** (1939), 122-128.

[Atk2] F. V. Atkinson, The mean value of the zeta-function on the critical line, *Proc. London Math. Soc.* **47** (1941), 174-200.

[Atk3] F. V. Atkinson, A mean value property of the Riemann zeta-function, *J. London Math. Soc.* **23** (1948), 128-135.

[Atk4] F. V. Atkinson, The mean value of the Riemann zeta-function, *Acta Math.* **81** (1949), 353-376.

[Bac] R. J. Backlund, Sur les zéros de la fonction $\zeta(s)$ de Riemann, *Comptes Rendus Acad. Sci.* **158** (1914), 1979-1982.

[Bet] S. Bettin, The second moment of the Riemann zeta-function with unbounded shifts, preprint available at arXiv:1111.0925.

[Boc] S. Bochner, On Riemann's functional equation with multiple gamma factors, *Annals of Math.* **67** (1958), 29-41.

[BoHe] E. Bombieri and D. Hejhal, Sur les zéros des fonctions zeta d'Epstein, *Comptes Rendus Acad. Sci. Paris* **304** (1987), 213-217.

[BoIw1] E. Bombieri and H. Iwaniec, On the order of $\zeta(\frac{1}{2}+it)$, *Ann. Scuola Norm. Sup. Pisa Cl. Sci.* (4)**13** (1986), 449-472.

[BoIw2] E. Bombieri and H. Iwaniec, Some mean value theorems for exponential sums, *Ann. Scuola Norm. Sup. Pisa Cl. Sci.* (4) **13** (1986), 473-486.

[BoGh] E. Bombieri and A. Ghosh, Around the Davenport-Heilbronn function, *Russian Math. Surveys* **66** (2011), 221-270, and *Uspekhi Mat. Nauk* **66** (2011), 15-66.

[Bom1] E. Bombieri, *Riemann Hypothesis. The Millennium Prize Problems*, Clay Math. Inst., Cambridge, MA, 2006, pp. 107-124.

[Bom2] E. Bombieri, The classical theory of zeta and L-functions, *Milan J. Math.* **78** (2010), 11-59.

[BCRW] P. Borwein, S. Choi, B. Rooney and A. Weirathmueller, *The Riemann Hypothesis, a Resource for the Afficionado and the Virtuoso Alike*, CMS Books in Mathematics, Canadian Math. Soc., 2008.

[BCY] H. Bui, B. Conrey and M. P. Young, More than 41% of the zeros of the zeta function are on the critical line, to appear in *Acta Arith.*, see arXiv:1002.4127.

[BhS1] G. Bhowmik and J.-C. Schlage-Puchta, Natural boundaries of Dirichlet series, *Func. Approx. Comment. Math.* **37** (2007), 17-29.

[BhS2] G. Bhowmik and J.-C. Schlage-Puchta, Essential singularities of Euler products, to appear, preprint available at arXiv:1001.1891.

[Bre] J. Bredberg, Large gaps between consecutive zeros on the critical line of the Riemann zeta-function, to appear, preprint available at arXiv:1101.3197.

[Bui] H. M. Bui, Large gaps between consecutive zeros of the Riemann zeta-function, *J. Number Theory* **131** (2011), 67-95.

[Cha] K. Chandrasekharan, *Arithmetical Functions*, Springer Verlag, Berlin-Heidelberg-New York, 1970.

[ChNa] K. Chandrasekharan and R. Narasimhan, Functional equations with multiple gamma factors and the average order of arithmetic functions, *Annals of Math.* **76** (1962), 93-136.

[ChSo] V. Chandee and K. Soundararajan, Bounding $|\zeta(\frac{1}{2}+it)|$ on the Riemann Hypothesis, *Bull. London Math. Soc.* (2011), 243-250.

[CMoP] E. Carletti, G. Monti Bragadin and A. Perelli, On general L-functions, *Acta Arith.* **66** (1994), 147-179.

[Con1] J. B. Conrey, The fourth moment of derivatives of the Riemann zeta-function, *Quarterly J. Math., Oxf. II. Ser.* **39** (1988), No. 153, 21-36.

[Con2] J. B. Conrey, More than two fifths of the zeros of the Riemann zeta function are on the critical line, *J. Reine Angew. Math.* **399** (1989), 1-26.

[Con3] J. B. Conrey, A note on the fourth power moment of the Riemann zeta-function, in: B. C. Berndt *et al.* (eds.) *Analytic Number Theory. Vol. 1. Proc. of a Conf. in Honor of Heini Halberstam, Urbana, 1995*, Birkhäuser, Prog. Math. 138 (1996), 225-230.

[Con4] J. B. Conrey, The Riemann hypothesis, *Notices Amer. Math. Soc.* **50** (2003), 341-353.

[CoGh1] J. B. Conrey and A. Ghosh, A mean value theorem for the Riemann zeta-function at its relative extrema on the critical line, *J. Lond. Math. Soc., II. Ser.* **32** (1985), 193-202.

[CoGh2] J. B. Conrey and A. Ghosh, A simpler proof of Levinson's theorem, *Math. Proc. Camb. Phil. Soc.* **97** (1985), 385-395.

[CoGh3] J. B.Conrey and A. Ghosh, On the Selberg class of Dirichlet series: small degrees, *Duke Math. J.* **72** (1993), 673-693.

[CGG1] J. B. Conrey, A. Ghosh and S. M. Gonek, A note on gaps between zeros of the zeta-function, *Bull. London Math. Soc.* **16** (1984), 421-424.

[CGG2] J. B. Conrey, A. Ghosh and S. M. Gonek, Large gaps between zeros of the zeta-function, *Mathematika* **33** (1986), 212-238.

[CGG3] J. B. Conrey, A. Ghosh and S. M. Gonek, Simple zeros of the Riemann zeta-function, *Proc. Lond. Math. Soc., III. Ser.* **76** (1998), No. 3, 497-522.

[CFKRS1] J. B. Conrey, D. W. Farmer, J. P. Keating, M. O. Rubinstein and N. C. Snaith, Integral moments of L-functions, *Proc. London Math. Soc.* (3) **91** (2005), 33-104.

[CFKRS2] J. B. Conrey, D. W. Farmer, J. P. Keating, M. O. Rubinstein and N. C. Snaith, Lower order terms in the full moment conjecture for the Riemann zeta function, *J. Number Theory* **128** (2008), 1516-1554.

[CoGo] J. B. Conrey and S. M. Gonek, High moments of the Riemann zeta-function, *Duke Math. J.* **107** (2001), 577-604.

[COSV] G. Csordas, A. M. Odlyzko, W. Smith and R. S. Varga, A new Lehmer pair of zeros and a new lower bound for the de Bruijn-Newman constant LAMBDA, *Electr. Trans. Num. Anal.* **1** (1993), 104-111.

[Dah] G. Dahlquist, On the analytic continuation of Eulerian products, *Ark. Mat.* **1** (1952), 533-554.

[DaHe] H. Davenport and H. Heilbronn, On the zeros of certain Dirichlet series I, II, *J. London Math. Soc.* **11** (1936), 181-185 and *ibid.*, 307-312.

[Del] H. Delange, Généralisation du théorème de Ikehara, *Annales scien. E.N.S.* **71** (1954), 213-242.

[Deli] P. Deligne, La conjecture de Weil. I., *Publications Mathématiques de l'IHÉS* **43** (1974), 273-307.

[DGG] A. Diaconu, P. Garrett and D. Goldfeld, Natural boundaries and a correct notion of integral moments of L-functions, preprint, 2009.

[DGH] A. Diaconu, D. Goldfeld and J. Hoffstein, Multiple Dirichlet series and moments of zeta and L-functions, *Compositio Math.* **139**, No. 3 (2003), 297-360.

[Edw] H. M. Edwards, *Riemann's Zeta-Function*, Academic Press, New York-London, 1974 (Dover Publications Inc., Mineola, NY, 2001. Reprint of the 1974 original).

[EMOT] A. Erdélyi, W. Magnus, F. Oberhettinger and F. G. Tricomi, *Higher Transcendental Functions*, Volume I, McGraw-Hill, 1953.

[Est1] T. Estermann, On certain functions represented by Dirichlet series, *Proc. London Math. Soc.* **27** (1928), 435-448.

[Est2] T. Estermann, *Introduction to Modern Prime Number Theory*, Cambridge University Press, Cambridge, 1969.

[Eul] L. Euler, Remarques sur un beau rapport entre les séries des puissances tout directes que réciproques, *Mém. Acad. Roy. Sci. Belles Lettres* **17** (1768), 83-106.

[Fen] S. Feng, Zeros of the Riemann zeta-function on the critical line, to appear, preprint available at arXiv:1003.0059.

[FGH] D. W. Farmer, S. M. Gonek and C. P. Hughes, The maximum size of L-functions. *J. Reine Angew. Math.* **609** (2007), 215-236.

[Fla] C. Flammer, *Spheroidal Wave Functions*, Stanford University Press, Stanford, CA, 1957.

[For1] K. Ford, Vinogradov's integral and bounds for the Riemann zeta-function, *Proc. London Math. Soc. III Ser.* **85** (2002), 565-633.

[For2] K. Ford, Zero-free regions for the Riemann zeta-function, in: M. A. Bennett *et al.* (ed.) *Number Theory for the Millennium II*. Proc. of the Conf.

on number theory, Urbana-Champaign, IL, USA, 2000. Natick, MA: A. K. Peters, 25-56 (2002).

[Fuj1] A. Fujii, On the distribution of zeros of the Riemann zeta function in short intervals, *Bull. Amer. Math. Soc.* **81** (1975), 139-142.

[Fuj2] A. Fujii, On the difference between r consecutive ordinates of the zeros of the Riemann zeta function, *Proc. Japan Acad.* **51** (1975), 741-743.

[Gab] W. Gabcke, Neue Herleitung und explizite Restabschätzung der Riemann-Siegel Formel, Univ. Göttingen, Ph.D. Dissertation, Göttingen, 1979, pp. 153.

[Gho] A. Ghosh, On the Riemann zeta function-mean value theorems and the distribution of $|S(T)|$, *J. Number Theory* **17** (1983), 93-102.

[Gol1] D. A. Goldston, Large differences between consecutive prime numbers, Ph.D. thesis, University of California, Berkeley, 1981.

[Gol2] D. A. Goldston, On the pair correlation conjecture for zeros of the Riemann zeta-function, *J. Reine Angew. Math.* **385** (1988), 24-40.

[Gol3] D. A. Goldston, Notes on pair correlation of zeros and prime numbers, in: F. Mezzadri *et al.* (eds.) *Recent Perspectives in Random Matrix Theory and Number Theory, Proc. "Random Matrix Approaches in Number Theory"*, London Mathematical Society Lecture Note Series, 322, Cambridge University Press, Cambridge (2005), 79-110.

[GoGo] D. A. Goldston and S. M. Gonek, A note on $S(t)$ and the zeros of the Riemann zeta-function, *Bull. London Math. Soc.* **39** (2007), 482-486.

[GoMo] D. A. Goldston and H. L Montgomery, *Pair Correlation and Primes in Short Intervals, Analytic Number Theory and Diophantine Problems*, Birkhäuser, Boston, Mass., 1987, pp. 187-203.

[Gon] S. M. Gonek, On negative moments of the Riemann zeta-function, *Mathematika* **36** (1989), 71-88.

[GrKo] S. W. Graham and G. Kolesnik, *Van der Corput's method for exponential sums*, London Mathematical Society Lecture Note Series, 126, Cambridge University Press, Cambridge, 1991.

[Gourd] X. Gourdon, The first 10^{13} zeros of the Riemann zeta-function and zeros computation at very large height, 2004, http://numbers.computation.free.fr/Constants/Miscellaneous.

[Gours] É. Goursat, *Cours d'Analyse Mathématique*, Dover, New York, 1959.

[Gra] J. P. Gram, Sur les zéros de la fonction de Riemann, *Acta Mathematica* **27** (1903), 289-304.

[Hal1] R. R. Hall, The behaviour of the Riemann zeta-function on the critical line, *Mathematica* **46** (1999), 281-313.

[Hal2] R. R. Hall, A Wirtinger type inequality and the spacing between the zeros of the Riemann zeta-function, *J. Number Theory* **93** (2002), 235-245.

[Hal3] R. R. Hall, On the extreme values of the Riemann zeta-function between its zeros on the critical line, *J. Reine Angew. Math.* **560** (2003), 29-41.

[Hal4] R. R. Hall, Generalized Wirtinger inequalities, random matrix theory, and the zeros of the Riemann zeta-function, *J. Number Theory* **97** (2002), 397-409.

[Hal5] R. R. Hall, On the stationary points of Hardy's function $Z(t)$, *Acta Arith.* **111** (2004), 125-140.

[Hal6] R. R. Hall, A new unconditional result about large spaces between zeta zeros, *Mathematika* **53** (2005), 101-113.

[Hal7] R. R. Hall, Extreme values of the Riemann zeta-function on short zero intervals, *Acta Arith.* **121** (2006), 259-273.

[Ham] H. Hamburger, Über die Riemmansche Funkionalgleichung der ζ-Funktion I, II, III, *Math. Zeit.* **10** (1921), 240-254, *ibid.* **11** (1922), 224-245, *ibid.* **13** (1922), 283-311.

[HaIv1] J. L. Hafner and A. Ivić, On some mean value results for the Riemann zeta-function, *Proceedings International Number Theory Conference Québec 1987*, Walter de Gruyter and Co., 1989, Berlin-New York, pp. 348-358.

[HaIv2] J. L. Hafner and A. Ivić, On the mean square of the Riemann zeta-function on the critical line, *J. Number Theory* **32** (1989), 151-191.

[Har] G. Harcos, Uniform approximate functional equation for principal L-functions, *Inter. Math. Research Notices* **18** (2002), 923-932.

[Har1] G. H. Hardy, On the zeros of Riemann's zeta-function, *Proc. London Math. Soc. Ser. 2* **13** (records of proceedings at meetings), March 1914.

[Har2] G. H. Hardy, Sur les zéros de la fonction $\zeta(s)$ de Riemann, *Comptes Rendus Acad. Sci. (Paris)* **158** (1914), 1012-1014.

[Har3] G. H. Hardy, *A Mathematician's Apology*, Cambridge University Press, Cambridge (1940), 2004 reissue.

[Har4] G. H. Hardy, *Ramanujan*, Cambridge University Press, London, 1940, reissue AMS Chelsea Pub., 1999.

[Har5] G. H. Hardy, *Divergent Series*, Clarendon Press, Oxford, 1949.

[Har6] G. H. Hardy, *A Course of Pure Mathematics* (10th edn). Cambridge University Press, Cambridge (1952) [1908], 2008 reissue.

[Har7] G. H. Hardy, *Collected Papers of G. H. Hardy; Including Joint papers with J.E. Littlewood and Others*, London Mathematical Society, 1966.

[HaLi1] G. H. Hardy and J. E. Littlewood, Contributions to the theory of the Riemann zeta-function and the distribution of primes, *Acta Math.* **41** (1918), 119-196.

[HaLi2] G. H. Hardy and J. E. Littlewood, The zeros of Riemann's zeta-function on the critical line, *Math. Zeitschrift* **10** (1921), no. 3-4, 283-317.

[HaLi3] G. H. Hardy and J. E. Littlewood, The approximate functional equation for $\zeta(s)$ and $\zeta^2(s)$, *Proc. London Math. Soc.* (2) **29** (1929), 81-97.

[HaRi] G. H. Hardy and M. Riesz, *The General Theory of Dirichlet Series*, Cambridge Tracts in Mathematics, Cambridge University Press, Cambridge, 1915.

[Has] C. B. Haselgrove, *Tables of the Riemann Zeta Function*, Cambridge University Press, Cambridge, 1960.

[HaWr] G. H. Hardy and E. M. Wright, *An Introduction to the Theory of Numbers* (6th edn), Oxford University Press, Oxford, 2008.

[Hea1] D. R. Heath-Brown, The twelfth power moment of the Riemann zeta-function, *Quart. J. Math.* (*Oxford*) **29** (1978), 443-462.

[Hea2] D. R. Heath-Brown, The mean value theorem for the Riemann zeta-function, *Mathematika* **25** (1978), 177-184.

[Hea3] D. R. Heath-Brown, The fourth power moment of the Riemann zeta function, *J. London Math. Soc.* (3)**38** (1979), 385-422.

[Hea4] D. R. Heath-Brown, Fractional moments of the Riemann zeta-function, *J. London Math. Soc.* **24** (1981), 65-78.

[Hej1] D. A. Hejhal, On the distribution of $|\log \zeta'(\frac{1}{2}+it)|$, in: K. E. Aubert *et al.* (eds.) *Number Theory, Trace Formulas and Discrete Groups*, Proceedings Selberg 1987 Symposium, Academic Press, 1989, 343-370.

[Hej2] D. A. Hejhal, On a result of Selberg concerning zeros of linear combinations of L-functions, *Int. Math. Res. Not.* **11** (2000), 551-577.

[HiOd1] G. A. Hiary and A. M. Odlyzko, The zeta function on the critical line: numerical evidence for moments and random matrix theory models, to appear, preprint available at arXiv:1008.2173.

[HiOd2] G. A. Hiary and A. M. Odlyzko, Numerical study of the derivative of the Riemann zeta-function at zeros, to appear, preprint available at arXiv:1105.4312.

[HKN] C. P. Hughes, J. P. Keating and O. Neil, Random matrix theory and the derivative of the Riemann zeta function, *Proc. R. Soc. Lond., Ser. A, Math. Phys. Eng. Sci.* **456** (2000), 2611-2627.

[Hut] J. I. Hutchinson, On the roots of the Riemann zeta-function, V. *Trans. Amer. Math. Soc.* **27** (1925), 27-49.

[Hux1] M. N. Huxley, *Area, Lattice Points and Exponential Sums*, Oxford Science Publications, Clarendon Press, Oxford, 1996.

[Hux2] M. N. Huxley, Exponential sums and the Riemann zeta function V, *Proc. London Math. Soc.* (3) **90** (2005), 1-41.

[HuIv] M. N. Huxley and A. Ivić, Subconvexity for the Riemann zeta-function and the divisor problem, Bulletin CXXXIV de l'Académie Serbe des Sciences et des Arts - 2007, *Classe des Sciences Mathématiques et Naturelles, Sciences Mathématiques* **32**, 13-32.

[Ing] A. E. Ingham, Mean-value theorems in the theory of the Riemann zeta-function, *Proc. London Math. Soc.* (2) **27** (1926), 273-300.

[Isr1] M. I. Israilov, Coefficients of the Laurent expansion of the Riemann zeta-function (Russian), *Dokl. Akad. Nauk SSSR* **12** (1979), 9-10.

[Isr2] M. I. Israilov, The Laurent expansion of the Riemann zeta-function (Russian), *Trudy Mat. Inst. Steklova* **158** (1981), 98-104.

[Iv1] A. Ivić, *The Riemann Zeta-Function*, John Wiley & Sons, New York, 1985 (reissue, Dover, Mineola, New York, 2003).

[Iv2] A. Ivić, On consecutive zeros of the Riemann zeta-function on the critical line, in: *Séminaire de Théorie des Nombres, Université de Bordeaux 1986/87*, Exposé no. **29**, 14 pp.

[Iv3] A. Ivić, On a problem connected with zeros of $\zeta(s)$ on the critical line, *Monatshefte Math.* **104** (1987), 17-27.

[Iv4] A. Ivić, *Mean Values of the Riemann Zeta-function*, LN's 82, Tata Inst. of Fundamental Research, Bombay, 1991 (Springer Verlag, Berlin).

[Iv5] A. Ivić, On the fourth moment of the Riemann zeta-function, *Publs. Inst. Math. (Belgrade)* **57**(71) (1995), 101-110.

[Iv6] A. Ivić, On sums of gaps between the zeros of $\zeta(s)$ on the critical line, *Univ. Beograd. Publ. Elektrotehn. Fak. Ser. Mat.* **6** (1995), 55-62.

[Iv7] A. Ivić, An approximate functional equation for a class of Dirichlet series, *Journal of Analysis* **3** (1995), 241-252.

[Iv8] A. Ivić, The Mellin transform and the Riemann zeta-function, in: W. G. Nowak and J. Schoißengeier Vienna (eds.) *Proceedings of the Conference on Elementary and Analytic Number Theory* (Vienna, July 18-20, 1996), Universität Wien & Universität für Bodenkultur, 1996, 112-127.

[Iv9] A. Ivić, On the error term for the fourth moment of the Riemann zeta-function, *J. London Math. Soc.* **60** (2)(1999), 21-32.

[Iv10] A. Ivić, The Laplace transform of the fourth moment of the zeta-function, *Univ. Beograd. Publ. Elektrotehn. Fak. Ser. Mat.* **11** (2000), 41-48.

[Iv11] A. Ivić, On some conjectures and results for the Riemann zeta-function, *Acta. Arith.* **99** (2001), 115-145.

[Iv12] A. Ivić, The Laplace transform of the fourth moment of of the zeta-function, *Univ. Beograd. Publ. Elektrotehn. Fak. Ser. Mat.* **11** (2000), 41-48.

[Iv13] A. Ivić, On small values of the Riemann zeta-function on the critical line and gaps between zeros, *Lietuvos Mat. Rinkinys* **42** (2002), 31-45.

[Iv14] A. Ivić, On the estimation of $\mathcal{Z}_2(s)$, in: *Anal. Probab. Methods Number Theory*, A. Dubickas *et al.* (eds.) TEV, Vilnius, 2002, 83-98.

[Iv15] A. Ivić, On the integral of Hardy's function, *Arch. Mathematik* **83** (2004), 41-47.

[Iv16] A. Ivić, The Mellin transform of the square of Riemann's zeta-function, *International J. of Number Theory* **1** (2005), 65-73.

[Iv17] A. Ivić, On the estimation of some Mellin transforms connected with the fourth moment of $|\zeta(\frac{1}{2}+it)|$, in: W. Schwarz and J. Steuding (eds.) *Elementare und Analytische Zahlentheorie (Tagungsband), Proceedings ELAZ-Conference May 24-28, 2004*, Franz Steiner Verlag, 2006, 77-88.

[Iv18] A. Ivić, On some reasons for doubting the Riemann Hypothesis, in: P. Borwein *et al.* (eds.) *The Riemann Hypothesis*, CMS Books in Mathematics, Springer, 2008.

[Iv19] A. Ivić, On the moments of the Riemann zeta-function in short intervals, *Hardy-Ramanujan Journal* **32** (2009), 4-23.

[Iv20] A. Ivić, On the Mellin transforms of powers of Hardy's function, *Hardy-Ramanujan Journal* **33** (2010), 32-58.

[Iv21] A. Ivić, On some problems involving Hardy's function, *Central European J. Math.* **8**(6) (2010), 1029-1040.

[IvJu] A. Ivić and M. Jutila, Gaps between consecutive zeros of the Riemann zeta-function, *Monatshefte Math.* **105** (1988), 59-73.

[IJM] A. Ivić, M. Jutila and Y. Motohashi, The Mellin transform of powers of the Riemann zeta-function, *Acta Arith.* **95** (2000), 305-342.

[IvMo1] A. Ivić and Y. Motohashi, A note on the mean value of the zeta and L-functions VII, *Proc. Japan Acad. Ser. A* **66** (1990), 150-152.

[IvMo2] A. Ivić and Y. Motohashi, The mean square of the error term for the fourth moment of the zeta-function, *Proc. London Math. Soc.* (3)**66** (1994), 309-329.

[IvMo3] A. Ivić and Y. Motohashi, The fourth moment of the Riemann zeta-function, *J. Number Theory* **51** (1995), 16-45.

[IvPe] A. Ivić and A. Perelli, Mean values of certain zeta-functions on the critical line, *Litovskij Mat. Sbornik* **29** (1989), 701-714.

[Iwa1] H. Iwaniec, *Introduction to the Spectral Theory of Automorphic Forms*, Bibl. de la Revista Iberoamericana, Madrid, 1995.

[Iwa2] H. Iwaniec, *Topics in Classical Automorphic Forms*, Graduate Studies in Mathematics 17, American Mathematical Society, Providence, RI, 1997.

[Iwa3] H. Iwaniec, *Spectral Methods of Automorphic Forms*, 2nd edn, Graduate Studies in Mathematics, 53, American Mathematical Society, Providence, RI, 1997.

[Joy] D. Joyner, *Distribution Theorems of L-functions*, Pitman Research Notes in Mathematics Series 142, Longman Scientific & Technical, Harlow, John Wiley & Sons, New York, 1986.

[Jut1] M. Jutila, On the value distribution of the zeta-function on the critical line, *Bull. London Math. Soc.* **15** (1983), 513-518.

[Jut2] M. Jutila, Mean values of Dirichlet series via Laplace transforms, in: Y. Motohashi (ed.) *Analytic Number Theory*, London Math. Soc. LNS 247, Cambridge University Press, Cambridge, 1997, 169-207.

[Jut3] M. Jutila, Atkinson's formula revisited, in: *Voronoï's Impact on Modern Science*, Book 1, Institute of Mathematics, National Academy of Sciences of Ukraine, Kyiv, 1998, 137-154.

[Jut4] M. Jutila, The Mellin transform of the square of Riemann's zeta-function, *Periodica Math. Hung.* **42** (2001), 179-190.

[Jut5] M. Jutila, The Mellin transform of the fourth power of the Riemann zeta-function, in: S. D Adhikari, *et al.* (eds.) *Number Theory*. Proc. Inter. Conf. on Analytic Number Theory with special emphasis on L-functions, held at the Inst. Math. Sc., Chennai, India, January 2002. Ramanujan Math. Soc. LNS 1 (2005), 15-29.

[Jut6] M. Jutila, Atkinson's formula for Hardy's function, *J. Number Theory* **129** (2009), 2853-2878.

[Jut7] M. Jutila, An estimate for the Mellin transform of Hardy's function, *Hardy-Ramanujan J.* **33** (2010), 23-31.

[Jut8] M. Jutila, The Mellin transform of Hardy's function is entire (in Russian), *Mat. Zametki* **88** (4) (2010), 635-639.

[Jut9] M. Jutila, An asymptotic formula for the primitive of Hardy's function, *Arkiv Mat.* **49**, No. 1, (2011), 97-107.

[Kac] J. Kaczorowski, Axiomatic theory of L-functions: the Selberg class, in: A. Perelli and C. Viola (eds.) *Analytic Number Theory*, Springer Verlag, Berlin-Heidelberg, 2006, 133-209.

[KaKo] A. A. Karatsuba and M. A. Korolev, The argument of the Riemann zeta function (English translation of Russian original) *Russ. Math. Surv.* **60** (2005), No. 3, 433-488; translation from *Usp. Mat. Nauk* **60** (2005), No. 3, 41-96.

[KaPe1] J. Kaczorowski and A. Perelli, The Selberg class: a survey, in: *Number Theory in Progress*, Vol. 2 (Zakopane-Koscielisko, 1997), de Gruyter, Berlin, 1999, 953-992.

[KaPe2] J. Kaczorowski and A. Perelli, On the structure of the Selberg class, I: $0 \leq d \leq 1$, *Acta Math.* **182** (1999), 207-241.

[KaPe3] J. Kaczorowski and A. Perelli, On the structure of the Selberg class, V: $1 < d < 5/3$, *Invent. Math.* **150** (2002), 485-516.

[KaPe4] J. Kaczorowski and A. Perelli, On the structure of the Selberg class, VI: non-linear twists, *Acta Arith.* **116** (2005), 315-341.

[KaPe5] J. Kaczorowski and A. Perelli, On the structure of the Selberg class, VII: $1 < d < 2$, *Annals of Math.* **173** (2011), 1397-1441.

[Kar1] A. A. Karatsuba, On the distance between adjacent zeros of the Riemann zeta-function lying on the critical line (Russian), *Trudy Mat. Inst. Steklova* **157** (1981), 49-63.

[Kar2] A. A. Karatsuba, On the zeros of the Davenport-Heilbronn function lying on the critical line (Russian), *Izv. Akad. Nauk SSSR ser. mat.* **54** no. 2 (1990), 303-315.

[KaS] J. Kalpokas and J. Steuding, On the value-distribution of the Riemann zeta-function on the critical line, *Moscow Journal of Combinatorics and Number Theory* **1** (2011), 26-42.

[KaVo] A. A. Karatsuba and S. M. Voronin, *The Riemann Zeta-Function*, Walter de Gruyter, Berlin-New York, 1992.

[Kea] J. Keating, The Riemann zeta-function and quantum chaology, *Proc. Internat. School of Phys. Enrico Fermi* **CXIX** (1993), 145-185.

[KeSn] J. P. Keating and N.C. Snaith, Random matrix theory and L-functions at $s = 1/2$, *Comm. Math. Phys.* **214** (2000), 57-89.

[Kob] H. Kober, Eine Mittelwertformel der Riemannschen Zetafunktion, *Compositio Math.* **3** (1936), 174-189.

[Kol] G. Kolesnik, On the estimation of multiple exponential sums, in: *Recent Progress in Analytic Number Theory*, Symp. Durham 1979 (Vol. 1), Academic Press, London, 1981, 231-246.

[Kore] J. Korevaar, *Tauberian Theory*, Grund. der math. Wissenschaften Vol. 329, Springer, Berlin, 2004.

[Kor1] M. A. Korolev, On the argument of the Riemann zeta function on the critical line. (English translation of Russian original) *Izv. Math.* **67** (2003), No. 2, 225-264; translation from *Izv. Ross. Akad. Nauk Ser. Mat.* **67** (2003), No. 2, 21-60.

[Kor2] M. A. Korolev, Sign change of the function $S(t)$ on short intervals (English translation of Russian original), *Izv. Math.* **69** (2005), No. 4, 719-731; translation from *Izv. Ross. Akad. Nauk, Ser. Mat.* **69** (2005), No. 4, 75-88.

[Kor3] M. A. Korolev, On the primitive of the Hardy function $Z(t)$, *Dokl. Math.* **75**, No. 2, 295-298 (2007); translation from *Dokl. Akad. Nauk, Ross. Akad. Nauk* **413**, No. 5, 599-602 (2007).

[Kor4] M. A. Korolev, On the integral of Hardy's function $Z(t)$, *Izv. Math.* **72**, No. 3, 429-478 (2008); translation from *Izv. Ross. Akad. Nauk, Ser. Mat.* **72**, No. 3, 19-68 (2008).

[Kor5] M. A. Korolev, Gram's law and the argument of the Riemann zeta-function, to appear, preprint available at arXiv:1106.0516.

[Kos] H. Kösters, On the occurrence of the sine kernel in connection with the shifted moments of the Riemann zeta function, *J. Number Theory* **130** (2005), 2596-2609.

[Kra] E. Krätzel, *Lattice Points, Mathematics and its Applications*: East European Series, 33. Dordrecht, Kluwer Academic Publishers; Berlin: VEB Deutscher Verlag der Wissenschaften, 1988.

[Lan] E. Landau, Euler und Functionalgleichung der Riemannschen Zeta-Funktion, *Biblio. Math.* (3) Bd. **7**, Leipzig, 1906, pp. 69-79.

[Lau1] A. Laurinčikas, On the zeta-function of Riemann on the critical line (Russian), *Litovskij Mat. Sbornik* **25** (1985), 114-118.

[Lau2] A. Laurinčikas, On the moments of the zeta-function of Riemann on the critical line (Russian), *Mat. Zametki* **39** (1986), 483-493.

[Lau3] A. Laurinčikas, The limit theorem for the Riemann zeta-function on the critical line I (Russian), *Litovskij Mat. Sbornik* **27** (1987), 113-132; II *ibid.* **27** (1987), 489-500.

[Lau4] A. Laurinčikas, A limit theorem for Dirichlet L-functions on the critical line (Russian), *Litovskij Mat. Sbornik* **27** (1987), 699-710.

[Lau5] A. Laurinčikas, *Limit Theorems for the Riemann Zeta-Function*, Kluwer, Dordrecht, 1996.

[Lau6] A. Laurinčikas, Limit theorems for the Mellin transforms of the Riemann zeta-function, *Fiz. Mat. Fak. Moksl. Semin. Darb.* **8** (2005), 63-75.

[Lav1] A. A. Lavrik, Uniform approximations and zeros of derivatives of Hardy's Z-function in short intervals (in Russian), *Analysis Mathem.* **17** (1991), 257-259.

[Lav2] A. A. Lavrik, Titchmarsh's problem in the discrete theory of the Riemann zeta-function. (English translation of Russian original) *Proc. Steklov Inst. Math.* **207** (1995), 179-209; translation from *Tr. Mat. Inst. Steklova* **207** (1994), 197-230.

[Lavr] A. F. Lavrik, On the principal term in the divisor problem and power series of the Riemann zeta-function in a neighborhood of its pole (Russian), *Trudy Mat. Inst. Steklova* **142** (1976), 165-173, reprinted in *Proc. Steklov Mat. Inst.* **3** (1979), 175-183.

[LaIE] A. F. Lavrik, M. I. Israilov and Ž. Edgorov, On integrals containing the error term in the divisor problem (Russian), *Acta Arith.* **37** (1980), 381-389.

[LaTs] Y.-K. Lau and K.-M. Tsang, Omega result for the mean square of the Riemann zeta function, *Manuscr. Math.* **117** (2005), 373-381.

[Leh1] D. H. Lehmer, On the roots of the Riemann zeta function, *Acta Math.* **95** (1956), 291-298.

[Leh2] D. H. Lehmer, Extended computation of the Riemann zeta-function, *Mathematika* **3** (1956), 102-108.

[Lev] N. Levinson, More than one third of the zeros of Riemann's zeta-function are on $\sigma = 1/2$, *Adv. Math.* **18** (1975), 383-346.

[Lin] E. Lindelöf, Quelques remarques sur la croissance de la fonction $\zeta(s)$, *Bull. Sci. Math.* **32** (1908), 341-356.

[Lit] J. E. Littlewood, On the zeros of the Riemann zeta-function, *Proc. Camb. Phil. Soc.* **22** (1924), 295-318.

[Luk] M. Lukkarinen, The Mellin transform of the square of Riemann's zeta-function and Atkinson's formula, Doctoral Dissertation, Annales Acad. Sci. Fennicae, No. 140, Helsinki, 2005, 74 pp.

[LRW] J. van de Lune, H. J. J. te Riele and D. T. Winter, On the zeros of the Riemann zeta function in the critical strip. IV, *Math. Comp.* **46** (1986), 667-681.

[Man] H. von Mangoldt, Zu Riemann's Abhandlung "Über die Anzahl ...,", *Crelle's J.* **114** (1895), 255-305.

[MaTa] K. Matsumoto and Y. Tanigawa, On the zeros of higher derivatives of Hardy's Z-function, *J. Number Theory* **75** (1999), 262-278.

[Meh] M. L. Mehta, *Random Matrices* (3rd edn), Pure and Applied Mathematics, 142. Elsevier/Academic Press, Amsterdam, 2004.

[Mil] M. B. Milinovich, Moments of the Riemann zeta-function at its relative extrema on the critical line, *Bull. London Math. Soc.* **43** (2011), 1119-1129.

[Mol] G. Molteni, A note on a result of Bochner and Conrey-Ghosh about the Selberg class, *Arch. Math.* **72** (1999), 219-222.

[Mon1] H. L. Montgomery, The pair correlation of zeros of the zeta-function, in: *Proc. Symp. Pure Math.* 24, AMS, Providence, RI, 1973, 181-193.

[Mon2] H. L. Montgomery, Extreme values of the Riemann zeta-function, *Comment. Math. Helv.* **52** (1977), 511-518.

[MOd] H. L. Montgomery and A. M. Odlyzko, Gaps between zeros of the zeta function, in: *Coll. Math. Soc. János Bolyai* 34, Topics in Classical Number Theory (Budapest, 1981), 1079-1106.

[Mos1] J. Moser, The proof of the Titchmarsh hypothesis in the theory of the Riemann zeta-function (Russian), *Acta Arith.* **36** (1980), 147-156.

[Mos2] J. Moser, On the order of a sum of E. C. Titchmarsh in the theory of the Riemann zeta-function (Russian), *Czech. Math. J.* **41**(116) (1991), 663-684.

[Mot1] Y. Motohashi, A note on the approximate functional equation for $\zeta^2(s)$, *Proc. Japan Acad.* **59A** (1983), 392-396, and II, *ibid.* **59A** (1983), 469-472.

[Mot2] Y. Motohashi, *Riemann-Siegel Formula*, Lecture Notes, University of Colorado, Boulder, 1987.

[Mot3] Y. Motohashi, An explicit formula for the fourth power mean of the Riemann zeta-function, *Acta Math.* **170** (1993), 181-220.

[Mot4] Y. Motohashi, A relation between the Riemann zeta-function and the hyperbolic Laplacian, *Annali Scuola Norm. Sup. Pisa, Cl. Sci. IV ser.* **22** (1995), 299-313.

[Mot5] Y. Motohashi, *Spectral Theory of the Riemann Zeta-Function*, Cambridge University Press, Cambridge, 1997.

[Mot6] Y. Motohashi, The Riemann zeta-function and Hecke congruence subgroups II, *Journal of Research Institute of Science and Technology, Tokyo*, 2009, to appear.

[MoVa] H. L. Montgomery and R. C. Vaughan, *Multiplicative Number Theory I. Classical Theory*, Cambridge studies in advanced mathematics, Cambridge University Press, Cambridge, 2007.

[Mue] J. Mueller, On the difference between consecutive zeros of the Riemann zeta-function, *J. Number Theory* **14** (1982), 327-331.

[NaSt] H. Nagoshi and J. Steuding, Universality for L-functions in the Selberg class, *Lithuanian Math. J.* **50** (2010), 293-311.

[Odl1] A. M. Odlyzko, On the distribution of spacings between zeros of the zeta function, *Math. Comp.* **48** (1987), 273-308.

[Odl2] A. M. Odlyzko, The 10^{22}-nd zero of the Riemann zeta function, in: M. van Frankenhuysen and M. L. Lapidus (eds.) *Dynamical, Spectral, and Arithmetic Zeta Functions*, Amer. Math. Soc., Contemporary Math. 290, 2001, 139-144.

[Odl3] A. M. Odlyzko, The 10^{20}-th zero of the Riemann zeta-function and 175 million of its neighbors, unpublished, see *www.dtc.umn.edu/~odlyzko/zeta_tables/index.html*.

[Per] A. Perelli, A survey of the Selberg class of L-functions, Part I, *Milan J. Math.* **73** (2005), 19-52, and Part II, *Riv. Mat. Univ. Parma* **7** (2004), 83-118.

[Pia] I. Piatetski-Shapiro, Multiplicity one theorems, in: A. Borel and W. Casselman (eds.), *Automorphic Forms, Representations and L-Functions*, Proc. Symp. Pure Math. 33, AMS Publications, 1979, 209-212.

[Pre] E. Preissmann, Sur la moyenne de la fonction zêta, in: K. Nagasaka (ed.) *Analytic Number Theory and Related Topics*, Proceedings of the Symposium, Tokyo, Japan, November 11-13, 1991, Singapore: World Scientific (1993), 119-125.

[Rad] M. Radziwill, The 4.36-th moment of the Riemann zeta-function, IMRN, to appear, preprint available at arXiv:1106.4806.

[RaSo] M. Radziwill and K. Soundararajan, Continuous lower bounds for moments of zeta and L-functions, preprint available at arXiv:1202.1351.

[Ram] K. Ramachandra, *On the mean-value and omega-theorems for the Riemann zeta-function*, LN's 85, Tata Inst. of Fundamental Research, Bombay, 1995 (Springer Verlag, Berlin).

[Ran] R. A. Rankin, Van der Corput's method and the theory of exponent pairs, *Quart. J. Math. Oxford* **6** (1955), 147-153.

[RaSa1] K. Ramachandra and A. Sankaranarayanan, Note on a paper by H.L. Montgomery-I, *Publ. Inst. Math.* **50** (64) (1991), 51-59.

[RaSa2] K. Ramachandra and A. Sankaranarayanan, On some theorems of Littlewood and Selberg I, *J. Number Theory* **44** (1993), 281-291.

[Ric] H.-E. Richert, Über Dirichlet Reihen mit Funktionalgleichung, *Publs. Inst. Math. Serbe Sci.* **1** (1957), 73-124.

[Rie] B. Riemann, Über die Anzahl der Primzahlen unter einer gegebener Grösse, *Monatshefte Preuss. Akad. Wiss.* (1859-1860), 671-680.

[RiLu] H. J. J. te Riele and J. van de Lune, Computational Number Theory at CWI in 1970-1994, *CWI Quarterly* **7(4)** (1994), 285-335.

[Riv1] T. Rivoal, La fonction zêta de Riemann prend une infinité de valeurs irrationnelles aux entiers impairs, *C. R. Acad. Sci., Paris, Sér. I, Math.* **331**, No. 4 (2000), 267-270.

[Riv2] T. Rivoal, Irrationalité d'au moins un des neuf nombres $\zeta(5)$, $\zeta(7), \ldots, \zeta(21)$, *Acta Arith.* **103**, No. 2 (2002), 157-167.

[RuYa] M. O. Rubinstein and S. Yamagishi, Computing the moment polynomials of the zeta function, to appear, preprint available at arXiv:1112.2201.

[Sar1] P. Sarnak, L-functions, in: *Proc. International Congress of Mathematicians, Vol. I* (Berlin, 1998), *Doc. Math.* 1998, Extra Vol. I, 453-465.

[Sar2] P. Sarnak, The grand Riemann hypothesis, *Milan J. Math.* **78** (2010), 61-63.

[Sel] A. Selberg, *Selected Papers, Vol. I*, Springer Verlag, Berlin 1989, and *Vol. II*, Springer Verlag, Berlin 1991.

[Sie] C. L. Siegel, Über Riemanns Nachlaß zur analytischen Zahlentheorie, *Quell. Stud. Gesch. Mat. Astr. Physik* **2** (1932), 45-80 (also in *Gesammelte Abhandlungen*, Band I, Springer Verlag, Berlin, 1966, 275-310).

[Sou1] K. Soundarajan, Omega results for the divisor and circle problems, *Int. Math. Res. Not.* **36** (2003), 1987-1998.

[Sou2] K. Soundarajan, Degree 1 elements of Selberg class, *Expo. Math.* **23** (2005), 65-70.

[Sou3] K. Soundarajan, Extreme values of zeta and L-functions, *Math. Annalen* **342** (2008), 467-486.

[Sou4] K. Soundarajan, Moments of the Riemann zeta function, *Ann. Math.* **170** (2010), 981-993.

[Spi] R. Spira, Some zeros of the Titchmarsh counterexample, *Math. Comp.* **63** (1994), 747-748.

[Sri] B. R. Srinivasan, Lattice point problems of many-dimensional hyperboloids II, *Acta Arith.* **8** (1963), 173-204, and II, *Math. Annalen* **160** (1965), 280-310.

[Ste] J. Steuding, *Value-distribution of L-functions*, Lecture Notes Math. 1877, Springer-Verlag, Berlin, 2007.

[Sti] T. J. Stieltjes, *Correspondance d'Hermite et de Stieltjes*, Tome 1, Gauthier-Villars, Paris, 1905.

[Sub] M. A. Subhankulov, *Tauberian Theorems with Remainder Terms* (Russian), Nauka, Moscow, 1976.

[Tit1] E. C. Titchmarsh, On van der Corput's method and the zeta-function of Riemann. IV, *Quart. J. Math.* **5** (1934), 98-105.

[Tit2] E. C. Titchmarsh, *Introduction to the Theory of Fourier Integrals* (2nd edn), Oxford University Press, Oxford, 1948.

[Tit3] E. C. Titchmarsh, *The Theory of the Riemann Zeta-Function* (2nd edn), Oxford University Press, Oxford, 1986.

[Tru1] T. S. Trudgian, Further results on Gram's law, DPhil Thesis, University of Oxford, Oxford, 2009, 95pp.

[Tru2] T. S. Trudgian, On the success and failure of Gram's law and the Rosser rule, *Acta. Arith.* **143** (2011), 225-256.

[Tru3] T. S. Trudgian, A modest improvement on the function $S(T)$, to appear in *Mathematics of Computation.*

[Tsa1] K.-M. Tsang, Some Ω-theorems for the Riemann zeta-function, *Acta Arith.* **46** (1985), 369-395.

[Tsa2] K.-M. Tsang, The large values of the Riemann zeta-function, Mathematika **40** (1993), 203-214.

[Vin1] I. M. Vinogradov, *Selected Works*, Springer-Verlag, Berlin-New York, 1985.

[Vin2] I. M. Vinogradov, *Method of Trigonometrical Sums in the Theory of Numbers*, Dover Publications, Mineola, NY, 2004.

[Vor] S. M. Voronin, Theorem on the "universality" of the Riemann zeta-function (Russian), *Izv. Akad. Nauk SSSR, Ser. Mat.* **39** (1975), no. 3, 475-486.

[Wats] G. N. Watson, *A Treatise on the Theory of Bessel Function* (2nd edn), Cambridge University Press, Cambridge, 1952.

[Watt] N. Watt, A note on the mean square of $|\zeta(\frac{1}{2}+it)|$, *J. London Math. Soc.* **82**(2) (2010), 279-294.

[Wie] R. Wiebelitz, Über approximative Funktionalgleichungen der Potenzen der Riemannschen Zeta-funktion, *Math. Nachr.* **6** (1951-1952), 263-270.

[Wil] J. R. Wilton, An approximate functional equation for the product of two ζ-functions, *Proc. London Math. Soc.* **31** (1930), 11-17.

[Zud] V. Zudilin, Irrationality of values of the Riemann zeta function, *Izv. Math.* **66**, No. 3 (2002), 489-542.

[You] M. P. Young, A short proof of Levinson's theorem, *Arch. Math.* **95** (2010), 539-548.

Author index

Subject index